W0044174

Modern Birkhäuser Classics

Many of the original research and survey monographs in pure and applied mathematics published by Birkhäuser in recent decades have been groundbreaking and have come to be regarded as foundational to the subject. Through the MBC Series, a select number of these modern classics, entirely uncorrected, are being re-released in paperback (and as eBooks) to ensure that these treasures remain accessible to new generations of students, scholars, and researchers.

Hans Triebel

Fractals and Spectra

Related to Fourier Analysis and Function
Spaces

Reprint of the 1997 Edition

 Birkhäuser

Hans Triebel
Mathematisches Institut
Friedrich-Schiller-Universität Jena
07737 Jena
Germany
hans.triebel@uni-jena.de

ISBN 978-3-0348-0033-4   e-ISBN 978-3-0348-0034-1
DOI 10.1007/978-3-0348-0034-1

Library of Congress Control Number: 2010936432

2000 Mathematics Subject Classification 46E35, 28A80, 46-02, 47G30, 47A10

© 1997 Birkhäuser Verlag
Originally published under the same title as volume 91 in the Monographs in Mathematics series by
Birkhäuser Verlag, Switzerland, ISBN 978-3-7643-5776-2
Reprinted 2011 by Springer Basel

Cover design: deblik, Berlin

Printed on acid-free paper

Springer Basel AG is part of Springer Science+Business Media

www.birkhauser-science.com

Contents

V Spectra of fractal pseudodifferential operators

Preface

This book deals with several aspects of fractal geometry in $\mathbb{R}^n$ which are closely connected with Fourier analysis, function spaces, and appropriate (pseudo)differential operators. It emerged quite recently that some modern techniques in the theory of function spaces are intimately related to methods in fractal geometry. Special attention is paid to spectral properties of fractal (pseudo)differential operators; in particular we shall play the drum with a fractal layer.

In some sense this book may be considered as the fractal twin of [ET96], where we developed adequate methods to handle spectral problems of degenerate pseudodifferential operators in $\mathbb{R}^n$ and in bounded domains. Besides a few special properties of function spaces we relied there on sharp estimates of entropy numbers of compact embeddings between these spaces and their relations to the distribution of eigenvalues. Some of the main assertions of the present book are based on just these techniques but now in a fractal setting. Since virtually nothing of these new methods is available in literature, a substantial part of what we have to say deals with recent developments in the theory of function spaces, also for their own sake. In this respect the book might also be considered as a continuation of [Tri83] and [Tri92].

We give a brief description of the contents of the book. Chapter I deals with fractals in $\mathbb{R}^n$. We present the basic material without proofs following closely [Fal85] and [Mat95]. Isotropic, anisotropic and nonisotropic d-sets and related self-affine fractals are discussed in greater details. Chapter II is devoted to entropy numbers in weighted l_p-spaces with a dyadic block structure. In Chapter III we introduce the function spaces $B_{pq}^s(\mathbb{R}^n)$ and $F_{pq}^s(\mathbb{R}^n)$ referring mainly to [Tri83, Tri92, ET96] for details as far as basic facts are concerned. On the other hand we give a new proof of atomic representations in these spaces. The method used allows us to atomize the atoms, which results in subatomic (or quarkonial) decompositions which prove to be useful in connection with entropy numbers. This method works even for vector-valued function spaces. One obtains a total decoupling and decomposition in elementary building blocks which resembles the Taylor expansion of analytic functions. The final section of this chapter deals with the Hausdorff dimension of the graph of a real function $f(x)$ belonging to the Hölder spaces $\mathcal{C}^s(\mathbb{R}^n)$ with $0 < s < 1$, demonstrating the close connection between atomic representations of functions and methods of fractal geometry. Chapter IV is devoted to function spaces on and of fractals. First we characterize the Hausdorff dimension $\dim_H \Gamma$ of a Borel set Γ in $\mathbb{R}^n$ with $\dim_H \Gamma < n$ in terms of distributions and function spaces. Furthermore spaces $L_p(\Gamma)$ and Besov spaces $\mathbb{B}_{pq}^s(\Gamma)$, $s > 0$, on compact fractals Γ are studied, especially their relations to the spaces $B_{pq}^\sigma(\mathbb{R}^n)$ and the asymptotics of the entropy numbers of compact embeddings between them. On that basis we introduce «regular» elliptic operators on compact d-sets via quadratic forms and study the distribution of their eigenvalues. The final Chapter V deals with spectra of fractal pseudodifferential operators of several types. Let

$$B = b\,a(\cdot,D), \quad a(\cdot,D) \in \Psi_{1,\varrho}^{-\varkappa}, \quad \varkappa > 0, \quad 0 \le \varrho \le 1,$$

(Hörmander classes) and

$$b \in \mathbb{B}_{pq}^{s}(\mathbb{R}^{n}), \quad s < 0, \quad supp\, b \quad \text{compact},$$

(pseudodifferential operators with fractal coefficients). These operators are compact in suitable spaces (under some restrictions on the parameters involved, in particular $\varkappa > 2|s|$) and one obtains for the eigenvalues μ_k (counted with respect to their algebraic multiplicities) the estimate

$$|\mu_k| \le c\, k^{-\frac{\varkappa - 2|s|}{n}}, \quad k \in \mathbb{N}.$$

For fractal pseudodifferential operators of type

$$B = b_1\, a(\cdot, D)\, b_2, \quad a(\cdot,D) \in \Psi_{1,\varrho}^{-\varkappa}, \quad \varkappa > 0, \quad 0 \le \varrho \le 1,$$

where $b_1 \in L_{r_1}(\Gamma)$, $b_2 \in L_{r_2}(\Gamma)$ and Γ is a compact d-set with $0 < d < n$, we obtain the estimate

$$|\mu_k| \le c\, k^{-\frac{\varkappa - n + d}{d}}, \quad k \in \mathbb{N},$$

for the distribution of the eigenvalues μ_k (counted with respect to their algebraic multiplicities) in suitable spaces (again under some restrictions on the parameters involved, in particular $\varkappa - n + d > d(\frac{1}{r_1} + \frac{1}{r_2})$). The exponent of k is sharp. In several (self-adjoint) cases one has even

$$c_1 k^{-\frac{\varkappa - n + d}{d}} \le |\mu_k| \le c_2 k^{-\frac{\varkappa - n + d}{d}}, \quad k \in \mathbb{N}, \tag{$*$}$$

for some c_1, c_2 with $0 < c_1 < c_2 < \infty$. Compared with the «Weyl exponent» $\frac{\varkappa}{n}$ one must replace n by d and one has the additional *fractal defect* $n - d$. Hence one obtains both sub-Weylian and super-Weylian behaviour. In particular one can play the (n-dimensional) drum, given by a smooth bounded domain Ω in $\mathbb{R}^n$, with a compact fractal layer Γ, where $\Gamma \subset \Omega$. Let Γ be an (isotropic, anisotropic, or nonisotropic) d-set; then the corresponding operator looks like

$$(-\Delta)^{-1} \circ tr^{\Gamma}$$

where tr^{Γ} is closely connected with the trace operator tr_{Γ} and $-\Delta$ stands for the Dirichlet Laplacian with respect to Ω. In case of isotropic d-sets one has $(*)$ with $\varkappa = 2$ provided that $n - 2 < d < n$. Even more interesting are anisotropic and nonisotropic d-sets in the plane (the music of the ferns). In that cases two-sided estimates for the μ_k's are given. At the end of this chapter we deal briefly with Schrödinger operators having fractal potentials and with some related nonlinear problems.

It is a pleasure to acknowledge the help I have received from Dorothee Haroske concerning the use of LaTeX. I am especially indebted to her and to Mircea Malarski for producing all the figures in this book. Last, but not least, I wish to thank David Edmunds who looked through the whole manuscript and offered many comments.

Jena, Winter 1996/7

Chapter I
Fractals

1 Basic notation in measure theory

1.1 Basic notation

Let $\mathbb{R}^n$ be euclidean n-space. We collect some basic notation and fundamental facts about measures on sets in $\mathbb{R}^n$. More details may be found in [Fal85] and [Mat95], and the references given there. Otherwise we assume that the reader is familiar with measure and integration theory. We follow [Mat95] (see also [Fed96], p. 53), by calling *measure* what is often called *outer measure*. The reader must be well aware of this in what follows.

Let $X \subset \mathbb{R}^n$ be an arbitrary set and let $[0, \infty]$ be the extended half-axis including 0 and ∞. A set function $\mu : \{A : A \subset X\} \to [0, \infty]$ on X is called a *measure* if

$$\mu(\emptyset) = 0,$$

$$\mu(A) \le \mu(B) \qquad \text{when} \qquad A \subset B \subset X,$$

$$\mu(\bigcup_{j=1}^{\infty} A_j) \le \sum_{j=1}^{\infty} \mu(A_j) \quad \text{when} \quad A_k \subset X \quad \text{for all} \quad k \in \mathbb{N}.$$

Here $\mathbb{N}$ stands for the *natural numbers*. As usual a set $A \subset X$ is called μ-*measurable* if

$$\mu(E) = \mu(E \cap A) + \mu(E \setminus A) \text{ for all } E \subset X.$$

The collection M of the μ-measurable sets is a σ-*algebra*; that means,

$$\emptyset \in M \quad \text{and} \quad X \in M,$$

$$\text{if} \quad A \in M \quad \text{then} \quad X \setminus A \in M,$$

$$\text{if} \quad A_k \in M \quad \text{for all} \quad k \in \mathbb{N}, \quad \text{then} \quad \bigcup_{j=1}^{\infty} A_j \in M.$$

Furthermore if $\mu(A) = 0$ for $A \subset X$, then $A \in M$. Recall the σ-*additivity*, that means

$$\mu(\bigcup_{j=1}^{\infty} A_j) = \sum_{j=1}^{\infty} \mu(A_j)$$

if the sets $A_j \in M$ are pairwise disjoint.

H. Triebel, *Fractals and Spectra*, Modern Birkhäuser Classics,
DOI 10.1007/978-3-0348-0034-1_1, © Birkhäuser Verlag 1997

Let $r > 0$ then

$$B(x,r) = \{y \in \mathbb{R}^n : |y - x| \le r\} \tag{1.1}$$

are the *closed balls* in $\mathbb{R}^n$ centred at $x \in \mathbb{R}^n$ and with radius r.

A measure μ on X is called *locally finite* if for every $x \in X$ and every $r > 0$

$$\mu(X \cap B(x,r)) < \infty.$$

1.2 Borel and Radon measures

Let X be a set in $\mathbb{R}^n$. A subset of X is called open (closed) if it is the intersection of an open (closed) set in $\mathbb{R}^n$ with X. As usual a set in X is called a *Borel set* if it belongs to the smallest σ-algebra containing the open subsets of X.

A measure μ on X is said to be a *Borel measure* if all Borel sets are μ-measurable.

A measure μ on X is called *Borel regular* if it is a Borel measure and if for any $A \subset X$ there is a Borel set $B \subset X$ such that $A \subset B$ and $\mu(A) = \mu(B)$.

A measure μ on X is called a *Radon measure* if it is a Borel measure and

$$\mu(K) < \infty \quad \text{for any compact set} \quad K \subset X,$$
$$\mu(V) = \sup\{\mu(K) : K \subset V, K \text{ compact}\} \quad \text{for open sets} \quad V \subset X,$$
$$\mu(A) = \inf\{\mu(V) : A \subset V \subset X, \ V \text{ open}\} \quad \text{for} \quad A \subset X.$$

1.3 Restrictions to subsets

Let X be a set in $\mathbb{R}^n$, let μ be a measure on X, and let $A \subset X$. The *restriction* of μ to A, denoted by $\mu|A$, is given by

$$(\mu|A)(B) = \mu(A \cap B) \quad \text{where} \quad B \subset X.$$

Then $\mu|A$ is a measure and any μ-measurable set is also $\mu|A$-measurable. If μ is Borel regular and if A is μ-measurable with $\mu(A) < \infty$, then $\mu|A$ is Borel regular.

1.4 Proposition *A measure μ on $\mathbb{R}^n$ is a Radon measure if, and only if, it is locally finite and Borel regular.*

As for a proof and further information see [Mat95], pp. 11, 12.

2   Hausdorff measures and Hausdorff dimensions

2.1 Coverings

Again we follow [Mat95], p. 54, closely. Let E be a non-empty set in $\mathbb{R}^n$; then its diameter is given by

$$diam\, E = \sup\{|x - y| : x \in E, y \in E\}.$$

Let $\mathscr{F}$ be a family of subsets of $\mathbb{R}^n$ and let ζ be a non-negative function on $\mathscr{F}$ with the properties:

(i) For every $\delta > 0$ there are $E_j \in \mathscr{F}$ such that $\mathbb{R}^n = \bigcup_{j=1}^{\infty} E_j$ and *diam* $E_j \leq \delta$.

(ii) For every $\delta > 0$ there exists an $E \in \mathscr{F}$ such that $\zeta(E) \leq \delta$ and *diam* $E \leq \delta$.

For $0 < \delta < \infty$ and $A \subset \mathbb{R}^n$ we define

$$\psi_\delta(A) = \inf\{\sum_{j=1}^{\infty} \zeta(E_j) : A \subset \bigcup_{j=1}^{\infty} E_j, \ diam\, E_j \leq \delta, \ E_j \in \mathscr{F}\}. \qquad (2.1)$$

Of course, $\psi_\delta(A)$ is monotone,

$$\psi_\delta(A) \leq \psi_\varepsilon(A), \quad \text{when} \quad 0 < \varepsilon < \delta < \infty,$$

and hence, $\psi = \psi(\mathscr{F}, \zeta)$, given by

$$\psi(A) = \lim_{\delta \to 0} \psi_\delta(A) = \sup_{\delta > 0} \psi_\delta(A), \quad A \subset \mathbb{R}^n, \qquad (2.2)$$

makes sense. We have $0 \leq \psi(A) \leq \infty$. By (ii) it follows that the countable coverings of A in the definition of $\psi_\delta(A)$ can be replaced by coverings which are *either* finite *or* countable.

2.2 Theorem

(i) ψ *is a Borel measure on* $\mathbb{R}^n$.

(ii) *If the members of* $\mathscr{F}$ *are Borel sets, then* ψ *is a Borel regular measure on* $\mathbb{R}^n$.

See [Mat95], p. 55, for a proof.

2.3 Theorem *Let the members of* $\mathscr{F}$ *be Borel sets. Let A be ψ-measurable with $\psi(A) < \infty$. Then $\psi|A$ is a Radon measure.*

Proof. This assertion follows immediately from Theorem 2.2, 1.3, and Proposition 1.4.

2.4 Hausdorff measures

The measures ψ constructed in 2.1–2.3 will be called *generalized Hausdorff measures*. Let $0 \leq d < \infty$. The *d-dimensional Hausdorff measure* $\mathcal{H}^d$ on $\mathbb{R}^n$ is given by

$$\mathscr{F} = \{E : E \subset \mathbb{R}^n \text{ closed}\},$$
$$\zeta(E) = (diam\, E)^d,$$

with $0^0 = 1$ and $(diam\, \emptyset)^d = 0$,

$$\mathcal{H}^d_\delta(A) = \inf\{\sum_j (diam\, E_j)^d : A \subset \bigcup E_j,\ diam\, E_j \leq \delta\} \qquad (2.3)$$

and

$$\mathcal{H}^d(A) = \lim_{\delta \to 0} \mathcal{H}^d_\delta(A) = \sup_{\delta > 0} \mathcal{H}^d_\delta(A), \quad A \subset \mathbb{R}^n. \qquad (2.4)$$

These are the classical Hausdorff measures. According to the end of 2.1 the covering $\bigcup E_j$ is either finite or countable. Extensive treatments may be found in [Fal85,90], [Mat95], and the references given there. We assume that the reader is familiar with the basic facts of this theory.

By Theorem 2.3 it follows that $\mathcal{H}^d|A$ is a Radon measure if A is $\mathcal{H}^d$-measurable and $\mathcal{H}^d(A) < \infty$.

2.5 Remark If one replaces $\mathscr{F}$ in 2.4 by the collection of all (open or closed) convex sets in $\mathbb{R}^n$ then the resulting measure is again $\mathcal{H}^d$. If one replaces $\mathscr{F}$ in 2.4 by the collection of all (open or closed) balls in $\mathbb{R}^n$ then one obtains the *d-dimensional spherical measure* $\mathcal{S}^d$, and

$$\mathcal{H}^d(A) \leq \mathcal{S}^d(A) \leq 2^d \mathcal{H}^d(A) \quad \text{for any} \quad A \subset \mathbb{R}^n.$$

In this book we are only interested in whether $\mathcal{H}^d(A)$ is 0, positive and finite, or infinite. Then we can identify $\mathscr{F}$ with the collection of all sets, all convex sets, all cubes, or all balls, either open or closed.

2.6 Definition *The Hausdorff dimension of a set* $X \subset \mathbb{R}^n$ *is*

$$\begin{aligned} \dim_H X &= \sup\{s : \mathcal{H}^s(X) > 0\} = \sup\{s : \mathcal{H}^s(X) = \infty\} \\ &= \inf\{t : \mathcal{H}^t(X) < \infty\} = \inf\{t : \mathcal{H}^t(X) = 0\}. \end{aligned} \qquad (2.5)$$

Obviously, the interpretation of (2.5) is that the two suprema and the two infima always coincide and that $\dim_H(X)$ is their common value. Recall that

$$0 \leq \dim_H X \leq n. \qquad (2.6)$$

Let $s < \dim_H X < t$; then

$$\mathcal{H}^s(X) = \infty \quad \text{and} \quad \mathcal{H}^t(X) = 0. \qquad (2.7)$$

2.7 Theorem *Let E be a Borel set in $\mathbb{R}^n$ with $\mathcal{H}^d(X) = \infty$. Then there exists a compact subset F of E such that $\mathcal{H}^d(F) > 0$ and*

$$\mathcal{H}^d(B(x,r) \cap F) \leq b\, r^d, \quad x \in \mathbb{R}^n, \quad r \leq 1, \qquad (2.8)$$

for some $b > 0$, where $B(x,r)$ is the ball given by (1.1).

This is a crucial theorem in fractal geometry and also in the context of this book. It is closely related to Frostman's lemma. The above formulation essentially coincides with [Fal85], pp. 69/70. See also [Mat95], p. 112.

3   *d*-sets

3.1 Definition *Let* $n \in \mathbb{N}$, *let* Γ *be a set in* $\mathbb{R}^n$, *and let* $0 \leq d \leq n$. *Then* Γ *is called a d-set if there exists a Borel measure* μ *in* $\mathbb{R}^n$ *with the following two properties:*

(i) *supp* $\mu = \Gamma$;
(ii) *there are two constants* $c_1 > 0$ *and* $c_2 > 0$ *such that for all* $\gamma \in \Gamma$ *and all* r *with* $0 < r < 1$,

$$c_1 r^d \leq \mu(B(\gamma, r) \cap \Gamma) \leq c_2 r^d. \qquad (3.1)$$

3.2 Remark Again $B(\gamma, r)$ is the ball (1.1) with γ in place of x. The set Γ is closed, μ is locally finite, by 2.3 (in connection with 3.4 below) it is Radon, and hence it can be interpreted by the Riesz representation theorem (see [Mat95], p. 15) as a tempered distribution in $\mathbb{R}^n$. Furthermore the supports *supp* μ of μ on the one hand as a measure and on the other hand as the related distribution coincide.

3.3 Remark Our notation of a d-set differs from that in fractal geometry, see [Fal85], p. 8, but it coincides with [JoW84]. See also [Mat95], p. 92, and the references given there. If Γ is a d-set then there is essentially only one Borel measure with the required properties. To explain what this means we first recall that two Borel measures μ_1 and μ_2 in $\mathbb{R}^n$ are called equivalent if there are two positive constants c_1 and c_2 such that

$$c_1 \mu_1(A) \leq \mu_2(A) \leq c_2 \mu_1(A) \quad \text{for any Borel set } A \subset \mathbb{R}^n. \qquad (3.2)$$

3.4 Theorem *Let* Γ *be a d-set according to Definition* 3.1 *with the Borel measure* μ. *Then* μ *is equivalent to* $\mathcal{H}^d|\Gamma$.

Proof.
Step 1. Let $\gamma \in \Gamma$, $0 < r < 1$, and $\Gamma(\gamma, r) = \Gamma \cap B(\gamma, r)$. Let $\Gamma(\gamma, r)$ be covered by balls $\{K_j\}_{j \in I}$ of radius r_j, where I is either finite or $I = \mathbb{N}$. Then we have

$$c_1 r^d \leq \mu(\Gamma(\gamma, r)) \leq \sum_j \mu(K_j) \leq c_2 \sum_j r_j^d.$$

By Remark 2.5 and a suitable choice of K_j we obtain

$$c_1 r^d \leq \mu(\Gamma(\gamma, r)) \leq c_3 \mathcal{H}^d(\Gamma(\gamma, r)),$$

where c_3 is independent of γ and r.

Step 2. We cover $\Gamma(\gamma, r)$ by finitely many balls K_j of radius ρ and centred at $\Gamma(\gamma, r)$ where $0 < \rho < r$. We may assume that any $y \in \mathbb{R}^n$ belongs to at most P such balls, where P is independent of ρ (but depends on n). Then we have by 2.4

$$\mathcal{H}_\rho^d(\Gamma(\gamma, r)) \leq \sum_j \rho^d \leq c_1 \sum_j \mu(K_j) \leq c_1 P \, \mu(\Gamma(\gamma, 2r)) \leq c_2 r^d,$$

where c_2 is independent of γ, r and ρ. In particular $\rho \to 0$ yields

$$\mathcal{H}^d(\Gamma(\gamma, r)) \leq c\, r^d.$$

Together with Step 1 and measure-theoretical arguments we obtain easily the desired equivalence according to (3.2).

3.5 Remark Up to equivalence constants which are independent of $\gamma \in \Gamma$ and $0 < r < 1$, the Hausdorff measure $\mathcal{H}^d(\Gamma(\gamma, r))$ can be obtained by optimal coverings of $\Gamma(\gamma, r)$ by balls of the same radius ρ and $\rho \to 0$. This follows from the theorem and Step 2. In other words, in the case of d-sets the *Minkowski content* is equivalent to the Hausdorff measure and, in particular, the *Minkowski dimension* coincides in that case with the Hausdorff dimension. As for details about Minkowski contents and Minkowski dimension we refer to [Mat95], pp. 76–81.

3.6 Corollary *Let Γ be a d-set according to Definition 3.1 with $0 < d < n$. Then $|\Gamma| = 0$, where $|\Gamma|$ stands for the Lebesgue measure,*

$$\dim_H \Gamma = d, \tag{3.3}$$

and μ is a Radon measure.

Proof. By Theorem 3.4 we may identify μ with $\mathcal{H}^d|\Gamma$. The assertion about the dimension follows from 2.6. The σ-additivity of measures and $d < n$ imply $|\Gamma| = 0$. The last statement follows from Theorem 2.3 and 1.2 with standard arguments in measure theory if $\mu(\Gamma) = \infty$.

3.7 Notation Let Γ be a d-set and let $0 < p \leq \infty$. Then $L_p(\Gamma)$ are the usual complex L_p-spaces with respect to $\mu = \mathcal{H}^d|\Gamma$, quasi-normed via

$$\|f|L_p(\Gamma)\| = \left(\int_\Gamma |f(\gamma)|^p\, \mu(d\gamma) \right)^{1/p}, \quad 0 < p < \infty,$$

(modification if $p = \infty$). We write μ instead of $\mu|\Gamma$ if it is clear from the context what is meant. Of course $L_p(\Gamma)$ are quasi-Banach spaces (Banach spaces if $p \geq 1$). Furthermore let $D|\Gamma$ be the restriction of $D(\mathbb{R}^n) = C_0^\infty(\mathbb{R}^n)$ to Γ. Here as usual, $D(\mathbb{R}^n)$ is the collection of all infinitely differentiable complex-valued functions on $\mathbb{R}^n$ with compact support.

3.8 Theorem *Let Γ be a d-set according to Definition 3.1 and let $0 < p < \infty$. Then $D|\Gamma$ is dense in $L_p(\Gamma)$.*

Proof. By standard arguments it is sufficient to approximate the characteristic function $\chi_\Lambda(\gamma)$ of a μ-measurable set $\Lambda \subset \Gamma$ with $\mu(\Lambda) < \infty$ in the desired way. We may assume that Λ is bounded in $\mathbb{R}^n$. By Corollary 3.6 the measure μ is Radon. By 1.2 and the approximation theorem for Radon measures (see [Mat95], Theorem 1.10 on p. 11) it follows that for any $\varepsilon > 0$ there exists a closed set $K(\varepsilon)$ and an open set $V(\varepsilon)$ with

$$K(\varepsilon) \subset \Lambda \subset V(\varepsilon) \subset \Gamma, \quad \mu(V(\varepsilon)) - \mu(K(\varepsilon)) \le \varepsilon.$$

Then the C^∞ functions φ_ε on $\mathbb{R}^n$ with $0 \le \varphi_\varepsilon \le 1$,

$$(\varphi_\varepsilon|\Gamma)(\gamma) = 1 \quad \text{if} \quad \gamma \in K(\varepsilon) \quad \text{and} \quad supp\, \varphi_\varepsilon|\Gamma \subset V(\varepsilon)$$

provide the desired approximation.

4 Self-affine fractals

4.1 Similarities and contractions
First we introduce some notation following again [Fal85] and [Mat95], see also the original paper by Hutchinson [Hut81]. A mapping $\psi : \mathbb{R}^n \to \mathbb{R}^n$ is called a *contraction* if

$$|\psi(x) - \psi(y)| \le r\,|x - y|, \quad x \in \mathbb{R}^n, \quad y \in \mathbb{R}^n, \quad \text{some } r,\ 0 < r < 1. \quad (4.1)$$

A contraction is called a *similarity* (similitude) if

$$|\psi(x) - \psi(y)| = r\,|x - y|, \quad x \in \mathbb{R}^n, \quad y \in \mathbb{R}^n, \quad \text{some } r,\ 0 < r < 1.$$

Any similarity is an affine map and can be written as

$$\psi(x) = y + r\,g(x), \quad x \in \mathbb{R}^n, \quad (4.2)$$

for some $y \in \mathbb{R}^n$ and $g \in O(n)$ (the group of all rotations in $\mathbb{R}^n$ and all rotations combined with a reflection on a suitable hyper-plane). To see this we put $g(x) = \frac{1}{r}(\psi(x) - \psi(0))$. We have $g(0) = 0$ and

$$|g(x) - g(y)| = |x - y|, \quad x \in \mathbb{R}^n, \quad y \in \mathbb{R}^n.$$

In particular, $g(x)$ is norm-preserving and by

$$
\begin{aligned}
2(g(x), g(y)) &= |g(x)|^2 + |g(y)|^2 - |g(x) - g(y)|^2 \\
&= |x|^2 + |y|^2 - |x - y|^2 \\
&= 2(x, y)
\end{aligned}
$$

also scalar product preserving. In particular if $\{x^j\}_{j=1}^n$ is orthonormal in $\mathbb{R}^n$, then $\{g(x^j)\}_{j=1}^n$ is also orthonormal and we have

$$
\begin{aligned}
g(x) &= \sum_{j=1}^n (g(x), g(x^j))\, g(x^j) \\
&= \sum_{j=1}^n (x, x^j)\, g(x^j),
\end{aligned}
$$

which proves that g is linear. Hence $g \in O(n)$ and ψ can be represented by (4.2).

Let $\{\psi_j\}_{j=1}^N$ be N contractions in $\mathbb{R}^n$ and let Γ be a non-empty compact set in $\mathbb{R}^n$. We put

$$
\psi(\Gamma) = \bigcup_{j=1}^N \psi_j(\Gamma), \quad (\psi^0(\Gamma) = \Gamma, \quad \psi^1 = \psi), \tag{4.3}
$$

$$
\psi^k(\Gamma) = \psi(\psi^{k-1}(\Gamma)) = \bigcup_{1 \le j_l \le N} \psi_{j_1} \circ \cdots \circ \psi_{j_k}(\Gamma), \quad k \in \mathbb{N}, \tag{4.4}
$$

and

$$
\psi^\infty(\Gamma) = \lim_{k \to \infty} \psi^k(\Gamma), \tag{4.5}
$$

where this limit stands for the collection of all points x for which there is a sequence $x^k \in \psi^k(\Gamma)$ with $x^k \to x$ if $k \to \infty$ or, what is the same, the limit element of $\psi^k(\Gamma)$ in the metric space of all non-empty compact sets in $\mathbb{R}^n$ equipped with the Hausdorff metric, see [Fal85], p. 37, for details.

4.2 Theorem *Let $\{\psi_j\}_{j=1}^N$ be N contractions in $\mathbb{R}^n$.*

(i) *There exists a unique non-empty compact set Γ such that*

$$
\Gamma = \psi(\Gamma), \quad \text{(the invariant set)}
$$

(ii) *Let Λ be an arbitrary non-empty compact set in $\mathbb{R}^n$. Then holds*

$$
\Gamma = \psi^\infty(\Lambda).
$$

4.3 Remark Both parts of the theorem follow easily from Schauder's fixed point theorem in the complete metric space of all non-empty compact sets in $\mathbb{R}^n$ equipped with the Hausdorff metric, see for details [Fal85], pp. 119–120. In particular, Γ can be generated if one starts with an arbitrary point $x \in \mathbb{R}^n$ and looks for the accumulation set of all its iterations $\psi_{j_1} \circ \cdots \circ \psi_{j_k}\{x\}$,

$$
\Gamma = \psi^\infty(\{x\}).
$$

4.4 Corollary *The invariant set* Γ *in Theorem* 4.2 *is the closure of the set of all fixed points of all maps* $\psi_{j_1} \circ \cdots \circ \psi_{j_k}$ *where* $k \in \mathbb{N}$ *and* $1 \le j_l \le N$.

Proof. This is an easy consequence of Remark 4.3. (Recall that by Banach's theorem any contraction has a unique fixed point.)

4.5 Definition *Let* $\{\psi_j\}_{j=1}^N$ *be similarities in* $\mathbb{R}^n$ *according to* 4.1 *with*

$$|\psi_j(x) - \psi_j(y)| = r_j\, |x - y|, \quad 0 < r_j < 1, \tag{4.6}$$

and let Γ *be the corresponding invariant set according to Theorem* 4.2 *with Hausdorff dimension* $d = \dim_H \Gamma$.

(i) *Then* Γ *is called self-similar if*

$$\mathcal{H}^d(\psi_j(\Gamma) \cap \psi_k(\Gamma)) = 0 \quad \text{for} \quad j \neq k.$$

(ii) $\{\psi_j\}_{j=1}^N$ *satisfies the open set condition if there is a non-empty open set* O *such that*

$$\psi(O) = \bigcup_{j=1}^N \psi_j(O) \subset O \quad \text{and} \quad \psi_j(O) \cap \psi_k(O) = \emptyset \quad \text{for} \quad j \neq k. \tag{4.7}$$

4.6 Remark The meaning of both parts of the definition is quite clear. One wishes to avoid too much overlapping if one applies different ψ_j's to Γ or to O. Whereas part (i) might be somewhat complicated to be checked, it turns out that part (ii) is easy to handle in concrete cases, even with the additional property $O \cap \Gamma = \emptyset$, which proves to be very useful for our purposes.

4.7 Theorem *Let* $\{\psi_j\}_{j=1}^N$ *be similarities satisfying the open set condition in Definition* 4.5(ii). *Let* $d \ge 0$ *be the unique number with*

$$\sum_{j=1}^N r_j^d = 1$$

(similarity dimension). Then the corresponding invariant set Γ *is self-similar and it is a compact* d*-set according to Definition* 3.1.

4.8 Remark This is one of the key assertions of the theory of self-similar fractals. We refer to [Fal85], pp. 118–124, [Mat95], pp. 65–69, [Hut81], and the references given there. In particular it follows by Corollary 3.6 that

$$d = \dim_H \Gamma,$$

including now $d = n$ and $d = 0$.

4.9 The Cantor set

To illustrate the definitions and theorems described so far we look at the Cantor set in $\mathbb{R}^n$. Let $Q = [0, 1]^n$ be the closed cube with side-length 1; each side will be trisected and the inner cube q with side-length $1/3$ will be taken out.

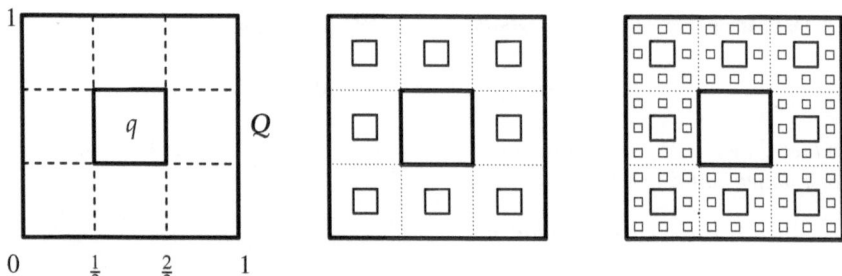

Fig. 4.1

The process will be repeated with the remaining $3^n - 1$ cubes with side-length $1/3$. Iteration ends up with the self-similar fractal Γ, see Fig. 4.1. This procedure can also be described via $3^n - 1 = N$ similarities ψ_j with the dilation factor $1/3$ transforming Q in the mentioned $3^n - 1$ cubes with side-length $1/3$. The open set condition is satisfied. One can take, for instance,

$$O_1 = \overset{\circ}{Q}, \quad \text{the interior of } Q, \tag{4.8}$$

or

$$O_2 = \bigcup_{k=0}^{\infty} \bigcup_{1 \le j_l \le N} \psi_{j_1} \circ \cdots \circ \psi_{j_k}(\overset{\circ}{q}), \tag{4.9}$$

where $\overset{\circ}{q}$ is the interior of the inner cube q in the first step of this procedure (or another open subset of q). The second choice has a big advantage for us later on, since

$$O_2 \cap \Gamma = \emptyset.$$

As for the Hausdorff dimension $d = \dim_H \Gamma$, according to 4.7 and 4.8 we have

$$(3^n - 1)\, 3^{-d} = 1, \quad \text{hence} \quad d = \frac{\log(3^n - 1)}{\log 3}. \tag{4.10}$$

4.10 Self-affine fractals: the set-up

So far we described in this Section 4 some basic assertions in fractal geometry. In the remaining subsections we discuss a few more special topics, especially anisotropic and nonisotropic fractals in the plane. Let Q be the closed unit square in $\mathbb{R}^2$, that is

$$Q = \{x = (x_1, x_2) : 0 \le x_1 \le 1, \, 0 \le x_2 \le 1\}, \tag{4.11}$$

and let $A = (A_l)_{l=1}^N$ be $N \geq 2$ affine maps of $\mathbb{R}^2$ onto itself, hence

$$A_l : x = (x_1, x_2) \mapsto (a_{1,1}^l x_1 + a_{1,2}^l x_2 + a_1^l, \ a_{2,1}^l x_1 + a_{2,2}^l x_2 + a_2^l), \qquad (4.12)$$

$$a_{1,1}^l a_{2,2}^l - a_{1,2}^l a_{2,1}^l \neq 0.$$

We assume that the maps A_l are contractions according to 4.1 and that

$$A_l Q \subset Q, \qquad (4.13)$$

$$A_l \overset{\circ}{Q} \cap A_m \overset{\circ}{Q} = \emptyset \quad \text{if} \quad l \neq m, \qquad (4.14)$$

where $\overset{\circ}{Q}$ is the interior of Q, the open unit square, and

$$\sum_{l=1}^N vol(A_l Q) < 1. \qquad (4.15)$$

The Cantor set in 4.9 is an isotropic example. Much as in 4.1 we put

$$A Q = (A Q)^1 = \bigcup_{l=1}^N A_l Q, \quad (A Q)^0 = Q,$$

$$(A Q)^k = A((A Q)^{k-1}) = \bigcup_{1 \leq j_l \leq N} A_{j_1} \cdots A_{j_k} Q, \quad k \in \mathbb{N}. \qquad (4.16)$$

This sequence of sets is monotonically decreasing. By Theorem 4.2 its limit

$$\Gamma = (A Q)^\infty = \bigcap_{k \in \mathbb{N}} (A Q)^k = \lim_{k \to \infty} (A Q)^k \qquad (4.17)$$

is the uniquely determined fractal generated by the contractions $(A_l)_{l=1}^N$. A typical example is shown in Fig. 4.2.

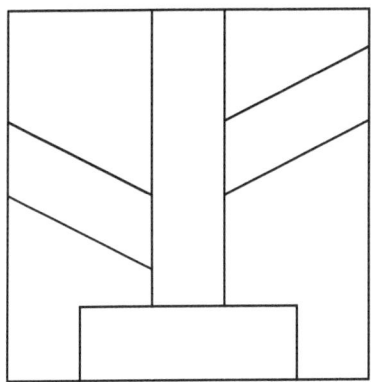 

Fig. 4.2

In particular it is clear that one can generate in that way fern-like fractals. There is a counterpart of the open set condition in 4.9 with $O_1 = \overset{\circ}{Q}$ according to Definition 4.5(ii). By (4.15) there is also a counterpart of O_2 in (4.9).

4.11 Remark: Ferns and their spores The self-affine fractals created in 4.10 look like ferns, at least if the maps involved are of the type as in Fig. 4.2, see also [Fal90], p. 133, for similar examples. Denoting the fixed points of all the maps $A_{j_1} \cdots A_{j_k}$ with $k \in \mathbb{N}$ and $1 \leq j_l \leq N$, as the *spores* of this *fern*, then it turns out that this

$$\text{fern is the closure of the set of its spores.} \tag{4.18}$$

This follows from Corollary 4.4.

4.12 Definition *Let* $A = (A_l)_{l=1}^N$ *be the* $N \geq 2$ *affine maps introduced in* 4.10. *The affine dimension of* $\Gamma = (A\,Q)^\infty$ *is the (uniquely determined) number* $d = \dim_A \Gamma$ *with*

$$\sum_{l=1}^N vol(A_l\,Q)^{\frac{d}{2}} = 1. \tag{4.19}$$

4.13 Proposition *Under the hypotheses of Definition* 4.12,

$$0 < \dim_A \Gamma < 2 \tag{4.20}$$

Proof. The assertion follows immediately from (4.15) and $N \geq 2$.

4.14 Remark If A_l is in addition a similarity according to (4.6), $l = 1, \ldots, N$, then $vol(A_l\,Q) = r_l^2$, and (4.19) coincides with the similarity dimension in Theorem 4.7 and hence, by Remark 4.8, we have

$$\dim_A \Gamma = \dim_H \Gamma \tag{4.21}$$

in this case. But in general this is not so and the two numbers $\dim_A \Gamma$ and $\dim_H \Gamma$ are completely unrelated, see 4.22.

4.15 Theorem *Let* $\Gamma = (A\,Q)^\infty$ *be the fractal introduced in* (4.17) *with the affine dimension* d *according* (4.19). *There is a uniquely determined Radon measure* μ *in* $\mathbb{R}^2$ *with* $supp\,\mu = \Gamma$ *and*

$$\mu(\Gamma \cap A_{l_1} \cdots A_{l_k}\,Q) = (vol\,A_{l_1} \cdots A_{l_k}\,Q)^{\frac{d}{2}}, \tag{4.22}$$

where $k \in \mathbb{N}$ *and* $1 \leq l_r \leq N$.

Proof. We use the following repeated mass distribution procedure. We start with $\mu(Q) = 1$ and distribute the mass via

$$\mu(A_l\,Q) = vol(A_l\,Q)^{\frac{d}{2}}, \quad l = 1, \ldots, N, \tag{4.23}$$

and subsequently

$$\mu(A_{l_1} \cdots A_{l_k}\,Q) = vol(A_{l_1} \cdots A_{l_k}\,Q)^{\frac{d}{2}}, \quad 1 \leq l_r \leq N. \tag{4.24}$$

Iteration of (4.19) yields

$$\sum_{l_1,\dots,l_k}^{1,\dots,N} vol(A_{l_1} \cdots A_{l_k} Q)^{\frac{d}{2}} = 1, \quad k \in \mathbb{N}. \tag{4.25}$$

Hence, by (4.16) and (4.17) the unit mass is finally distributed on Γ. We refer to [Fal90], pp. 13/14, in particular Proposition 1.7, as far as this technique is concerned. The result is the desired Radon measure μ. Without going in detail we outline the main idea. On the set

$$\{f \geq 0, f \text{ continuous on the compact set } \Gamma\}$$

we construct a positive linear functional Lf by following the above limit process in the same way as one introduces the Riemann integral via partial sums. By the Riesz representation theorem there is a uniquely determined Radon measure μ such that

$$Lf = \int_\Gamma f(\gamma) \mu(d\gamma),$$

see [Mat95], p. 15. It has the property (4.22).

4.16 A discussion
Let Ω be a bounded smooth domain in $\mathbb{R}^2$ and let $\Gamma = (A Q)^\infty \subset \Omega$ be the fractal according to Theorem 4.15 with the measure μ. Let $-\Delta$ be the Dirichlet Laplacian in Ω and let tr^μ, roughly speaking, be the pointwise multiplication operator with the above measure μ interpreted as a (singular) distribution. One of the aims of this book is the study of spectral properties of operators of the type

$$(-\Delta)^{-1} \circ tr^\mu \tag{4.26}$$

in suitable function spaces, then in an $\mathbb{R}^n$-setting (see Chapter V for details). Taking for granted the necessary explanation what is meant by operators of type (4.26) one can imagine that the different natures of the isotropic operator $-\Delta$ and the (in general, nonisotropic) measure μ cause additional difficulties. Our way to handle this problem is to cover $(A Q)^\infty$ or $(A Q)^k$ according to (4.17) and (4.16) by cubes (balls) of side-length (radius) 2^{-j} with $j \to \infty$ in an optimal way. This makes clear why we shall pay special attention in 4.17 to maps A_l according to (4.12)–(4.17) where $A_l Q$ are rectangles with sides parallel to the axes of coordinates creating PXT-fractals Γ.

Looking for other possibilities and having these difficulties in mind one might be tempted to replace the square in (4.11) by an (equilateral or slightly distorted) triangle or a circle, preserving the related counterparts of (4.13)–(4.15). The outcome is quite different from our point of view. Replacing Q in (4.11) by a circle

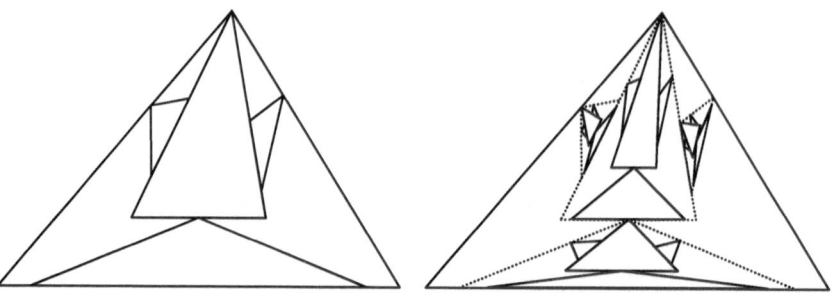

Fig. 4.3

one arrives at fractals which fit in our scheme. We return to this case in 4.19. On the other hand Fig. 4.3 shows the typical situation in the case of a triangle T.

One can imagine that (without additional assumptions about the maps A_l) an effective covering of $\Gamma = (AT)^\infty$ or $(AT)^k$ with squares of side-length 2^{-j}, $j \to \infty$, causes many difficulties.

In the subsections 4.17 and 4.19 we discuss those two cases of anisotropic and nonisotropic fractals which are of interest for us later on.

4.17 The anisotropic case: PXT-fractals
Let again Q be the unit square in (4.11). We specify the affine maps in (4.12) by

$$A_l : x = (x_1, x_2) \mapsto (\eta_1^l \, 2^{-a_1^l \, \varkappa_l} \, x_1, \; \eta_2^l \, 2^{-a_2^l \, \varkappa_l} \, x_2) + x^l, \qquad (4.27)$$

where again $l = 1, \dots, N$ with $N \geq 2$,

$$a_1^l > 0, \, a_2^l > 0 \quad \text{with} \quad a_1^l + a_2^l = 2, \quad \text{and} \quad 0 < \varkappa_l < \infty. \qquad (4.28)$$

Furthermore η_1^l and/or η_2^l are 1 or -1 (indicating possible reflections). This is the usual normalization in the theory of anisotropic function spaces, where $a^l = (a_1^l, a_2^l)$ with $a_1^l + a_2^l = 2$ is called the *anisotropicity*, see [SchT87], 4.2.1, p. 197. Otherwise we assume that (4.13), (4.14), and (4.15) hold, where the latter now reads as follows,

$$\sum_{l=1}^{N} 2^{-2 \varkappa_l} < 1. \qquad (4.29)$$

Let $\Gamma = (AQ)^\infty$, given by (4.17), be the resulting *anisotropic fractal*. We reserve the word *anisotropic* for use when the generating affine maps A_l have the special structure (4.27) (otherwise the resulting fractal will be called *nonisotropic*). In our understanding *isotropic fractals* are special *anisotropic fractals* and *anisotropic fractals* are special *nonisotropic fractals*. Fig. 4.4 is a specification of Fig. 4.2. We call Γ a PXT-fractal:

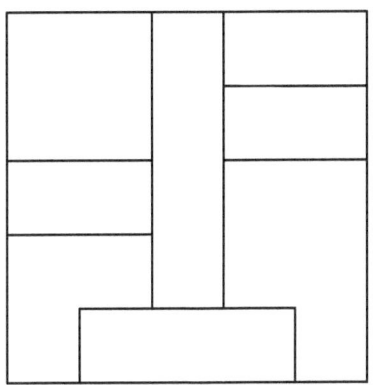 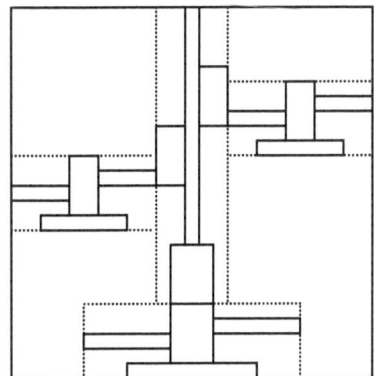

Fig. 4.4 PXT-fractal

it looks like a $\underline{X}$mas-$\underline{T}$ree, somewhat cubistic, hence $\underline{P}$ stands for Picasso. By
(4.19) the affine dimension $d = \dim_A \Gamma$ is now given by

$$\sum_{l=1}^{N} 2^{-d \varkappa_l} = 1. \tag{4.30}$$

4.18 Regular anisotropic fractals

Let $n_1 \in \mathbb{N}$ and $n_2 \in \mathbb{N}$ with $n_1 \geq 2$ and $n_2 \geq 2$; and let log be taken with respect
to the base 2. We specify (4.27)–(4.29) by

$$d_1^l = 2 \frac{\log n_1}{\log n_1 n_2}, \quad d_2^l = 2 \frac{\log n_2}{\log n_1 n_2}, \quad \varkappa_l = \frac{1}{2} \log n_1 n_2, \tag{4.31}$$

for $l = 1, \ldots, N$ with $N \geq 2$, and x^l in (4.27) is chosen such that we have
the situation as depicted in Fig. 4.5. The resulting fractal Γ is called *regular*

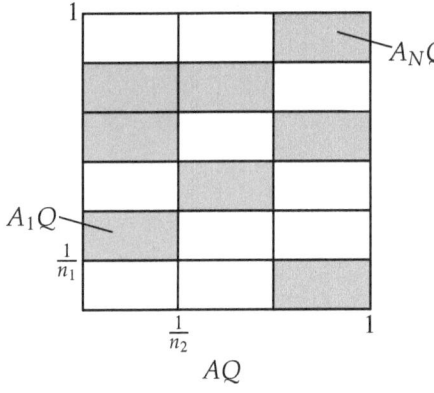 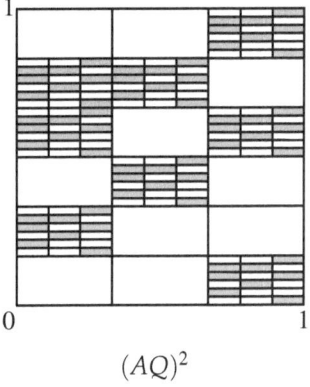

Fig. 4.5 Regular anisotropic fractal

$$T_1 = \begin{pmatrix} 0.60000 & 0.00000 \\ 0.00000 & 0.20000 \end{pmatrix} \times \begin{pmatrix} x \\ y \end{pmatrix} + \begin{pmatrix} 0.20000 \\ 0.00000 \end{pmatrix}$$

$$T_2 = \begin{pmatrix} 0.20000 & 0.00000 \\ 0.00000 & 0.80000 \end{pmatrix} \times \begin{pmatrix} x \\ y \end{pmatrix} + \begin{pmatrix} 0.40000 \\ 0.20000 \end{pmatrix}$$

$$T_3 = \begin{pmatrix} 0.00000 & -0.40000 \\ 0.20000 & 0.20000 \end{pmatrix} \times \begin{pmatrix} x \\ y \end{pmatrix} + \begin{pmatrix} 0.40000 \\ 0.30000 \end{pmatrix}$$

$$T_4 = \begin{pmatrix} 0.00000 & 0.40000 \\ -0.20000 & 0.20000 \end{pmatrix} \times \begin{pmatrix} x \\ y \end{pmatrix} + \begin{pmatrix} 0.60000 \\ 0.70000 \end{pmatrix}$$

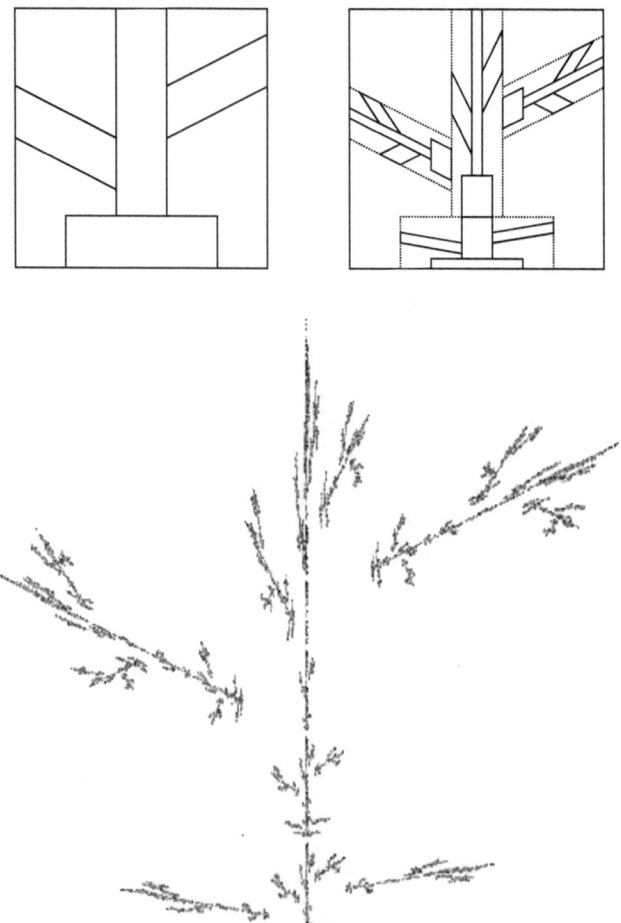

Fig. 4.2 (p. 11)

$$T_1 = \begin{pmatrix} 0.83333 & -0.10417 \\ 0.00000 & 0.25000 \end{pmatrix} \times \begin{pmatrix} x \\ y \end{pmatrix} + \begin{pmatrix} 0.10000 \\ 0.00000 \end{pmatrix}$$

$$T_2 = \begin{pmatrix} 0.33333 & 0.08333 \\ 0.00000 & 0.75000 \end{pmatrix} \times \begin{pmatrix} x \\ y \end{pmatrix} + \begin{pmatrix} 0.40000 \\ 0.20000 \end{pmatrix}$$

$$T_3 = \begin{pmatrix} 0.10000 & -0.08750 \\ 0.20000 & 0.07500 \end{pmatrix} \times \begin{pmatrix} x \\ y \end{pmatrix} + \begin{pmatrix} 0.46000 \\ 0.32000 \end{pmatrix}$$

$$T_4 = \begin{pmatrix} 0.01667 & 0.11042 \\ -0.16667 & 0.24583 \end{pmatrix} \times \begin{pmatrix} x \\ y \end{pmatrix} + \begin{pmatrix} 0.76000 \\ 0.48000 \end{pmatrix}$$

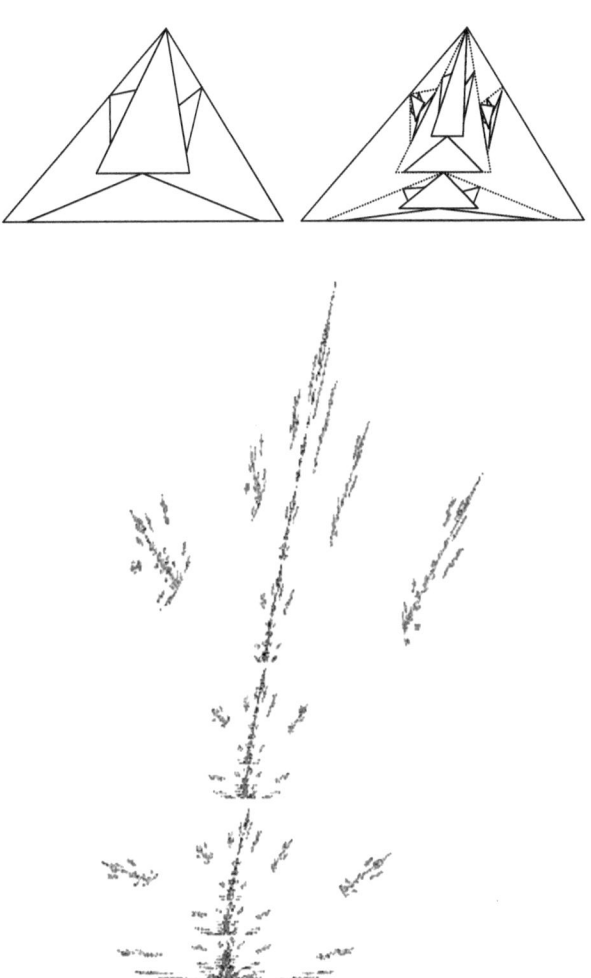

Fig. 4.3 (p. 14)

anisotropic. Of course $N < n_1 n_2$ by (4.29) and

$$N (n_1 n_2)^{-\frac{d}{2}} = 1 \tag{4.32}$$

by (4.30) for the affine dimension $d = \dim_A \Gamma$. In the isotropic case, that means in addition $n = n_1 = n_2 \geq 2$, we have $N n^{-d} = 1$. Hence by Theorem 4.7 and Remark 4.8 we obtain in that case

$$\dim_H \Gamma = \dim_A \Gamma = d. \tag{4.33}$$

But it turns out that for general regular anisotropic fractals the situation is quite different. We return to this problem in 4.21 and 4.22. However the outcome is even worse. As mentioned in [Fal90], pp. 128/129, $\dim_H \Gamma$ depends for the above general regular anisotropic case not only on the number N of rectangles, but also on their relative position. *This sheds some additional light on the difficulties in estimating the distribution of eigenvalues of operators of type (4.26) in those cases.*

4.19 The nonisotropic case: OF-fractals
At the end of our discussion in 4.16 we discarded fractals as depicted in Fig. 4.3 and favoured the anisotropic fractals described in 4.17 and 4.18. However there is a second, now genuine, nonisotropic case in $\mathbb{R}^2$ which fits in our scheme outlined in 4.16. About the use of the words *isotropic, anisotropic* and *nonisotropic* fractal or related d-set we refer to 4.17, and also to 5.2 and 5.3 below. For that purpose we replace the unit square Q in (4.11) by the circle

$$B = \{x \in \mathbb{R}^2 : |x| \leq \frac{1}{2}\}. \tag{4.34}$$

In particular $vol\, B < 1$. Let again $A = (A_l)_{l=1}^{N}$ be $N \geq 2$ contractive affine maps of $\mathbb{R}^2$ onto itself as described in (4.12). The counterparts of (4.13) and (4.14) are now given by

$$A_l B \subset B \tag{4.35}$$

and

$$A_l \overset{\circ}{B} \cap A_m \overset{\circ}{B} = \emptyset \quad \text{if} \quad l \neq m. \tag{4.36}$$

We construct in the same way as in (4.16) and (4.17) with B in place of Q the *nonisotropic fractal* $\Gamma = (A B)^\infty$. Fig. 4.6 is the counterpart of Fig. 4.2.

Of course, $A_l B$ are ellipses with diameters less than 1. By (4.35), (4.36), and $N \geq 2$ they do not fill B, which is the natural counterpart of (4.15). As in 4.12 one may introduce the affine dimension $\dim_A \Gamma$ of $\Gamma = (A B)^\infty$ as the (uniquely determined) number $d = \dim_A \Gamma$ with

$$\sum_{l=1}^{N} vol(A_l B)^{\frac{d}{2}} = (vol\, B)^{\frac{d}{2}}. \tag{4.37}$$

By the above observation we have again

$$0 < \dim_A \Gamma < 2. \tag{4.38}$$

We call Γ an OF-fractal (<u>O</u>val-<u>F</u>erny-fractal), where «Oval» reminds on the procedure and «Ferny» on the result. But maybe it looks more like a *cactus*.

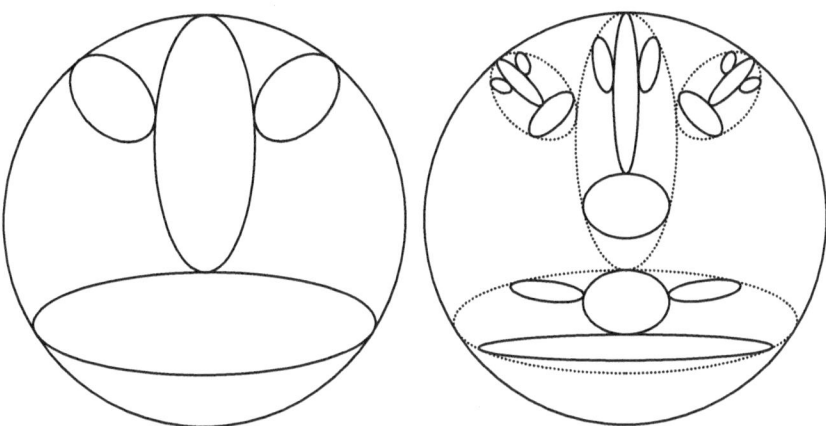

Fig. 4.6

4.20 Remark We return to our somewhat heuristic discussion in 4.16. One can imagine that both PXT-fractals and OF-fractals may be covered in an optimal way by squares (circles) of side-length (radius) 2^{-j} where $j \to \infty$. To do the same in the triangular case depicted in Fig. 4.3 one might first encircle the triangle T by a circle or an ellipse and follow this circle or ellipse in the course of the repeated mappings. But then it is quite clear that there is an uncontrollable overlapping. This clarifies the difference between the triangular case and the above OF-fractals. The OF-fractals are rather *airy* sets whereas the fractals generated in the triangular case are more massive.

4.21 Anisotropic curves
We specify the regular anisotropic fractals discussed in 4.18. As in (4.27), (4.28), and (4.31) we assume

$$a_1 > 0, \quad a_2 > 0 \quad \text{with} \quad a_1 + a_2 = 2, \quad \varkappa > 0, \tag{4.39}$$

such that

$$1 = n_1 \, 2^{-\varkappa a_1} = n_2 \, 2^{-\varkappa a_2} \quad \text{for some } n_1 \in \mathbb{N} \text{ and } n_2 \in \mathbb{N}. \tag{4.40}$$

In specification of the situation as shown in Fig. 4.5 we suppose now that in each column precisely one rectangle $A_l Q$ is located. It turns out that the two cases $0 < a_1 < 1$ and $1 < a_1 < 2$ are rather different, not to speak about $a_1 = 1 = a_2$. In addition, in the case $1 < a_1 < 2$, let

$$2^{\varkappa(a_1 - a_2)} = \frac{n_1}{n_2} = 2n + 1 \quad \text{for some } n \in \mathbb{N}, \tag{4.41}$$

be an odd natural number and let the rectangles $A_l Q$ be arranged as depicted in Fig. 4.7, where we always choose $\eta_2^l = 1$ in (4.27), and $\eta_1^l = 1$ in the first n_2

$$T_1 = \begin{pmatrix} 0.60000 & 0.00000 \\ 0.00000 & 0.20000 \end{pmatrix} \times \begin{pmatrix} x \\ y \end{pmatrix} + \begin{pmatrix} 0.20000 \\ 0.00000 \end{pmatrix}$$

$$T_2 = \begin{pmatrix} 0.20000 & 0.00000 \\ 0.00000 & 0.80000 \end{pmatrix} \times \begin{pmatrix} x \\ y \end{pmatrix} + \begin{pmatrix} 0.40000 \\ 0.20000 \end{pmatrix}$$

$$T_3 = \begin{pmatrix} 0.40000 & 0.00000 \\ 0.00000 & 0.20000 \end{pmatrix} \times \begin{pmatrix} x \\ y \end{pmatrix} + \begin{pmatrix} 0.00000 \\ 0.40000 \end{pmatrix}$$

$$T_4 = \begin{pmatrix} 0.40000 & 0.00000 \\ 0.00000 & 0.20000 \end{pmatrix} \times \begin{pmatrix} x \\ y \end{pmatrix} + \begin{pmatrix} 0.60000 \\ 0.60000 \end{pmatrix}$$

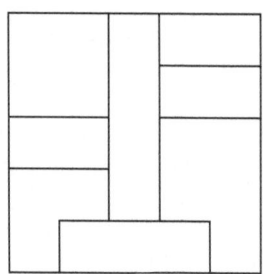 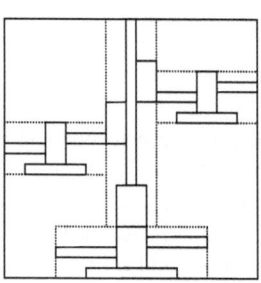

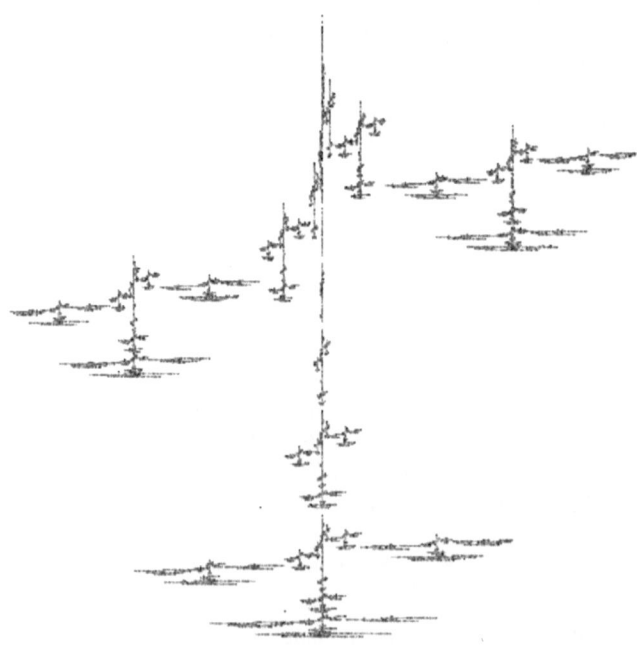

Fig. 4.4 (p. 15)

$$T_1 = \begin{pmatrix} 0.85233 & 0.00000 \\ 0.00000 & 0.25000 \end{pmatrix} \times \begin{pmatrix} x \\ y \end{pmatrix} + \begin{pmatrix} 0.00000 \\ -0.50000 \end{pmatrix}$$

$$T_2 = \begin{pmatrix} 0.25000 & 0.00000 \\ 0.00000 & 0.62500 \end{pmatrix} \times \begin{pmatrix} x \\ y \end{pmatrix} + \begin{pmatrix} 0.00000 \\ 0.37500 \end{pmatrix}$$

$$T_3 = \frac{\sqrt{2}}{2} \begin{pmatrix} -0.16667 & 0.25000 \\ 0.16667 & 0.25000 \end{pmatrix} \times \begin{pmatrix} x \\ y \end{pmatrix} + \begin{pmatrix} 0.46000 \\ 0.59000 \end{pmatrix}$$

$$T_4 = \frac{\sqrt{2}}{2} \begin{pmatrix} 0.16667 & -0.25000 \\ 0.16667 & 0.25000 \end{pmatrix} \times \begin{pmatrix} x \\ y \end{pmatrix} + \begin{pmatrix} -0.46000 \\ 0.59000 \end{pmatrix}$$

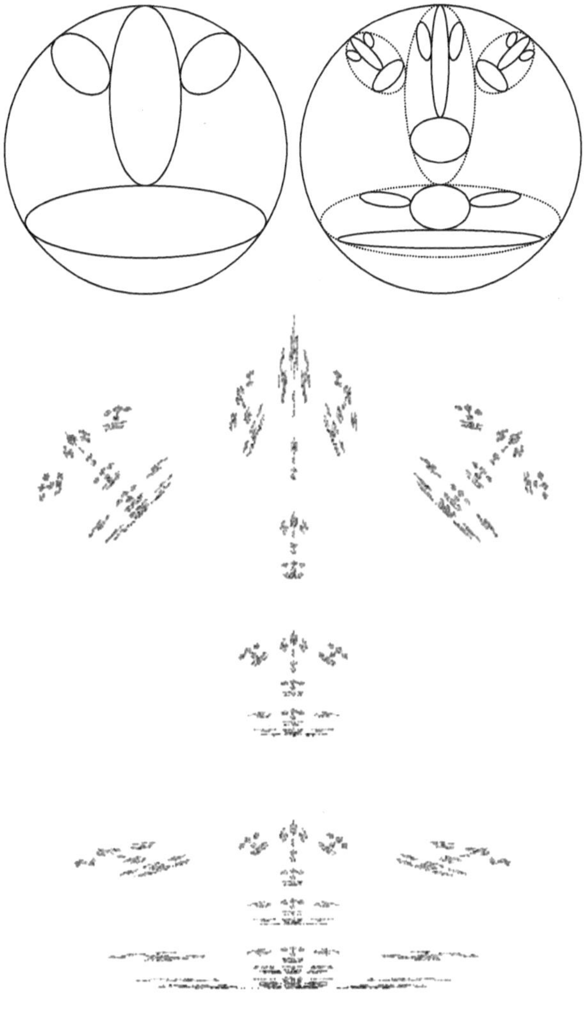

Fig. 4.6 (p. 19)

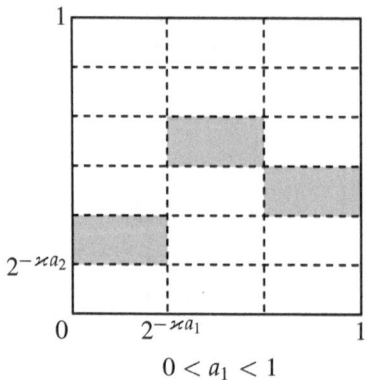
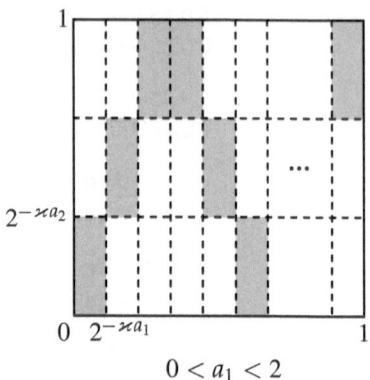

Fig. 4.7

columns, $\eta_1^l = -1$ in the second n_2 columns (additional reflection), then again $\eta_1^l = 1$ in the third n_2 columns and so on. Under these assumptions the resulting regular anisotropic fractal Γ is the graph of a continuous function f,

$$\Gamma = \{(x_1, f(x_1)) : 0 \le x_1 \le 1\} = graph\, f. \tag{4.42}$$

Let $0 < s < 1$; then $\mathscr{C}^s[0,1]$ denotes the Hölder space of all (complex) continuous functions $g(t)$ on $[0,1]$ such that

$$|g(t_1) - g(t_2)| \le c\,(t_2 - t_1)^s, \quad 0 \le t_1 \le t_2 \le 1, \tag{4.43}$$

for some $c \ge 0$. We claim

$$f \in \mathscr{C}^s[0,1] \quad \text{with} \quad s = \frac{a_2}{a_1} < 1, \tag{4.44}$$

where f is the function in (4.42). To prove (4.44) we first remark that $\Gamma \subset (A\,Q)^k$, see (4.16). We look at a rectangle belonging to $(A\,Q)^k$ with side lengths $2^{-k\,\varkappa a_1}$

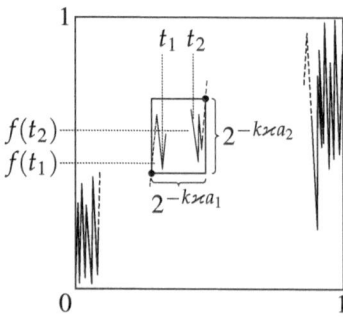

Fig. 4.8

and $2^{-k \varkappa a_2}$, see Fig. 4.8. The continuity of the curve when entering or leaving this rectangle is clear by construction (large dots, maybe). Hence we may assume that t_1 and t_2 are as indicated and that

$$c\, 2^{-k \varkappa a_1} \le t_2 - t_1 \le 2^{-k \varkappa a_1} \tag{4.45}$$

where $c > 0$ is independent of t_1, t_2, and k. Then we have

$$|f(t_2) - f(t_1)| \le 2^{-k \varkappa a_2} \le c'(t_2 - t_1)^{\frac{a_2}{a_1}}. \tag{4.46}$$

This completes the proof of (4.44).

Next we compare for the above fractal Γ the affine dimension $\dim_A \Gamma$ introduced in Definition 4.12 with the usual Hausdorff dimension $\dim_H \Gamma$, see also Remark 4.14 for a first discussion.

4.22 Theorem *Let $\Gamma = (A\,Q)^\infty$ be the anisotropic curve constructed in 4.21 (as a special case of the regular anisotropic fractals introduced in 4.18), where a_1 and a_2 are given by (4.39) and (4.40). Then*

$$\dim_A \Gamma = a_1 \quad and \quad \dim_H \Gamma = 2 - \min(1, \frac{a_2}{a_1}). \tag{4.47}$$

Proof.
Step 1. The first assertion follows immediately from (4.32) with $N = 2^{\varkappa a_1}$ and $n_1 n_2 = 2^{\varkappa (a_1 + a_2)} = 2^{2\varkappa}$.

Step 2. If $a_1 = a_2 = 1$ then we have $\dim_H \Gamma = 1$ by (4.21). We calculate $\dim_H \Gamma = 1$ when $0 < a_1 < 1$. By Definition 2.6 and (2.3) with $\delta = 2^{-\varkappa a_1 J}, J \in \mathbb{N}$, we cover Γ in an optimal way by squares (with sides parallel to the axes of the coordinates) with side-length $2^{-\varkappa a_1 j}$, $j \ge J$. We may assume that these squares fit in the division scheme according to Fig. 4.7. We always calculate modulo unimportant positive factors what is justified by (2.5). In particular, since Γ is compact, we may assume that the number of the squares involved is finite, hence $J \le j \le J'$. For one of these squares q one arrives at the situation shown in Fig. 4.9 with

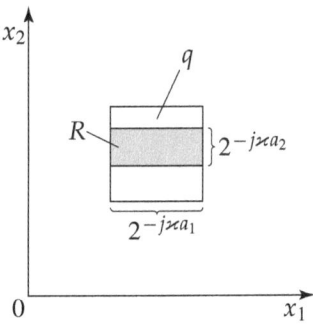

Fig. 4.9

$$\Gamma \cap q \subset R, \quad vol\, q = 2^{-2j\varkappa a_1}, \quad vol\, R = 2^{-2j\varkappa}.$$

By Theorem 4.15 with $d = a_1$ and the underlying mass distribution procedure we have $\mu(R) = 2^{-j\varkappa a_1}$ and hence

$$\sum_q (vol\, q)^{\frac{1}{2}} = \sum_q \mu(q) = \sum_R \mu(R) = 1, \tag{4.48}$$

where we sum over the involved squares: Although the diverse squares q may have different side-lengths, and hence the related rectangles R are also of different size, the mass distribution procedure described in the proof of Theorem 4.15, still works. One has to refine the level j to the highest level needed, maybe the above number J'. Now by Definition 2.6 and (2.3), (2.4) it follows that $\dim_H \Gamma = 1$.

Step 3. Let $1 < a_1 < 2$. The situation is now slightly different. The beginning is the same as in Step 2: covering of Γ with finitely many squares q of side-length $2^{-\varkappa a_1 j}$, $j \geq J$. Instead of Fig. 4.9 we have now Fig. 4.10. By (4.41) and iteration there are $(2n + 1)^k$ rectangles belonging to $(A\,Q)^k$ in each row. A corresponding

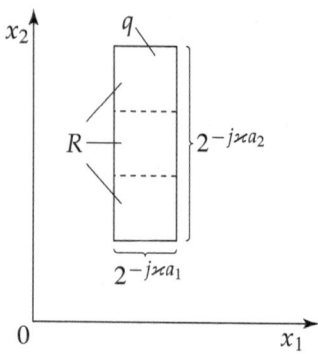

Fig. 4.10

assertion holds for any admissible rectangle R. The mass distribution procedure described in the proof of Theorem 4.15 applies to R. But by the property just described we may assume that each of the $2^{j\varkappa(a_1-a_2)} = (2n+1)^j$ squares of side-length $2^{-j\varkappa a_1}$ in which R can be subdivided receives $2^{-j\varkappa(a_1-a_2)}\mu(R)$ of the mass in the course of the mass distribution. Hence we have

$$\mu(q) \sim 2^{-j\varkappa(a_1-a_2)}\mu(R) = 2^{-j\varkappa(a_1-a_2)}\,2^{-j\varkappa a_1} = 2^{-j\varkappa(2a_1-a_2)} = (vol\,q)^{\frac{D}{2}} \tag{4.49}$$

with $D = 2 - \frac{a_2}{a_1}$, where we used (4.24) with $d = \dim_A \Gamma = a_1$. Now by the same argument as in (4.48) we obtain $\dim_H \Gamma = D = 2 - \frac{a_2}{a_1}$. The proof is complete.

4.23 Corollary *Let $\Gamma = (A\,Q)^\infty = graph\, f$ be the curve constructed in 4.21 with $1 < a_1 < 2$ in (4.39). Then Γ is an (isotropic) d-set according to Definition 3.1 with $d = 2 - \frac{a_2}{a_1}$.*

Proof. The proof follows from the construction in Step 3 of the proof of Theorem 4.22, in particular (4.49).

4.24 Remark Changing slightly the square $Q = [0, 1]^2$ in Fig. 4.7 one is starting from, then any a_1 in (4.39) with $1 < a_1 < 2$ can be admitted and any d with $1 < d < 2$,

$$d = 2 - \frac{a_2}{a_1} = 3 - \frac{2}{a_1}, \quad 1 < a_1 < 2, \tag{4.50}$$

can be obtained in this way, whereas (4.47) remains unchanged. This sheds new light on the distribution of eigenvalues of (pseudo)differential operators mentioned in the Preface and discussed in detail in Chapter V. In case of (isotropic) d-sets we obtain sharp assertions for the exponents as in $(*)$ in the Preface. For anisotropic fractals of the above type we shall prove two-sided estimates with different exponents where $\dim_A \Gamma$ plays a decisive role. However by the above observations one cannot expect to obtain sharp exponents since some of the considered anisotropic fractals, say, with $\dim_A \Gamma$ are also (isotropic) d-sets with $d \neq \dim_A \Gamma$.

4.25 The anisotropic defect
For the curves discussed in Theorem 4.22, Corollary 4.23, and Remark 4.24 we have the anisotropic defect

$$\triangle(a_1) = \dim_H \Gamma - \dim_A \Gamma = 3 - \frac{2}{a_1} - a_1, \quad 1 < a_1 < 2, \tag{4.51}$$

which is shown in Fig. 4.11.

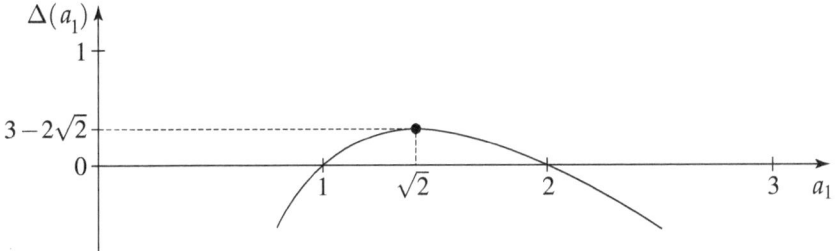

Fig. 4.11

5   Nonisotropic *d*-sets

5.1 Preliminaries
In Definition 3.1 we introduced d-sets in $\mathbb{R}^n$ in qualitative terms. On the other hand, Section 4 dealt in a more constructive way with self-affine fractals of diverse types. In case of similarities one obtains self-similar fractals which are special d-sets, see Theorem 4.7 and 4.9. As for the self-affine fractals introduced in 4.10,

the situation is different. They need not be d-sets although there is a related (nonisotropic) Radon measure, see Theorem 4.15. In two distinguished cases, the PXT-fractals in 4.17, and the OF-fractals in 4.19, we have even more information. The main aim of this section is to combine these two approaches, the qualitative one and the constructive one, in a suitable way. To avoid additional technical complications we restrict ourselves to fractals in the plane $\mathbb{R}^2$. But it is quite clear how to generalize this type of fractal to $\mathbb{R}^n$.

We introduce a few notation. We deal with sets of open rectangles in $\mathbb{R}^2$,

$$\{R_l^j : l = 1, \ldots, N_j\}, \quad j \in \mathbb{N}_0 = \mathbb{N} \cup \{0\}, \tag{5.1}$$

generalizing the PXT-fractals introduced in 4.17 (anisotropic case), and sets of open ellipses in $\mathbb{R}^2$,

$$\{E_l^j : l = 1, \ldots, N_j\}, \quad j \in \mathbb{N}_0, \tag{5.2}$$

generalizing the OF-fractals introduced in 4.19 (nonisotropic case). We assume $N_0 = 1$ and $R_1^0 = \overset{\circ}{Q}$ (open square) and $E_1^0 = \overset{\circ}{B}$ (open circle), where their closures Q and B are given by (4.11) and (4.34), respectively. The basic idea is to cover the resulting fractal Γ for any fixed $j \in \mathbb{N}$ with N_j rectangles R_l^j or ellipses E_l^j, see Fig. 5.1. The rectangles R_l^j have sides parallel to the axes of the coordinates

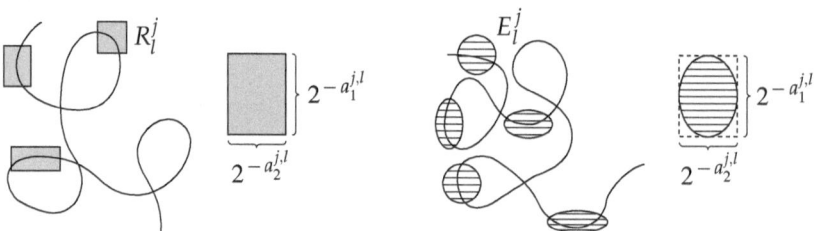

Fig. 5.1

with side-lengths

$$2^{-a_1^{j,l}} \text{ and } 2^{-a_2^{j,l}} \quad \text{where} \quad 0 < a_1^{j,l} \le a_2^{j,l} \tag{5.3}$$

and

$$c\,2^{-2j} \le \operatorname{vol} R_l^j \le 2^{-2j}, \quad j \in \mathbb{N}_0, \tag{5.4}$$

for some c, $0 < c < 1$, independent of j and l. In case of the ellipses no assumptions are made about whether or not the main axes of these ellipses are parallel to the axes of coordinates or not. Otherwise we have (5.3) with respect to the lengths of the main axes of the ellipses E_l^j and

$$c\,2^{-2j} \le \operatorname{vol} E_l^j \le 2^{-2j}, \quad j \in \mathbb{N}_0. \tag{5.5}$$

Finally we compare the above set-up with the construction of the PXT-fractals in 4.17 and the OF-fractals in 4.19. By (4.27) we have a good notational agreement. But the differences are also quite clear. The resulting fractal Γ in (4.17) is constructed via a sequence of sets $(A Q)^j$ given by (4.16) (with j in place of k). In contrast to this procedure we have on the jth level now (5.1) or (5.2), organized by a volume argument given by (5.4) or (5.5), respectively. But the difference is not so big as it may seem at first glance: The creation of PXT-fractals and OF-fractals can be re-organized so that they fit in the above setting. We even used this possibility at the end of Step 2 of the proof of Theorem 4.22.

5.2 Definition *Let d, c_1, and c_2 be positive numbers and let $N_j \in \mathbb{N}$ where $j \in \mathbb{N}_0$ with $N_0 = 1$ and*

$$0 < d < 2, \quad c_1 \, 2^{jd} \leq N_j \leq c_2 \, 2^{jd}. \tag{5.6}$$

A compact set Γ in $\mathbb{R}^2$ is called an anisotropic d-set if there exists a sequence of open rectangles R_l^j according to (5.1) with $R_1^0 = \overset{\circ}{Q}$, (5.3), and (5.4) with the following properties:

(i)
$$R_l^j \cap R_m^j = \emptyset \quad \text{for} \quad l \neq m, \quad (j \in \mathbb{N}). \tag{5.7}$$

(ii) *For any rectangle R_l^{j+1} there is a rectangle R_m^j, $m = m(l)$, such that*

$$R_l^{j+1} \subset R_m^j, \quad j \in \mathbb{N}_0. \tag{5.8}$$

(iii)
$$(vol \, R_l^j)^{\frac{d}{2}} = \sum_{R_m^{j+1} \subset R_l^j} (vol \, R_m^{j+1})^{\frac{d}{2}} \tag{5.9}$$

for any $j \in \mathbb{N}_0$ and $l = 1, \ldots, N_j$.

(iv) *There is a number a, called the (anisotropic) deviation with $0 \leq a \leq 1$ and a number $J \in \mathbb{N}$ such that*

$$(1 - a)j \leq a_1^{j,l} \leq a_2^{j,l} \leq (1 + a)j; \quad j \geq J; \quad l = 1, \ldots, N_j. \tag{5.10}$$

(v)
$$\Gamma = \bigcap_j \left(\bigcup_{l=1}^{N_j} \overline{R_l^j} \right). \tag{5.11}$$

5.3 Definition *Let d, c_1 and c_2 be positive numbers and let $N_j \in \mathbb{N}$ where $j \in \mathbb{N}_0$ with (5.6) and $N_0 = 1$. A compact set Γ in $\mathbb{R}^2$ is called a nonisotropic d-set if there exists a sequence of open ellipses E_l^j according to (5.2) with $E_1^0 = \overset{\circ}{B}$, (5.3), and (5.5) with the properties (i)–(v) of Definition 5.2 with E_l^j in place of R_l^j and the (nonisotropic) deviation a.*

5.4 Discussion

Of course, $\bigcup\limits_{l=1}^{N_j}\overline{R_l^j}$ is a decreasing sequence of compact sets and hence

$$\Gamma = \lim_{j \to \infty} \bigcup_{l=1}^{N_j} \overline{R_l^j}. \tag{5.12}$$

This is the direct counterpart of (4.17). We comment on the role of the (anisotropic or nonisotropic) *deviation a*. At the first glance (5.10) seems to be very restrictive. But this is not the case. Assume that we replaced (5.10) by the apparently weaker version

$$0 < a_1^{j,l} \le a_2^{j,l} \le (1+a)j; \quad j \ge J; \quad l = 1,\dots,N_j; \tag{5.13}$$

for some $a \ge 0$. By (5.4) and (5.5) we have

$$a_1^{j,l} \ge 2j - a_2^{j,l} \ge j(1-a), \tag{5.14}$$

what coincides with the left-hand side of (5.10). Also the restriction $a \le 1$ is rather natural. Assume that $a_2^{j_r,l_r} \ge (1+a)j_r$ for some $a > 1$ and for suitable sequences j_r, l_r holds, then we have a contradiction to the left-hand sides of (5.4) and (5.5). It might be reasonable to modify the notation of the (anisotropic or nonisotropic) deviation somewhat. It is quite clear that $j \to \infty$ in Definitions 5.2.and 5.3 is of interest. Hence one might be tempted to call a' with

$$1 + a' = \overline{\lim_{j \to \infty}} \frac{a_2^{j,l}}{j} \quad (< \infty) \tag{5.15}$$

the (anisotropic or nonisotropic) *deviation*. Of course $a' \le a$. But for sake of simplicity we stick at (5.10). Since $J \in \mathbb{N}$ may be chosen as large as one wants there is essentially no difference.

By our normalization $\overline{R_1^0} = Q$ and (5.9) we have

$$\sum_{l=1}^{N_j} \left(vol\, R_l^j\right)^{\frac{d}{2}} = 1, \quad j \in \mathbb{N}_0, \tag{5.16}$$

what is related to (4.25). Even more interesting is the counterpart of (4.22).

5.5 Theorem *Let Γ be the anisotropic (nonisotropic) d-set introduced in Definition 5.2 (5.3). Let U_l^j be either R_l^j, anisotropic case, or $(E_l^j$, nonisotropic case). Then there is a uniquely determined Radon measure μ in $\mathbb{R}^2$ with $supp\,\mu = \Gamma$ and*

$$\mu\left(\Gamma \cap \overline{U_l^j}\right) = \left(vol\, U_l^j\right)^{\frac{d}{2}}, \quad j \in \mathbb{N}_0, \quad l = 1,\dots,N_j. \tag{5.17}$$

Proof. With obvious modifications we can follow the proof of Theorem 4.15.

5.6 Proposition

(i) *The PXT-fractal introduced in* 4.17 *is an anisotropic d-set with* $d = \dim_A \Gamma$ *(affine dimension).*

(ii) *The OF-fractal introduced in* 4.19 *is a nonisotropic d-set with* $d = \dim_A \Gamma$.

Proof. We prove (i). The affine dimension has been introduced in Definition 4.12. Let A_l be the affine maps introduced in (4.27). Let c in (5.4) be given by

$$c = \min\{2^{-2\varkappa_l}; l = 1, \ldots, N\}. \tag{5.18}$$

We construct the sequence of rectangles $\{R_l^j : l = 1, \ldots, N_j\}$ by induction. We choose $R_1^0 = \overset{\circ}{Q}$ and assume that $\{R_l^j : l = 1, \ldots, N_j\}$ is known where

$$R_l^j = A_{j_1} \cdots A_{j_k} \overset{\circ}{Q} \tag{5.19}$$

for some $j_1, \ldots, j_k$,

$$\bigcup \overline{R_l^j} \supset \Gamma \quad \text{and} \quad c\, 2^{-2j} \leq vol\, R_l^j \leq 2^{-2j}. \tag{5.20}$$

If $vol\, R_l^j \leq 2^{-2(j+1)}$ for some l, then we take that R_l^j as a new R_m^{j+1}. Let $vol\, R_l^j > 2^{-2(j+1)}$. Then we examine the rectangles $\{A_k R_l^j : k = 1, \ldots, N\}$. We have by (4.27) and (5.18)

$$vol\, A_k R_l^j = 2^{-2\varkappa_k} vol\, R_l^j \geq c\, 2^{-2(j+1)}. \tag{5.21}$$

If $vol\, A_k R_l^j \leq 2^{-2(j+1)}$ then we choose $A_k R_l^j$ as a new rectangle R_m^{j+1}. Otherwise we repeat this procedure. By (5.21) we arrive at $\{R_l^{j+1} : l = 1, \ldots, N_{j+1}\}$ with (5.20), now with $j + 1$ in place of j. By (4.25) we have

$$\sum_{l_1,\ldots,l_k}^{1,\ldots,N} vol(A_{m_1} \cdots A_{m_r} A_{l_1} \cdots A_{l_k} Q)^{\frac{d}{2}} = vol(A_{m_1} \cdots A_{m_r} Q)^{\frac{d}{2}} \tag{5.22}$$

where $d = \dim_A \Gamma$ is the affine dimension. We identify R_l^j with some $A_{m_1} \cdots A_{m_r} \overset{\circ}{Q}$. To prove (5.9) we assume $R_m^{j+1} \subset R_l^j$ which is given by some $A_{m_1} \cdots A_{m_r} A_{l_1} \cdots A_{l_k} \overset{\circ}{Q}$ with different values of k. Let K be the largest of these involved k's. At that level K one has (5.22). Summing up by using appropriate counterparts of (5.22) with R_m^{j+1} on the right-hand side one obtains (5.9). However by (4.20), (5.4) with the constant c given by (5.18), and (5.16) we obtain (5.6). Since (5.7), (5.8), and (5.11) are clear by construction it remains to prove (5.10). Let again $R_l^j = A_{m_1} \cdots A_{m_r} Q$, then

$$\varkappa_{m_1} + \cdots + \varkappa_{m_r} \sim j \tag{5.23}$$

and hence $r \sim j$. By (4.27), (4.28), and 5.4 it is now clear that one finds a number $1 > a > 0$ with (5.10). This completes the proof of part (i). The proof of (ii) is the same.

5.7 Remark If Γ is an anisotropic or nonisotropic d-set then it might happen that it is also an anisotropic or nonisotropic d'-set with $d \neq d$' or even an (isotropic) d-set according to Definition 3.1. Examples are mentioned at the end of Remark 4.24.

5.8 Remark We compare Definitions 5.2 and 5.3 and Theorem 5.5 with previous constructions. As mentioned in Remark 3.5 in case of (isotropic) d-sets the Hausdorff measure $\mathcal{H}^d|\Gamma$ coincides (up to equivalence constants) with the related Minkowski contents. This supports the above constructions which are in some sense *anisotropic* and *nonisotropic Minkowski contents*. Secondly one may compare the above constructions with the set-up described in 2.1. If one chooses ζ in (2.1) in an appropriate way then the Theorems 2.2 and 2.3 are closely related to the above Theorem 5.5. But they do not coincide at the first glance since the criterion for selecting admitted sets is *diam* E_j in (2.1) compared with volumes in (5.3), (5.4) or (5.5).

5.9 Anisotropic versus nonisotropic
In Definition 5.2 we did not stress the point that the rectangles R_i^j have sides parallel to the axes of coordinates. This is unimportant both in this section and also in the later considerations in Sections 18 and 22 in connection with function spaces. We compare later on the resulting fractal Γ in (5.11) with some isotropic spaces $B_{pq}^s(\mathbb{R}^n)$. But we have the feeling that this is hardly the last word and that well adapted anisotropic spaces as treated in [SchT87], Ch. 4, might be the better choice. But it was not only for this vague perspective that we stuck to the view that the rectangles in Definition 5.2 should have sides parallel to the axes of coordinates, and to call that case *anisotropic*. If one looks at the *nonisotropic* case in the Fig.'s 4.6 and 5.1 constructed by means of ellipses, one gets the feeling that the created fractal is rather airy, also in comparison with other constructions, as, for instance, in Fig. 4.3. Replacing in Fig. 4.6 and also in Definition 5.3 the ellipses E_i^j by corresponding rectangles with sides which are not necessarily parallel to the axes of coordinates one creates an even more airy and somewhat disconnected fractal. That the elliptic case as described in Fig. 4.6 is denser than a so-obtained arbitrary rectangular case will become clearer later on at the end of Section 18 when we try to embed spaces $L_p(\Gamma)$ in (isotropic) spaces $B_{pq}^s(\mathbb{R}^n)$. This might justify distinguishing between *anisotropic* and *nonisotropic* d-sets as given by the Definitions 5.2. and 5.3, respectively.

5.10 Proper d-sets
By Definition 3.1 of an (isotropic) d-set the measure μ is rather evenly distributed. If $\varkappa$ with $0 < \varkappa < 1$ then we have by (3.1)

$$\mu\left(B(\gamma, \varkappa r) \cap \Gamma\right) \sim \mu\left(B(\gamma, r) \cap \Gamma\right) \sim r^d \tag{5.24}$$

where the equivalence constants depend on $\varkappa$, but not on $\gamma \in \Gamma$ and $0 < r < 1$. For anisotropic and nonisotropic d-sets in $\mathbb{R}^2$ we have (5.17) for the generated measure μ, but no counterpart of (5.24). In connection with the spectral theory of fractals developed in Chapter V at least a weak version of (5.24) would be desirable in the anisotropic and nonisotropic case. Let

$$R = \{(x_1, x_2) \in \mathbb{R}^2, \ |x_1 - x_1^0| < r_1, |x_2 - x_2^0| < r_2\} \qquad (5.25)$$

be a rectangle in $\mathbb{R}^2$ with sides parallel to the axes of the coordinates. Let $0 < \varkappa < 1$ then

$$\varkappa R = \{(x_1, x_2) \in \mathbb{R}^2, \quad |x_1 - x_1^0| < \varkappa r_1, |x_2 - x_2^0| < \varkappa r_2\} \qquad (5.26)$$

is the rectangle with the same centre and dilated sides. Similarly let E be an ellipse in $\mathbb{R}^2$ with the main axes r_1 and r_2 in the directions y^1 and y^2, respectively, and let $0 < \varkappa < 1$, then $\varkappa E$ is the ellipse with the same centre and the main axes $\varkappa r_1$ and $\varkappa r_2$ in the directions y^1 and y^2, respectively.

5.11 Definition *The anisotropic (nonisotropic) d-set Γ introduced in Definition 5.2 (5.3) and equipped with the measure μ according to Theorem 5.5, is called proper if there are two numbers $\varkappa$ and c, $0 < \varkappa < 1$ and $0 < c \leq 1$, such that*

$$\mu(\Gamma \cap \varkappa \overline{U_l^j}) \geq c \, (vol \, U_l^j)^{\frac{d}{2}}, \quad j \in \mathbb{N}_0, \ l = 1, \ldots, N_j, \qquad (5.27)$$

where U_l^j are the rectangles R_l^j in the anisotropic case, (ellipses E_l^j in the non-isotropic case).

5.12 Remark By (5.27) and (5.17) we have a modest counterpart of (5.24). But this is completely sufficient for our later purposes. On the other hand (5.27) is an additional assumption. Take, for example, the case $0 < a_1 < 1$ for the anisotropic curves in 4.21 and in Fig. 4.7, where the shaded rectangles fill the bottom row. In that case we arrive at $\Gamma = \{(x_1, 0), 0 \leq x_1 \leq 1\}$ and the left-hand side of (5.27) is always zero. If we apply (5.27) to the PXT-fractals introduced in 4.17 and to the OF-fractals introduced in 4.19 then it turns out that we exclude just these pathological phenomena. Recall that

$$\overset{\circ}{Q} = \{x = (x_1, x_2) : \quad 0 < x_1 < 1, \ 0 < x_2 < 1\} \qquad (5.28)$$

and

$$\overset{\circ}{B} = \{x \in \mathbb{R}^2 : |x| < \frac{1}{2}\} \qquad (5.29)$$

are the open interiors of Q and B given by (4.11) and (4.34), respectively. Now we complement Proposition 5.6.

5.13 Proposition

(i) *Let Γ be the PXT-fractal introduced in 4.17 and let*

$$\Gamma \cap \overset{\circ}{Q} \neq \emptyset. \qquad (5.30)$$

   Then Γ is a proper anisotropic d-set.

(ii) *Let Γ be the OF-fractal introduced in 4.19 and let*

$$\Gamma \cap \overset{\circ}{B} \neq \emptyset. \qquad (5.31)$$

   Then Γ is a proper nonisotropic d-set.

Proof. By Proposition 5.6 it remains to prove (5.27). By (5.30) we have for $\gamma \in \Gamma \cap \overset{\circ}{Q}$,

$$\gamma \in A_{m_1} \cdots A_{m_M} Q \subset \varkappa Q \tag{5.32}$$

for some $0 < \varkappa < 1$ and some $A_{m_1}, \ldots, A_{m_M}$. Applying to (5.32) an arbitrary map $A_{j_1} \cdots A_{j_k}$ we obtain

$$A_{j_1} \cdots A_{j_k}(A_{m_1} \cdots A_{m_M} Q) \subset \varkappa A_{j_1} \cdots A_{j_k} Q. \tag{5.33}$$

However by the mass distribution procedure as described in the proof of Theorem 4.15 we get (5.27). The proof for the nonisotropic case is the same.

5.14 Remark The conditions (5.30) and (5.31) are rather natural. It simply means that we exclude the not very attractive cases where $\Gamma \subset \partial Q$ or $\Gamma \subset \partial B$.

Chapter II
ℓ_p-spaces

6 Entropy numbers and eigenvalues

6.1 Preliminaries and notation

The main aim of Chapter II is to study entropy numbers in (weighted) ℓ_p-spaces. This will be done in the Sections 7–9. In the present section we describe briefly the necessary abstract background without proofs. We follow closely [ET96] where proofs, further details, explanations, and more references are given.

A *quasi-norm* on a complex linear space B is a map $\| \cdot |B\|$ from B to the non-negative reals $\mathbb{R}_+$ such that

$$\|x|B\| = 0 \text{ if, and only if, } x = 0, \tag{6.1}$$

$$\|\lambda x|B\| = |\lambda|\,\|x|B\| \quad \text{for all scalars } \lambda \in \mathbb{C} \text{ and all } x \in B, \tag{6.2}$$

there is a constant C such that for all $x \in B$ and $y \in B$

$$\|x + y|B\| \le C(\|x|B\| + \|y|B\|). \tag{6.3}$$

Plainly $C \ge 1$; if $C = 1$ is allowed then $\| \cdot |B\|$ is a norm in B. As usual, B is called a *quasi-Banach space* if every Cauchy sequence with respect to $\| \cdot |B\|$ is a convergent sequence.

Given any $p \in (0, 1]$, a *p-norm* on a complex linear space B is a map $\| \cdot |B\| \to \mathbb{R}_+$ which satisfies (6.1), (6.2), and instead of (6.3),

$$\|x + y|B\|^p \le \|x|B\|^p + \|y|B\|^p \quad \text{for } x, y \in B. \tag{6.4}$$

Two quasi-norms or p-norms $\| \cdot |B\|_1$ and $\| \cdot |B\|_2$ are said to be *equivalent* if there is a constant $c \ge 1$ such that for all $x \in B$,

$$c^{-1}\|x|B\|_1 \le \|x|B\|_2 \le c\,\|x|B\|_1. \tag{6.5}$$

It can be shown (see [Kön86], p. 47 or [DeVL93], p. 20) that if $\| \cdot |B\|_1$ is a quasi-norm on B then there exists $p \in (0, 1]$ and a p-norm $\| \cdot |B\|_2$ on B which is equivalent to $\| \cdot |B\|_1$.

Let A, B be quasi-Banach spaces and let $T : A \to B$ linear. Just as for the Banach space case, T will be called *bounded* or *continuous* if

$$\|T\| = \sup\{\|Ta|B\| : a \in A, \|a|A\| \le 1\} < \infty. \tag{6.6}$$

H. Triebel, *Fractals and Spectra*, Modern Birkhäuser Classics,
DOI 10.1007/978-3-0348-0034-1_2, © Birkhäuser Verlag 1997

The family of all such T will be denoted by $L(A, B)$ or $L(A)$ if $A = B$. Otherwise terminology which is standard in the context of Banach spaces will be taken without further comment to quasi-Banach spaces. In particular if $T \in L(B)$ then $\sigma(T)$ stands for its spectrum.

In [ET96], pp. 3–7, we developed a Riesz theory for compact operators $T \in L(B)$ in quasi-Banach spaces B parallel to the well-known assertions in the Banach spaces case. Especially, if

$T \in L(B)$ is compact, then $\sigma(T) \setminus \{0\}$ consists of an at most countably infinite number of eigenvalues of finite algebraic multiplicity which may accumulate only at the origin.

If B is a quasi-Banach space then $U_B = \{b \in B : \|b|B\| \le 1\}$ stands for the unit ball in B.

6.2 Definition *Let A, B be quasi-Banach spaces and let $T \in L(A, B)$. Then for all $k \in \mathbb{N}$, the kth entropy number $e_k(T)$ of T is defined by*

$$e_k(T) =$$
$$\inf \left\{ \varepsilon > 0 : T(U_A) \subset \bigcup_{j=1}^{2^{k-1}} (b_j + \varepsilon U_B) \text{ for some } b_1, \ldots, b_{2^{k-1}} \in B \right\}. \qquad (6.7)$$

6.3 Remark This formulation coincides with the definition given in [ET96], p. 7, which simply generalizes to quasi-Banach spaces what has been done before for Banach spaces. Further comments and some discussions may be found in [ET96], pp. 7–9, and, in greater detail, in [CaS90] and [EEv87].

6.4 Proposition *Let A, B, C be quasi-Banach spaces, let $S, T \in L(A, B)$ and suppose that $R \in L(B, C)$.*
  (i) *$\|T\| \ge e_1(T) \ge e_2(T) \ge \ldots ; e_1(T) = \|T\|$ if B is a Banach space.*
  (ii) *For all $k, l \in \mathbb{N}$*

$$e_{k+l-1}(R \circ S) \le e_k(R)\, e_l(S). \qquad (6.8)$$

  (iii) *If B is a p-Banach space, where $0 < p \le 1$, then for all $k, l \in \mathbb{N}$*

$$e_{k+l-1}^p(S + T) \le e_k^p(S) + e_l^p(T). \qquad (6.9)$$

6.5 Remark This formulation coincides with Lemma 1 in [ET96], pp. 7,8, where also a simple proof may be found. In case of quasi-Banach spaces it may happen that $\|T\| > e_1(T)$, see [ET96], Remark 4 on p. 9.

6.6 Compact operators
Recall that $T \in L(B)$ is compact if, and only if, for every $\varepsilon > 0$ there is a finite ε-net covering $T(U_B)$. By (6.7) this is the same as

$$T \in L(B) \text{ is compact if, and only if, } e_k(T) \to 0 \text{ for } k \to \infty. \qquad (6.10)$$

6.7 Interpolation properties
The entropy numbers behave very well with respect to real interpolation of quasi-Banach spaces. We gave in [ET96], pp. 13–15, a rather careful treatment of this subject, which in turn was based on [HaT94a]. Further properties in the context of Banach spaces and historical comments may be found in [Tri78], 1.16.2, and [Pie80], 12.1.

6.8 Eigenvalues
Again let B be a (complex) quasi-Banach space and let $T \in L(B)$ be compact. As we mentioned at the end of 6.1 the spectrum of T, apart from the point 0, consists solely of eigenvalues of finite algebraic multiplicity: let $\{\mu_k(T)\}_{k\in\mathbb{N}}$ be the sequence of all non-zero eigenvalues of T, repeated according to algebraic multiplicity and ordered so that

$$|\mu_1(T)| \geq |\mu_2(T)| \geq \ldots \to 0. \qquad (6.11)$$

If T has only $m(< \infty)$ distinct eigenvalues and M is the sum of their algebraic multiplicities we put $\mu_n(T) = 0$ for all $n > M$.

6.9 Theorem *Let T and $\{\mu_k(T)\}_{k\in\mathbb{N}}$ be as in 6.8. Then*

$$\left(\prod_{m=1}^{k} |\mu_m(T)|\right)^{\frac{1}{k}} \leq \inf_{n\in\mathbb{N}} 2^{\frac{n}{2k}} e_n(T), \quad k \in \mathbb{N}. \qquad (6.12)$$

6.10 Corollary *For all $k \in \mathbb{N}$*

$$|\mu_k(T)| \leq \sqrt{2}\, e_k(T). \qquad (6.13)$$

6.11 Remark This is Carl's famous inequality which connects spectral properties of compact operators with the geometry of the map T described in terms of entropy numbers. (6.13) in the context of Banach spaces was proved by Carl in [Carl81]. In [ET96], pp. 18–20, we gave a proof of (6.12) which generalizes the proof given in [CaT80] from Banach spaces to quasi-Banach spaces. Plainly, (6.13) follows from (6.11) and (6.12) with $n = k$.

6.12 Remark Further results, comments, references and, in particular comparisons of entropy numbers with other geometric quantities, especially approximation numbers, may be found in [ET96], [CaS90], [EEv87], [Kön86], [Pie87], and [LGM96].

7   The spaces ℓ_p^M

7.1   Preliminaries and notation

We follow again [ET96], p. 97. Let $M \in \mathbb{N}$ and let $0 < p \leq \infty$. By ℓ_p^M we shall mean the linear space of all complex M-tuples $y = (y_j)$, endowed with the quasi-norm

$$\|y|\ell_p^M\| = \left(\sum_{j=1}^{M} |y_j|^p \right)^{\frac{1}{p}}, \quad \text{if} \quad 0 < p < \infty, \tag{7.1}$$

and

$$\|y|\ell_\infty^M\| = \sup_j |y_j|, \quad \text{if} \quad p = \infty. \tag{7.2}$$

Let

$$U_p^M = \{y \in \ell_p^M : \|y|\ell_p^M\| \leq 1\} \tag{7.3}$$

be the closed unit ball in ℓ_p^M. Since $\mathbb{C}^M$ may be identified with $\mathbb{R}^{2M}$, we shall understand by the volume of U_p^M the Lebesgue measure of

$$\left\{ (x_1, \ldots, x_{2M}) \in \mathbb{R}^{2M} : \sum_{j=1}^{M} (x_{2j-1}^2 + x_{2j}^2)^{\frac{p}{2}} \leq 1 \right\}. \tag{7.4}$$

Let $p \in (0, \infty]$ be given. There are two positive constants c_1 and c_2 (which may depend on p) such that for all $M \in \mathbb{N}$

$$c_1 M^{-\frac{1}{p}} \leq \left(vol \, U_p^M \right)^{\frac{1}{2M}} \leq c_2 M^{-\frac{1}{p}}. \tag{7.5}$$

This follows from the Proposition in [ET96], p. 97, and the end of the proof on p. 98.

Plainly the identity from $\ell_{p_1}^M$ in $\ell_{p_2}^M$ is a compact operator. Our aim is to estimate the corresponding entropy numbers according to Definition 6.2. In what follows we assume that $\log = \log_2$ is taken with respect to the base 2. First we complement the results in [ET96], p. 98.

7.2 Proposition *Let* $0 < p_1 \leq \infty$, $0 < p_2 \leq \infty$ *and for each* $k \in \mathbb{N}$ *let* e_k *be the entropy numbers of the embedding*

$$id : \ell_{p_1}^M \to \ell_{p_2}^M.$$

Then

$$e_k \geq c \quad if \quad 1 \leq k \leq \log(2M), \tag{7.6}$$

and

$$e_k \geq c2^{-\frac{k}{2M}} (2M)^{\frac{1}{p_2} - \frac{1}{p_1}} \ if \ k \in \mathbb{N}, \tag{7.7}$$

where c *is a positive constant which is independent of* M *(and* k*) but may depend upon* p_1 *and* p_2.

Proof.

Step 1. We prove (7.6). Let $y = (y_j) \in \ell_p^M$ for some p where all components y_j are zero with exception of one component which is either 1 or -1. There are 2M such elements belonging to $U_{p_1}^M$ and $U_{p_2}^M$. Let y^1 and y^2 be two such points and assume that they belong to the same $\ell_{p_2}^M$-ball of radius ε, hence

$$y^1 \in x + \varepsilon U_{p_2}^M \quad \text{and} \quad y^2 \in x + \varepsilon U_{p_2}^M \quad \text{for some} \quad x \in \ell_{p_2}^M. \qquad (7.8)$$

For some c which is independent of M and $\overline{p_2} = \min(p_2, 1)$ we have

$$c \leq \|y^1 - y^2 |\ell_{p_2}^M\|^{\overline{p_2}} \leq$$

$$\|y^1 - x |\ell_{p_2}^M\|^{\overline{p_2}} + \|y^2 - x |\ell_{p_2}^M\|^{\overline{p_2}} \leq 2\varepsilon^{\overline{p_2}}. \qquad (7.9)$$

Now (7.6) follows from (7.9), $2^{k-1} \leq M < 2M$ and (6.7).

Step 2. We prove (7.7). We cover $U_{p_1}^M$ with 2^{k-1} balls in $\ell_{p_2}^M$ of radius ε chosen in an appropriate way. Then we have by the interpretation (7.4)

$$vol\, U_{p_1}^M \leq 2^{k-1} \varepsilon^{2M} \, vol\, U_{p_2}^M$$
$$\leq 2^k e_k^{2M} \, vol\, U_{p_2}^M. \qquad (7.10)$$

Now (7.7) follows from (7.10) and (7.5).

7.3 Theorem *Let $0 < p_1 \leq p_2 \leq \infty$ and for each $k \in \mathbb{N}$ let e_k be the kth entropy number of the embedding*

$$id : \ell_{p_1}^M \to \ell_{p_2}^M.$$

Then

$$c_1 \leq e_k \leq c_2 \quad if \quad 1 \leq k \leq \log(2M), \qquad (7.11)$$

$$e_k \leq c_2 \left(k^{-1} \log(1 + \frac{2M}{k}) \right)^{\frac{1}{p_1} - \frac{1}{p_2}} \quad if \quad \log(2M) \leq k \leq 2M, \qquad (7.12)$$

$$c_1 2^{-\frac{k}{2M}} (2M)^{\frac{1}{p_2} - \frac{1}{p_1}} \leq e_k$$
$$\leq c_2 2^{-\frac{k}{2M}} (2M)^{\frac{1}{p_2} - \frac{1}{p_1}} \quad if \quad k \geq 2M, \qquad (7.13)$$

where c_1 and c_2 are positive constants which are independent of M (and k) but may depend upon p_1 and p_2.

7.4 Remark The estimate from below is covered by Proposition 7.2. A proof of the estimate from above may be found in [ET96], pp. 98–101.

7.5 Remark If $1 \leq p_1 < p_2 \leq \infty$, then also the estimate (7.12) is an equivalence as in the two other cases: see [Schü84] and [Kön86], 3.c.8, pp. 190–191.

8 Weighted ℓ_p-spaces

8.1 Preliminaries and notation

Let $d > 0$, $\delta \geq 0$ and $(M_j)_{j \in \mathbb{N}_0}$ be a sequence of natural numbers. We always assume that there are two positive numbers c_1 and c_2 with

$$c_1 \leq M_j \, 2^{-jd} \leq c_2 \quad \text{for every} \quad j \in \mathbb{N}_0. \tag{8.1}$$

Let $0 < p \leq \infty$ and $0 < q \leq \infty$. Then by $\ell_q(2^{j\delta} \ell_p^{M_j})$ we shall mean the linear space of all complex sequences $x = (x_{j,l} : j \in \mathbb{N}_0; \; l = 1, \dots, M_j)$ endowed with the quasi-norm

$$\|x|\ell_q(2^{j\delta} \ell_p^{M_j})\| \;=\; \left(\sum_{j=0}^{\infty} \left(\sum_{l=1}^{M_j} 2^{j\delta p} |x_{j,l}|^p \right)^{\frac{q}{p}} \right)^{\frac{1}{q}} \tag{8.2}$$

with the obvious modifications according to (7.2) if $p = \infty$ and/or $q = \infty$. In case of $\delta = 0$ we write $\ell_q(\ell_p^{M_j})$ and if, in addition $p = q$, then we have the ℓ_p-spaces with the components ordered in the given way. Plainly, $\ell_q(2^{j\delta} \ell_p^{M_j})$ consists of dyadic blocks of spaces $\ell_p^{M_j}$ as introduced in 7.1 clipped together via the weights $2^{j\delta}$. We are interested in the counterpart of Theorem 7.3. Let $d > 0$, $\delta > 0$ and

$$0 < p_1 \leq p_2 \leq \infty, \quad 0 < q_1 \leq \infty, \quad 0 < q_2 \leq \infty. \tag{8.3}$$

Then the identity map

$$id: \quad \ell_{q_1}(2^{j\delta} \ell_{p_1}^{M_j}) \to \ell_{q_2}(\ell_{p_2}^{M_j}) \tag{8.4}$$

is compact, where M_j is restricted by (8.1). To prove this claim we use the decomposition

$$id = \sum_{j=0}^{\infty} id_j \tag{8.5}$$

where

$$\begin{aligned} id_j \, x &= (\delta_{jk} \, x_{k,l} : k \in \mathbb{N}_0; \, l = 1, \dots, M_k) \\ &= (0, \dots, 0, x_{j,1}, \dots, x_{j,M_j}, 0, 0, \dots) \end{aligned} \tag{8.6}$$

selects the jth block. We have

$$\begin{aligned} \|id_j \, x|\ell_{q_2}(\ell_{p_2}^{M_k})\| &= \|(x_{j,l})| \, \ell_{p_2}^{M_j}\| \\ &\leq \|(x_{j,l})| \, \ell_{p_1}^{M_j}\| \\ &\leq 2^{-j\delta} \|x| \, \ell_{q_1}(2^{k\delta} \ell_{p_1}^{M_k})\|. \end{aligned} \tag{8.7}$$

Now by (8.5) and (8.7) it follows that id is compact.

8.2 Theorem *Let $d > 0$, $\delta > 0$, and $M_j \in \mathbb{N}$ with (8.1). Let p_1, p_2, q_1, q_2 be given by (8.3). Let e_k be the entropy numbers of the compact operator id in (8.4) according to Definition 6.2. There are two positive numbers c and C such that*

$$c\, k^{-\frac{\delta}{d}+\frac{1}{p_2}-\frac{1}{p_1}} \le e_k \le C\, k^{-\frac{\delta}{d}+\frac{1}{p_2}-\frac{1}{p_1}}, \quad k \in \mathbb{N}. \tag{8.8}$$

Proof.
Step 1. First we prove the left-hand side of (8.8). In the commutative diagram

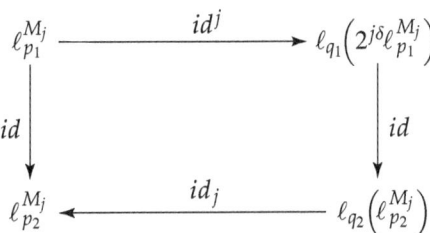

the operator id_j is given as in (8.6), now acting in the indicated slightly modified way, whereas id^j maps $\ell_{p_1}^{M_j}$ identically onto $2^{j\delta}\,\ell_{p_1}^{M_j}$ interpreted as a dyadic block of $\ell_{q_1}(2^{j\delta}\ell_{p_1}^{M_j})$. In what follows we reserve id for the identity given by (8.4), otherwise we indicate the spaces involved. Hence

$$id\left(\ell_{p_1}^{M_j} \to \ell_{p_2}^{M_j}\right) = id_j \circ id \circ id^j, \quad j \in \mathbb{N}. \tag{8.9}$$

Plainly,

$$\|id^j\| = 2^{j\delta} \quad \text{and} \quad \|id_j\| = 1, \tag{8.10}$$

and consequently by (6.8)

$$e_k\left(id : \ell_{p_1}^{M_j} \to \ell_{p_2}^{M_j}\right) \le 2^{j\delta} e_k, \quad k \in \mathbb{N},\, j \in \mathbb{N}. \tag{8.11}$$

By (7.13) with $k = 2M_j$ we obtain

$$e_{2M_j} \ge c\, 2^{-j\delta}\, 2^{jd(\frac{1}{p_2}-\frac{1}{p_1})}, \quad j \in \mathbb{N}. \tag{8.12}$$

By (8.1) and the monotonicity properties of the entropy numbers described in Proposition 6.4(i) it follows that

$$e_k \ge c\, k^{-\frac{\delta}{d}+\frac{1}{p_2}-\frac{1}{p_1}}, \quad k \in \mathbb{N}, \tag{8.13}$$

for some $c > 0$.

Step 2. The estimate from above is more complicated. Let $J \in \mathbb{N}$ and

$$L\delta = J\delta + Jd\left(\frac{1}{p_1} - \frac{1}{p_2}\right); \tag{8.14}$$

in particular $L \geq J$. We put $e_t = e_{[t]}$ if $t \geq 1$ and assume $L \in \mathbb{N}$. We wish to prove

$$e_{2^{Jd}} \leq c\, 2^{-J\delta + Jd\left(\frac{1}{p_2} - \frac{1}{p_1}\right)}, \quad J \in \mathbb{N}. \tag{8.15}$$

This is equivalent to the estimate from above in (8.8). We split the sum in (8.5) in three parts,

$$id = \sum_{j=0}^{J} id_j + \sum_{j=J+1}^{L} id_j + \sum_{j=L+1}^{\infty} id_j. \tag{8.16}$$

Of course here id_j is considered as a map between the two spaces in (8.4) according to (8.6) in contrast to id_j in the above diagram in Step 1. There is no danger of confusion. In particular by (8.7) we have

$$\left\| \sum_{j=L+1}^{\infty} id_j \right\| \leq c\, 2^{-L\delta}, \tag{8.17}$$

which by (8.14) coincides with the right-hand side of (8.15). Let $\varrho = \min(1, p_2, q_2)$. It is easy to see that $\ell_{q_2}(\ell_{p_2}^{M_j})$ is a ϱ-Banach space. Then we obtain by (6.9), (8.16), and (8.17)

$$e_k^{\varrho} \leq c\, 2^{-L\delta\varrho} + \sum_{j=0}^{J} e_{k_j}^{\varrho}(id_j) + \sum_{j=J+1}^{L} e_{k_j}^{\varrho}(id_j) \tag{8.18}$$

where $k = \sum_{j=0}^{L} k_j$. By (8.7) we have

$$e_{k_j}(id_j) = 2^{-j\delta} e_{k_j}\left(id : \ell_{p_1}^{M_j} \rightarrow \ell_{p_2}^{M_j}\right), \tag{8.19}$$

and hence by Theorem 7.3

$$e_{k_j}(id_j) \leq$$

$$c\, 2^{-j\delta} \left[k_j^{-1} \log\left(c\, 2^{jd}\, k_j^{-1}\right) \right]^{\frac{1}{p_1} - \frac{1}{p_2}} \quad \text{if} \quad k_j \leq 2M_j, \tag{8.20}$$

which covers also (7.11) and

$$e_{k_j}(id_j) \leq c\, 2^{-j\delta}\, 2^{-\frac{k_j}{2M_j}} (2M_j)^{\frac{1}{p_2} - \frac{1}{p_1}} \quad \text{if} \quad k_j > 2M_j. \tag{8.21}$$

Now we choose
$$k_j = 2^{Jd}\, 2^{-(J-j)\varepsilon} \quad \text{if} \quad j = 0,\dots,J \tag{8.22}$$
and
$$k_j = 2^{Jd}\, 2^{-(j-J)\varkappa} \quad \text{if} \quad j = J+1,\dots,L, \tag{8.23}$$
where ε and $\varkappa$ are positive numbers which will be chosen later on. We obtain
$$k \sim 2^{Jd}, \tag{8.24}$$
where «$\sim$» indicates equivalences (two-sided estimates up to unimportant positive constants which are independent of J). We deal with the two sums in (8.18) separately.

Step 3. Let $j = 0,\dots,J$. By (8.22) we have
$$k_j = 2^{jd}\, 2^{(J-j)(d-\varepsilon)} \ge 2^{jd}, \tag{8.25}$$
where we choose $0 < \varepsilon < d$. By (8.1) and (8.20), (8.21) it follows that we can always apply (8.21) in that case. Then we obtain
$$e_{k_j}(id_j) \le c\, 2^{\lambda_{j,J}} \tag{8.26}$$
with
$$\lambda_{j,J} = -J\delta + Jd(\tfrac{1}{p_2} - \tfrac{1}{p_1}) + (J-j)\left(\delta - d(\tfrac{1}{p_2} - \tfrac{1}{p_1})\right) - c\, 2^{(J-j)(d-\varepsilon)} \tag{8.27}$$
and consequently
$$\sum_{j=0}^{J} e_{k_j}^{\varrho}(id_j) \le c\, 2^{-J\delta\varrho + J\varrho d(\tfrac{1}{p_2} - \tfrac{1}{p_1})}. \tag{8.28}$$

Step 4. Let $j = J+1,\dots,L$. By (8.23) we have
$$k_j \le c\, 2^{jd}. \tag{8.29}$$
Hence we can always apply (8.20) and obtain
$$e_{k_j}(id_j) \le c\, 2^{-j\delta}\left[2^{-Jd+(j-J)\varkappa}\log\left(c\, 2^{(j-J)d}\, 2^{(j-J)\varkappa}\right)\right]^{\tfrac{1}{p_1} - \tfrac{1}{p_2}}$$
$$\le c\, 2^{-J\delta + Jd(\tfrac{1}{p_2} - \tfrac{1}{p_1})}\, 2^{(J-j)[\delta + \varkappa(\tfrac{1}{p_2} - \tfrac{1}{p_1})]}\,[(d+\varkappa)(j-J)]^{\tfrac{1}{p_1} - \tfrac{1}{p_2}}. \tag{8.30}$$
We choose $\varkappa > 0$ such that
$$\varkappa\left(\tfrac{1}{p_1} - \tfrac{1}{p_2}\right) < \delta \tag{8.31}$$
and obtain
$$\sum_{j=J+1}^{L} e_{k_j}^{\varrho}(id_j) \le c\, 2^{-J\delta\varrho + J d\varrho(\tfrac{1}{p_2} - \tfrac{1}{p_1})}. \tag{8.32}$$

Step 5. By (8.24), (8.14), (8.18), (8.28), and (8.32) we have
$$e_{c\, 2^{Jd}} \le c'\, 2^{-J\delta + Jd(\tfrac{1}{p_2} - \tfrac{1}{p_1})}, \quad J \in \mathbb{N}, \tag{8.33}$$
where c and c' are appropriate positive constants. This coincides essentially with (8.15) and completes the proof of the right-hand side of (8.8).

8.3 Remark In the case of Banach spaces, which means that the numbers p_1, p_2, q_1 and q_2 in (8.3) are larger than or equal to 1, the above theorem is more or less known, see [Kühn84]. In that paper the proof is based on interpolation properties of entropy numbers and entropy numbers of diagonal operators in ℓ_p-spaces due to Carl, see [CaS90].

8.4 Estimates of constants

Theorem 8.2 is the basis for the study of entropy numbers of embedding operators in function spaces. Usually, in non-limiting situations, all parameters p_1, p_2, q_1, q_2, d and δ are fixed and there is no need to have additional information on the dependence of c and C in (8.8) on these parameters. However in some limiting cases we deal with a sequence of target spaces and we have to know how C in (8.8) depends on p_2, q_2 and δ, whereas the dependence of C on p_1, q_1 and d is not so interesting for our later purposes. To facilitate the estimates we assume in addition

$$ 1 \le p_2 \le \infty, \quad 1 \le q_2 \le \infty, \quad \text{and} \quad 0 < \delta \le 1. \tag{8.34} $$

This is not really necessary, but sufficient for our later purposes.

8.5 Corollary *Under the hypotheses of Theorem 8.2, complemented by (8.34), we have*

$$ e_k \le c\,\delta^{-1-2(\frac{1}{p_1}-\frac{1}{p_2})}\, k^{-\frac{\delta}{d}+\frac{1}{p_2}-\frac{1}{p_1}}, \quad k \in \mathbb{N}, \tag{8.35} $$

for some positive constant c which is independent of p_2, q_2, and δ (but may depend on p_1, q_1, and d).

Proof. We follow the arguments in the Steps 2–5 of the proof of Theorem 8.2. By (8.34) we have $\varrho = 1$ in (8.18). We estimate the constant c in the first term on the right-hand side of (8.18). By (8.7) and (8.17) we have

$$ \left\| \sum_{j=L+1}^{\infty} id_j \right\| \le \sum_{j=L+1}^{\infty} \|id_j\| \le \sum_{j=L+1}^{\infty} 2^{-j\delta} \le c\,\delta^{-1}\,2^{-L\delta}, \tag{8.36} $$

where c is independent of δ. Next we remark that we may assume that the constant c_2 in (7.11)–(7.13) is independent of p_2. We refer to [ET96], Remark 2 on p. 101. But this is not a deep result. It follows immediately from Theorem 7.3 with $p_2 = \infty$ and the interpolation properties of the entropy numbers mentioned in 6.7. Having this in mind it follows that the constants c in (8.20) and (8.21) are independent of p_2, q_2 and δ. We may choose $\varepsilon = \frac{d}{2}$ in (8.22) and (8.25). Hence ε is not of interest for us. As for $\varkappa$ in (8.23) and (8.31) we may choose $\varkappa = \frac{\delta p_1}{2}$. Then (8.24) must be substituted now by

$$ 2^{Jd} \le k \le \frac{c}{\delta}\,2^{Jd}, \tag{8.37} $$

where c is independent of δ (and p_2 and q_2). Now by the above remarks about the constants in Theorem 7.3 and $\delta \le 1$ it follows from (8.26) and (8.27) that the constant c in (8.28) (now with $\varrho = 1$) is independent of p_2, q_2 and δ. We estimate the constant c in (8.32) (again with $\varrho = 1$). By the above choice of $\varkappa$ it follows from (8.30)

$$\sum_{j=J+1}^{L} e_{k_j}(id_j) \le c_1 \, 2^{-J\delta + Jd(\frac{1}{p_2} - \frac{1}{p_1})} \, \delta^{\frac{1}{p_2} - \frac{1}{p_1}} \sum_{l=1}^{L-J} 2^{-c_2 l\delta} \, (\delta l)^{\frac{1}{p_1} - \frac{1}{p_2}} \tag{8.38}$$

where c_1 and c_2 are independent of δ, p_2 and q_2. The last factor can be estimated from above by

$$\int_0^\infty e^{-c_3\delta t} \, (\delta t)^{\frac{1}{p_1} - \frac{1}{p_2}} \, dt \le c_4 \, \delta^{-1}. \tag{8.39}$$

Now by (8.18), (8.36), (8.28) and (8.38) with (8.39) we obtain

$$e_k \le c \, \delta^{-1 - \frac{1}{p_1} + \frac{1}{p_2}} \, 2^{-J\delta + Jd(\frac{1}{p_2} - \frac{1}{p_1})}, \tag{8.40}$$

where c is independent of δ, p_2, and q_2, and k is given by (8.37). With $2^{Jd} \sim k\delta$ in (8.40) we have

$$e_k \le c \, \delta^{-1 - 2(\frac{1}{p_1} - \frac{1}{p_2})} \, k^{-\frac{\delta}{d} + \frac{1}{p_2} - \frac{1}{p_1}}, \quad k \in \mathbb{N}. \tag{8.41}$$

The proof of (8.35) is complete.

8.6 Remark The restrictions $p_2 \ge 1$ and $q_2 \ge 1$ are unimportant. Otherwise one has $\varrho < 1$ in (8.18). There is no problem to follow the above reasoning in this more general case.

8.7 Comparison The estimates for entropy numbers of compact embeddings between function spaces will be based in non-limiting cases on (8.8), whereas in some limiting cases we need the additional information given in the above corollary. Although the context is slightly different (so far) one can compare the exponent $1 + 2(\frac{1}{p_1} - \frac{1}{p_2})$ of δ in (8.35) with the exponents $1 + \frac{2}{p}$ in [ET96], p. 130, formula (8), and $-\frac{2s}{n} - \varepsilon$ in [ET96], p. 139, formula (3), where any $\varepsilon > 0$ is admitted. It comes out that $1 + \frac{2}{p}$ originates precisely from $1 + 2(\frac{1}{p_1} - \frac{1}{p_2})$ (restricted to the treated case), whereas $-\frac{2s}{n} - \varepsilon$ is somewhat better ($\varepsilon = 1$ would be the direct counterpart). On the other hand the situation considered in [ET96], p. 139, is more special. We return in 23.5 to these comparisons in greater detail and shed more light upon these admittedly somewhat cryptical remarks.

8.8 A digression: Matrix operators

It is not our aim to discuss the spectral theory of compact operators acting in ℓ_p-spaces. This has been done in great detail by A. Pietsch, B. Carl and other mathematicians. We refer to [Pie87], esp. pp. 230–231, [Kön86], esp. pp. 150–151, and [CaS90]. Our intention here is simply to demonstrate the power of Theorem 8.2. For that purpose we estimate the distribution of eigenvalues of some matrix operators in ℓ_p-spaces. We avoid any technical complications and we are far from the most general case which can be treated in that way. In this sense we leave it to the interested reader to compare the results obtained here with the more systematic treatments in the above-mentioned books. Let $0 < p \leq \infty$; recall that ℓ_p is the linear space of all complex sequences $x = (x_k : k \in \mathbb{N})$ endowed with the quasi-norm

$$\|x|\ell_p\| = \left(\sum_{k=1}^{\infty} |x_k|^p \right)^{\frac{1}{p}} \tag{8.42}$$

with the obvious modification if $p = \infty$. Let

$$A = (a_{lk} : l \in \mathbb{N}, \ k \in \mathbb{N}), \quad a_{lk} \in \mathbb{C}, \tag{8.43}$$

and as usual let

$$Ax = \left(\sum_{k=1}^{\infty} a_{lk} x_k : \ l \in \mathbb{N} \right) \quad \text{for} \quad x = (x_k : k \in \mathbb{N}). \tag{8.44}$$

Let $d > 0$ and $\delta > 0$, and $M_j \sim 2^{jd}$ according to (8.1) with $j \in \mathbb{N}_0$. We put

$$M^j = \sum_{m=0}^{j-1} M_m \sim 2^{jd}, \quad j \in \mathbb{N}; \quad \text{and} \quad M^0 = 0, \tag{8.45}$$

and assume that the entries a_{lk} can be represented as

$$a_{lk} = 2^{-j\delta} b_{lm}^j, \quad l \in \mathbb{N}, \tag{8.46}$$

$$k = M^j + m \quad \text{for} \quad j \in \mathbb{N}_0 \quad \text{and} \quad m = 1, \ldots, M_j, \tag{8.47}$$

and

$$\sum_{l=0}^{\infty} \left(\sum_{j=0}^{\infty} \sum_{m=1}^{M_j} |b_{lm}^j| \right)^p < \infty, \quad 0 < p \leq \infty, \tag{8.48}$$

with the obvious modification if $p = \infty$. In other words, for fixed l we sum first the absolute values of the entries in the lth row. Hence,

$$B = (b_{lk} = b_{lm}^j : l \in \mathbb{N}, \ k \text{ given by } (8.47)) \tag{8.49}$$

is a so-called *Hille-Tamarkin matrix*, see the above references. In particular, A can be decomposed by

$$A = B \circ D, \tag{8.50}$$

where D is a diagonal matrix with the entries $d_k = 2^{-j\delta}$, where k is given by (8.47). As we shall see, A generates a compact operator in ℓ_p. Then we can apply the Riesz theory mentioned in 6.1, 6.6 and 6.8. In particular, the non-zero eigenvalues $\mu_k(A)$ of A, repeated according to algebraic multiplicity, can be ordered as in (6.11).

8.9 Proposition *Let $0 < p \le \infty$ and let A be the above operator. Then there is a positive constant c such that*

$$|\mu_k(A)| \le c \, k^{-\frac{\delta}{d} - \frac{1}{p}}, \quad k \in \mathbb{N}. \tag{8.51}$$

Proof. We decompose A as

$$A = B \circ id \circ D, \tag{8.52}$$

where D and B have the above meaning. We claim

$$\begin{aligned}
D &: \ell_p \rightarrow \ell_p \left(2^{j\delta} \ell_p^{M_j}\right), \\
id &: \ell_p \left(2^{j\delta} \ell_p^{M_j}\right) \rightarrow \ell_\infty, \\
B &: \ell_\infty \rightarrow \ell_p.
\end{aligned} \tag{8.53}$$

The first line is obvious where we used the notation introduced in (8.2). By Theorem 8.2 the operator id is compact and

$$e_k(id) \le C \, k^{-\frac{\delta}{d} - \frac{1}{p}}, \quad k \in \mathbb{N}. \tag{8.54}$$

Let $x = (x_k : k$ given by (8.47)$) \in \ell_\infty$. Then by (8.49)

$$\begin{aligned}
|(Bx)_l| &= \left| \sum_{j=0}^{\infty} \sum_{m=1}^{M_j} b_{lm}^j x_{M^j + m} \right| \\
&\le \|x|\ell_\infty\| \sum_{j=0}^{\infty} \sum_{m=1}^{M_j} |b_{lm}^j|.
\end{aligned} \tag{8.55}$$

Now the last line in (8.53) is a consequence of (8.48) and (8.55). Since D and B are bounded, (8.52), (8.54), and (6.8) prove

$$e_k(A) \le c \, k^{-\frac{\delta}{d} - \frac{1}{p}}, \quad k \in \mathbb{N}. \tag{8.56}$$

Finally, (8.51) is a consequence of Corollary 6.10.

8.10 Remark It can be easily seen that the exponent in (8.51) is sharp: Let A be a diagonal operator, $a_{kl} = 0$ if $k \neq l$ and

$$a_{kk} = k^{-\frac{\delta}{d} - \frac{1}{p}} (\log k)^\alpha. \qquad (8.57)$$

By (8.47) and (8.48) with (8.45) we have

$$\sum_{l=0}^{\infty} \left(\sum_{j=0}^{\infty} \sum_{m=1}^{M_j} |b_{lm}^j| \right)^p \sim \sum_{j=0}^{\infty} 2^{-jd} j^{\alpha p} 2^{jd} < \infty \qquad (8.58)$$

if $\alpha < -\frac{1}{p}$. In that case A has the required properties. On the other hand, a_{kk} are the eigenvalues of A. Hence the exponent in (8.51) is the best possible.

9   Weighted ℓ_p-spaces: a generalization

9.1 Preliminaries and notation

Unfortunately Theorem 8.2 and Corollary 8.5 are not completely sufficient for our later purposes. We need something like an ℓ_u-version of these two assertions. Fortunately it comes out that these generalizations are nothing more than a technical appendix to the results just mentioned. We use the same notation as in 8.1. In particular, let $d > 0$, $\delta \geq 0$ and $(M_j)_{j \in \mathbb{N}_0}$ be a sequence of natural numbers with (8.1) for some positive numbers c_1 and c_2. Let again $\ell_q(2^{j\delta} \ell_p^{M_j})$ with $0 < p \leq \infty$ and $0 < q \leq \infty$ be the quasi-Banach space introduced in 8.1 and quasi-normed by (8.2). Let, in addition, $\mu \geq 0$ and $0 < u \leq \infty$. Then by

$$\ell_u \left[2^{\mu m} \ell_q \left(2^{j\delta} \ell_p^{M_j} \right) \right]$$

we shall mean the linear space of all $\ell_q(2^{j\delta} \ell_p^{M_j})$-valued sequences $x = (x^m : m \in \mathbb{N}_0)$ endowed with the quasi-norm

$$\left\| x \mid \ell_u \left[2^{\mu m} \ell_q \left(2^{j\delta} \ell_p^{M_j} \right) \right] \right\| = \left(\sum_{m=0}^{\infty} 2^{\mu m u} \left\| x^m \mid \ell_q (2^{j\delta} \ell_p^{M_j}) \right\|^u \right)^{\frac{1}{u}} \qquad (9.1)$$

with the obvious modification according to the vector-valued version of (7.2) if $u = \infty$. In case of $\mu = \delta = 0$ we write $\ell_u [\ell_q(\ell_p^{M_j})]$ in accordance with the notation introduced in 8.1. We are interested in an extension of Theorem 8.2. Let $d > 0$, $\delta > 0$, $\mu > 0$,

$$0 < p_1 \leq p_2 \leq \infty \qquad (9.2)$$

and

$$0 < q_1 \leq \infty,\ 0 < q_2 \leq \infty,\ 0 < u_1 \leq \infty,\ 0 < u_2 \leq \infty. \qquad (9.3)$$

Then the identity map

$$id : \quad \ell_{u_1}\left[2^{\mu m}\,\ell_{q_1}\left(2^{j\delta}\,\ell_{p_1}^{M_j}\right)\right] \;\to\; \ell_{u_2}\left[\ell_{q_2}\left(\ell_{p_2}^{M_j}\right)\right] \tag{9.4}$$

is compact. This is simply the extension of what had been said in 8.1, see (8.4)–(8.7), from the scalar case to the ℓ_u-valued case. However this generalization is an immediate consequence of $\mu > 0$. Now Theorem 8.2 can be rather easily extended to the vector-valued case.

9.2 Theorem *Let $d > 0$, $\delta > 0$, $\mu > 0$, and $M_j \in \mathbb{N}$ with (8.1). Let p_1, p_2, q_1, q_2, u_1, u_2 be given by (9.2) and (9.3). Let e_k be the entropy numbers of the compact operator id according to (9.4). There are two positive numbers c and C such that*

$$c\,k^{-\frac{\delta}{d}+\frac{1}{p_2}-\frac{1}{p_1}} \;\le\; e_k \;\le\; C\,k^{-\frac{\delta}{d}+\frac{1}{p_2}-\frac{1}{p_1}}, \quad k \in \mathbb{N}. \tag{9.5}$$

Proof.
Step 1. The estimate from below is covered by Step 1 of the proof of Theorem 8.2.

Step 2. We reduce the estimate from above to the corresponding scalar case in Theorem 8.2. Let

$$id_m : \quad x \mapsto x^m, \quad \text{where} \quad x = (x^l)_{l \in \mathbb{N}_0} \tag{9.6}$$

has the same meaning as in (9.1). Then we have

$$id = \sum_{m=0}^{\infty} id_m. \tag{9.7}$$

Let, for brevity, $a = \frac{\delta}{d} + \frac{1}{p_1} - \frac{1}{p_2}$, and let $J \in \mathbb{N}$ and

$$L = [\frac{a}{\mu}J] \in \mathbb{N}_0. \tag{9.8}$$

Of course $a > 0$. Then it follows that

$$\left\|\sum_{l=L+1}^{\infty} id_l\right\| \;\le\; c\,2^{-\mu L} \;\le\; c'\,2^{-aJ}. \tag{9.9}$$

Let

$$k_l = 2^J\,2^{-l\varepsilon} \quad \text{where } l = 0, \ldots, L \quad \text{and} \quad \varepsilon a < \mu. \tag{9.10}$$

Then we have

$$k = \sum_{l=0}^{L} k_l \sim 2^J \tag{9.11}$$

and by (8.8)

$$e_{k_l}(id_l) \le c\,2^{-\mu l}\,2^{-aJ}\,2^{la\varepsilon}; \quad l = 0, \ldots, L. \tag{9.12}$$

Now by (9.9)–(9.11) and (6.9) it follows that

$$e_{c_1\,2^J}(id) \le c_2\,2^{-aJ}, \quad J \in \mathbb{N}, \tag{9.13}$$

for some $c_1 > 0$ and $c_2 > 0$. This proves the right-hand side of (9.5).

9.3 Remark We are also interested in an extension of Corollary 8.5 to the vector-valued case. In accordance with (8.34) we assume in addition

$$1 \leq p_2 \leq \infty, \ 1 \leq q_2 \leq \infty, \ 1 \leq u_2 \leq \infty, \ \text{and} \ 0 < \delta \leq 1. \tag{9.14}$$

Again, these conditions are not really necessary, but sufficient for our later purposes.

9.4 Corollary *Under the hypotheses of Theorem 9.2 complemented by (9.14) we have*

$$e_k \leq c \, \delta^{-1-2\left(\frac{1}{p_1}-\frac{1}{p_2}\right)} k^{-\frac{\delta}{d}+\frac{1}{p_2}-\frac{1}{p_1}}, \quad k \in \mathbb{N}, \tag{9.15}$$

for some positive constant c which is independent of p_2, q_2, u_2, and δ (but may depend on p_1, q_1, u_1, d and μ).

Proof. By slight modification we may assume that L in (9.8) and ε in (9.10) are chosen independently of the indicated numbers. Then we have the same situation in (9.9) and (9.11) by a similar argument as in (8.36). Replacing the constant c in (9.12) by the corresponding constant on the right-hand side of (8.35) we obtain (9.13) with the desired constant. This proves (9.15).

9.5 Remark As for comments we refer to 8.6 and 8.7.

Chapter III
Function spaces on $\mathbb{R}^n$

10 The spaces B^s_{pq} and F^s_{pq}

10.1 Introduction

This chapter deals mainly with the function spaces B^s_{pq} and F^s_{pq} and their special cases on $\mathbb{R}^n$. These spaces have been treated in detail in [Tri83], [Tri92] and, as far as some more recent results are concerned, in [ET96] and [RuS96]. The aim of this chapter is twofold. Firstly, to make this and the following chapters self-contained we give all the necessary definitions and collect those assertions which are needed later on. Secondly, beginning with Section 12, we discuss and establish some new properties in detail. This applies especially to so-called harmonic and local characterizations, as well as atomic and subatomic representations both of scalar-valued and vector-valued function spaces of the above type. We need later on only the subatomic (or quarkonial) decompositions in the scalar case, but its extension to the vector-valued case (that means with values in arbitrary complex Banach spaces) is even more striking. So we include these considerations as a digression which may be skipped. These results might be of some use in the theory of linear and nonlinear parabolic and evolutionary equations. But these possible applications are not the subject of this book.

10.2 Basic notation

Let $\mathbb{R}^n$ be euclidean n-space, put $\mathbb{R} = \mathbb{R}^1$. Let $\mathbb{N}$ be the collection of all natural numbers and $\mathbb{N}_0 = \mathbb{N} \cup \{0\}$. Let $S(\mathbb{R}^n)$ be the Schwartz space of all complex-valued, rapidly decreasing, infinitely differentiable functions on $\mathbb{R}^n$. By $S'(\mathbb{R}^n)$ we denote its topological dual, the space of all tempered distributions on $\mathbb{R}^n$. Furthermore, $L_p(\mathbb{R}^n)$ with $0 < p \leq \infty$, is the usual quasi-Banach space with respect to the Lebesgue measure, quasi-normed by

$$\|f|L_p(\mathbb{R}^n)\| = \left(\int_{\mathbb{R}^n} |f(x)|^p \, dx \right)^{\frac{1}{p}} \tag{10.1}$$

with the usual modification if $p = \infty$. If $\varphi \in S(\mathbb{R}^n)$ then

$$\hat{\varphi}(\xi) = (F\varphi)(\xi) = (2\pi)^{-\frac{n}{2}} \int_{\mathbb{R}^n} e^{-i\xi x} \varphi(x) \, dx, \quad \xi \in \mathbb{R}^n, \tag{10.2}$$

denotes the Fourier transform of φ. As usual, $F^{-1}\varphi$ or $\check{\varphi}$, stands for the inverse Fourier transform, given by the right-hand side of (10.2) with i in place of $-i$. Whether we use $F\varphi$, $F^{-1}\varphi$ or $\hat{\varphi}$, $\check{\varphi}$ will be a matter of convenience. Of course,

H. Triebel, *Fractals and Spectra*, Modern Birkhäuser Classics,
DOI 10.1007/978-3-0348-0034-1_3, © Birkhäuser Verlag 1997

ξx denotes the scalar product in $\mathbb{R}^n$. Both F and F^{-1} are extended to $S'(\mathbb{R}^n)$ in the standard way. Let $\varphi \in S(\mathbb{R}^n)$ with

$$\varphi(x) = 1 \quad \text{if} \quad |x| \leq 1 \quad \text{and} \quad \varphi(x) = 0 \quad \text{if} \quad |x| \geq \frac{3}{2}. \tag{10.3}$$

We put $\varphi_0 = \varphi$, $\varphi_1(x) = \varphi(\frac{x}{2}) - \varphi(x)$ and

$$\varphi_k(x) = \varphi_1(2^{-k+1}x), \quad x \in \mathbb{R}^n, \ k \in \mathbb{N}. \tag{10.4}$$

Then, since

$$1 = \sum_{k=0}^{\infty} \varphi_k(x) \quad \text{for all } x \in \mathbb{R}^n, \tag{10.5}$$

the φ_k form a dyadic resolution of unity. Recall that $(\varphi_k \hat{f})^{\vee}$ is an entire analytic function on $\mathbb{R}^n$ for any $f \in S'(\mathbb{R}^n)$. In particular, $(\varphi_k \hat{f})^{\vee}(x)$ makes sense pointwise.

10.3 Definition

(i) *Let $s \in \mathbb{R}$, $0 < p \leq \infty$ and $0 < q \leq \infty$. Then $B_{pq}^s(\mathbb{R}^n)$ is the collection of all $f \in S'(\mathbb{R}^n)$ such that*

$$\|f \mid B_{pq}^s(\mathbb{R}^n)\|_{\varphi} = \left(\sum_{j=0}^{\infty} 2^{jsq} \left\| (\varphi_j \hat{f})^{\vee} \mid L_p(\mathbb{R}^n) \right\|^q \right)^{\frac{1}{q}} \tag{10.6}$$

(with the usual modification if $q = \infty$) is finite.

(ii) *Let $s \in \mathbb{R}$, $0 < p < \infty$ and $0 < q \leq \infty$. Then $F_{pq}^s(\mathbb{R}^n)$ is the collection of all $f \in S'(\mathbb{R}^n)$ such that*

$$\|f \mid F_{pq}^s(\mathbb{R}^n)\|_{\varphi} = \left\| \left(\sum_{j=0}^{\infty} 2^{jsq} |(\varphi_j \hat{f})^{\vee}(\cdot)|^q \right)^{\frac{1}{q}} \mid L_p(\mathbb{R}^n) \right\| \tag{10.7}$$

(with the usual modification if $q = \infty$) is finite.

10.4 Remark These spaces, including their forerunners and special cases, have a long history, see also 10.5. The theory of these spaces in its full extent has been developed in [Tri83] and [Tri92]. As for more recent topics we refer to [ET96] and [RuS96]. In particular, both $B_{pq}^s(\mathbb{R}^n)$ and $F_{pq}^s(\mathbb{R}^n)$ are quasi-Banach spaces which are independent of the function φ chosen according to (10.3). This justifies our omission of the subscript φ in (10.6) and (10.7) in what follows. If $p \geq 1$ and $q \geq 1$ then both $B_{pq}^s(\mathbb{R}^n)$ and $F_{pq}^s(\mathbb{R}^n)$ are Banach spaces. For sake of clarity we shall occasionally write $B_{p,q}^s(\mathbb{R}^n)$ or $F_{p,q}^s(\mathbb{R}^n)$ instead of $B_{pq}^s(\mathbb{R}^n)$ or $F_{pq}^s(\mathbb{R}^n)$, especially when p and q are concrete numbers.

10.5 Concrete spaces and lifts

Following [ET96], pp. 25–27, we list some special cases of the above spaces $B^s_{pq}(\mathbb{R}^n)$ and $F^s_{pq}(\mathbb{R}^n)$ without further comments. Details may be found in [Tri83], especially 2.2.2, p. 35, and [Tri92], especially Ch. 1.

(i) Let $1 < p < \infty$; then

$$F^0_{p,2}(\mathbb{R}^n) = L_p(\mathbb{R}^n). \tag{10.8}$$

This is a Paley-Littlewood theorem, see [Tri83], 2.5.6, p. 87.

(ii) Let $1 < p < \infty$ and $s \in \mathbb{N}_0$; then

$$F^s_{p,2}(\mathbb{R}^n) = W^s_p(\mathbb{R}^n) \tag{10.9}$$

are the *classical Sobolev spaces*, usually normed by

$$\|f \mid W^s_p(\mathbb{R}^n)\| = \sum_{|\alpha| \le s} \|D^\alpha f \mid L_p(\mathbb{R}^n)\|. \tag{10.10}$$

This generalizes assertion (i), see again [Tri83], 2.5.6, p. 87.

(iii) Let $\sigma \in \mathbb{R}$; then

$$I_\sigma : \quad f \mapsto (\langle \xi \rangle^\sigma \hat{f})^\vee, \tag{10.11}$$

with $\langle \xi \rangle = (1 + |\xi|^2)^{\frac{1}{2}}$, is an one-to-one map of $S(\mathbb{R}^n)$ onto itself and of $S'(\mathbb{R}^n)$ onto itself. As for the spaces $B^s_{pq}(\mathbb{R}^n)$ and $F^s_{pq}(\mathbb{R}^n)$ with $s \in \mathbb{R}, 0 < p \le \infty$ ($p < \infty$ for the F scale), $0 < q \le \infty$, I_σ acts as a lift (equivalent quasi-norms):

$$I_\sigma B^s_{pq}(\mathbb{R}^n) = B^{s-\sigma}_{pq}(\mathbb{R}^n) \quad \text{and} \quad I_\sigma F^s_{pq}(\mathbb{R}^n) = F^{s-\sigma}_{pq}(\mathbb{R}^n), \tag{10.12}$$

see [Tri83], 2.3.8, p. 58. In particular, let

$$H^s_p(\mathbb{R}^n) = I_{-s} L_p(\mathbb{R}^n), \quad s \in \mathbb{R}, \quad 1 < p < \infty, \tag{10.13}$$

be the (fractional) *Sobolev spaces*. Then (10.8), (10.9), and (10.12) yield

$$H^s_p(\mathbb{R}^n) = F^s_{p,2}(\mathbb{R}^n), \quad s \in \mathbb{R}, \ 1 < p < \infty, \tag{10.14}$$

and

$$H^s_p(\mathbb{R}^n) = W^s_p(\mathbb{R}^n) \quad \text{if} \quad s \in \mathbb{N}_0 \quad \text{and} \quad 1 < p < \infty. \tag{10.15}$$

As for notations we follow the recent custom and refer to the members of the *full scale* $H^s_p(\mathbb{R}^n)$ given by (10.13) *as Sobolev spaces*.

(iv) Denote

$$\mathcal{C}^s(\mathbb{R}^n) = B^s_{\infty,\infty}(\mathbb{R}^n), \quad s \in \mathbb{R}, \tag{10.16}$$

as the *Hölder-Zygmund spaces.* Let

$$(\Delta_h^1 f)(x) \;=\; f(x+h) - f(x), \quad (\Delta_h^{l+1} f)(x) \;=\; \Delta_h^1 (\Delta_h^l f)(x), \quad (10.17)$$
$$x \in \mathbb{R}^n, \quad h \in \mathbb{R}^n, \quad l \in \mathbb{N},$$

be the iterated differences in $\mathbb{R}^n$. Let $0 < s < m \in \mathbb{N}$; then

$$\|f \,|\, \mathscr{C}^s(\mathbb{R}^n)\|_m \;=\; \sup_{x \in \mathbb{R}^n} |f(x)| \;+\; \sup |h|^{-s} |\Delta_h^m f(x)|, \qquad (10.18)$$

where the second supremum is taken over all $x \in \mathbb{R}^n$ and $h \in \mathbb{R}^n$ with $0 < |h| \le 1$, are equivalent norms on $\mathscr{C}^s(\mathbb{R}^n)$. Some differences in (10.18) can be replaced by some derivatives. We refer to [Tri83], 2.2.2, 2.5.12. Again we have extended the Hölder-Zygmund spaces to $s \le 0$.

(v) Assertion (iv) can be generalized as follows. Once more let $0 < s < m \in \mathbb{N}$ and $1 \le p \le \infty$, $1 \le q \le \infty$. Then

$$\|f \,|\, L_p(\mathbb{R}^n)\| \;+\; \left(\int_{|h| \le 1} |h|^{-sq} \, \|\Delta_h^m f \,|\, L_p(\mathbb{R}^n)\|^q \, \frac{dh}{|h|^n} \right)^{\frac{1}{q}} \qquad (10.19)$$

(with the usual modification if $q = \infty$) are equivalent norms on $B_{pq}^s(\mathbb{R}^n)$; see again [Tri83], 2.2.2, 2.5.12. These are the *classical Besov spaces.* Again some differences in (10.19) can be replaced by some derivatives.

(vi) The *inhomogeneous Hardy spaces* $h_p(\mathbb{R}^n)$ and the space $bmo\,(\mathbb{R}^n)$ of all functions with bounded mean oscillation can also be included in the scales introduced in 10.3 (in the latter case after some modifications). We shall hardly need these spaces in this book. So we refer to [ET96], p. 27, for a short description and to the literature quoted there.

10.6 Equivalent quasi-norms

The list in 10.5 makes the point that many classical function spaces, with their own historical roots, are special cases of the two scales B_{pq}^s and F_{pq}^s. It is possible to obtain some of these equivalent norms, for example (10.18) and (10.19), by (10.6) if φ in (10.3) and (10.4) is modified in a suitable way. We studied in [Tri92], 2.4.1 and 2.5.1 under which modified conditions on φ in (10.3) and (10.4), the right-hand sides of (10.7) and (10.6) are equivalent quasi-norms on $F_{pq}^s(\mathbb{R}^n)$ and $B_{pq}^s(\mathbb{R}^n)$, respectively. Of crucial interest for us is the possibility to obtain in that way characterizations of the above spaces in terms of harmonic functions

$$u(x,t) \;=\; (e^{-t|\xi|} \, \hat{f}(\xi))^{\vee}(x), \quad x \in \mathbb{R}^n, t > 0, \qquad (10.20)$$

on $\mathbb{R}_+^{n+1}$, see [Tri92], pp. 152–153, but we return to this point later on in a modified form in great detail.

11 Properties

11.1 Some properties
In this section we collect a few properties of the spaces $B_{pq}^s(\mathbb{R}^n)$ and $F_{pq}^s(\mathbb{R}^n)$ needed later on and which are proved in [Tri83], [Tri92], and [ET96].

11.2 Local means
We follow [Tri92], 2.4.6 and 2.5.3, pp. 122–124 and 138–139, respectively. Let k^0 and k_0 be two C^∞ functions in $\mathbb{R}^n$ with compact support and

$$\widehat{k_0}(0) \neq 0, \quad \text{and} \quad \widehat{k^0}(0) \neq 0. \tag{11.1}$$

Let

$$k_N(y) = \Delta^N k^0(y) = \left(\sum_{j=1}^n \frac{\partial^2}{\partial y_j^2} \right)^N k^0(y) \quad \text{for} \quad N \in \mathbb{N}. \tag{11.2}$$

Then the *local means*

$$k_M(t,f)(x) = \int_{\mathbb{R}^n} k_M(y) f(x+ty)\, dy$$

$$= t^{-n} \int_{\mathbb{R}^n} k_M\left(\frac{y-x}{t} \right) f(y)\, dy, \tag{11.3}$$

$$x \in \mathbb{R}^n,\ t > 0,\ M \in \mathbb{N}_0,$$

make sense for any $f \in S'(\mathbb{R}^n)$ (appropriately interpreted). Let

$$\sigma_p = n \left(\frac{1}{p} - 1 \right)_+ \quad \text{where} \quad 0 < p \leq \infty. \tag{11.4}$$

We have the following two assertions:

(i) *Let* $0 < p < \infty$, $0 < q \leq \infty$ *and* $s \in \mathbb{R}$. *Let* $N \in \mathbb{N}$ *with*

$$2N > \max(s, \sigma_p). \tag{11.5}$$

Then

$$\|k_0(1,f)\,|\,L_p(\mathbb{R}^n)\| + \left\| \left(\sum_{j=1}^\infty 2^{jsq} |k_N(2^{-j},f)(\cdot)|^q \right)^{\frac{1}{q}} |\,L_p(\mathbb{R}^n) \right\| \tag{11.6}$$

and

$$\|k_0(1,f)\,|\,L_p(\mathbb{R}^n)\| + \left\| \left(\int_0^1 t^{-sq} |k_N(t,f)(\cdot)|^q \frac{dt}{t} \right)^{\frac{1}{q}} |\,L_p(\mathbb{R}^n) \right\| \tag{11.7}$$

(modification if $q = \infty$*) are equivalent quasi-norms in* $F_{pq}^s(\mathbb{R}^n)$.

(ii) *Let* $0 < p \leq \infty$, $0 < q \leq \infty$ *and* $s \in \mathbb{R}$. *Let* $N \in \mathbb{N}$ *with (11.5). Then*

$$\|k_0(1,f) \,|\, L_p(\mathbb{R}^n)\| + \left(\sum_{j=1}^{\infty} 2^{jsq} \, \|k_N(2^{-j},f) \,|\, L_p(\mathbb{R}^n)\|^q \right)^{\frac{1}{q}} \tag{11.8}$$

and

$$\|k_0(1,f) \,|\, L_p(\mathbb{R}^n)\| + \left(\int_0^1 t^{-sq} \, \|k_N(t,f) \,|\, L_p(\mathbb{R}^n)\|^q \, \frac{dt}{t} \right)^{\frac{1}{q}} \tag{11.9}$$

(modification if $q = \infty$) *are equivalent quasi-norms in* $B_{pq}^s(\mathbb{R}^n)$.

The question arises whether each $f \in S'(\mathbb{R}^n)$ such that the quasi-norm in, say, (11.6) is finite belongs to $F_{pq}^s(\mathbb{R}^n)$. We discussed problems of this type in [Tri92], 2.4.2, in some detail and in greater generality. See the recent papers [BPT96a,b] where questions of that type are solved in a rather final way. The following special case is covered by the arguments developed in [Tri92], 2.4.2:

(iii) *Let* $1 \leq p \leq \infty$, $0 < q \leq \infty$, *and* $s \in \mathbb{R}$. *Let* $N \in \mathbb{N}$ *with (11.5). Then*

$$B_{pq}^s(\mathbb{R}^n) = \{f \in S'(\mathbb{R}^n) : \text{the quasi-norm in (11.8) is finite}\}. \tag{11.10}$$

We return to problems of this type related to local means in 12.8 and 12.9.

11.3 Localization principle
Let $F_{pq}^s(\mathbb{R}^n)$ and $B_{pq}^s(\mathbb{R}^n)$ be the spaces introduced in Definition 10.3. Let ψ be a compactly supported C^∞ function in $\mathbb{R}^n$ and let

$$\psi_k(x) = \psi(x - k), \quad k \in \mathbb{Z}^n, \quad x \in \mathbb{R}^n, \tag{11.11}$$

where again $\mathbb{Z}^n$ stands for the lattice of all points $x = (x_1, \ldots, x_n) \in \mathbb{R}^n$ with integer-valued components x_j. We assume that

$$\sum_{k \in \mathbb{Z}^n} \psi_k(x) = 1, \quad x \in \mathbb{R}^n, \tag{11.12}$$

is a resolution of unity. We have the following assertion:

Let $0 < p < \infty$, $0 < q \leq \infty$, *and* $s \in \mathbb{R}$. Then

$$\left(\sum_{k \in \mathbb{Z}^n} \|\psi_k f \,|\, F_{pq}^s(\mathbb{R}^n)\|^p \right)^{\frac{1}{p}} \tag{11.13}$$

is an equivalent quasi-norm in $F_{pq}^s(\mathbb{R}^n)$.

A corresponding assertion for the spaces $B_{pq}^s(\mathbb{R}^n)$ holds only if $p = q$, that means in the case $B_{pp}^s(\mathbb{R}^n) = F_{pp}^s(\mathbb{R}^n)$. We refer for proofs and details to [Tri92], 2.4.7, pp. 124–129.

11.4 Some embeddings

Embedding theorems for the spaces $B_{pq}^s(\mathbb{R}^n)$ and $F_{pq}^s(\mathbb{R}^n)$ introduced in 10.3 and their special cases described in 10.5 have a long history. We studied assertions of that type in [Tri83] and more recently in [SicT95]. A short survey of some of these results may also be found in [ET96], 2.3.3, p. 43–45. We refer in this context also to [RuS96]. We select only a few of these assertions which are useful for our purposes here omitting any details and further references which may be found in the just quoted literature. Recall that «$\subset$» in connection with two quasi-Banach spaces always means that there is a continuous linear embedding.

(i) *Let* $0 < p \le \infty,\ 0 < q_0 \le \infty,\ 0 < q_1 \le \infty$ *and* $-\infty < s_0 < s_1 < \infty$; *then*

$$B_{pq_1}^{s_1}(\mathbb{R}^n) \subset B_{pq_0}^{s_0}(\mathbb{R}^n) \tag{11.14}$$

and a corresponding assertion for the F*-spaces (now with* $p < \infty$*). This follows immediately from Definition 10.3. The following assertions are less obvious.*

(ii) *Let* $s \in \mathbb{R}, 0 < p < \infty,\ 0 < q \le \infty,\ 0 < u \le \infty,\ 0 < v \le \infty$*. Then*

$$B_{pu}^s(\mathbb{R}^n) \subset F_{pq}^s(\mathbb{R}^n) \subset B_{pv}^s(\mathbb{R}^n) \tag{11.15}$$

if, and only if, $0 < u \le \min(p,q)$ *and* $\max(p,q) \le v \le \infty$.

(iii) *Let*

$$s \in \mathbb{R}, \quad 0 < p_0 < p < p_1 \le \infty, \quad s_0 - \frac{n}{p_0} = s - \frac{n}{p} = s_1 - \frac{n}{p_1} \tag{11.16}$$

and suppose that $0 < q \le \infty,\ 0 < u \le \infty,\ 0 < v \le \infty$*. Then*

$$B_{p_0 u}^{s_0}(\mathbb{R}^n) \subset F_{pq}^s(\mathbb{R}^n) \subset B_{p_1 v}^{s_1}(\mathbb{R}^n) \tag{11.17}$$

if, and only if, $0 < u \le p \le v \le \infty$.

(iv) *Let*

$$s_0 \in \mathbb{R},\ 0 < p_0 < p_1 < \infty,\ s_0 - \frac{n}{p_0} = s_1 - \frac{n}{p_1},\ 0 < u \le \infty,\ 0 < v \le \infty. \tag{11.18}$$

Then

$$F_{p_0 u}^{s_0}(\mathbb{R}^n) \subset F_{p_1 v}^{s_1}(\mathbb{R}^n). \tag{11.19}$$

In particular, by (10.14) holds

$$H_{p_0}^{s_0}(\mathbb{R}^n) \subset H_{p_1}^{s_1}(\mathbb{R}^n), \quad 0 < p_0 < p_1 < \infty,\ s_0 - \frac{n}{p_0} = s_1 - \frac{n}{p_1}. \tag{11.20}$$

We took over the non-trivial embedding assertions (ii), (iii), and (iv) from [ET96], pp. 44–45, where one finds further results of that type, especially for limiting situations, and detailed references. In particular by (11.16) and (11.18) it is quite clear that the number $s - \frac{n}{p}$ plays a crucial role for the spaces $B_{pq}^s(\mathbb{R}^n)$ and $F_{pq}^s(\mathbb{R}^n)$. It is sometimes denoted as the *differential dimension* of the related spaces. In an $(\frac{1}{p}, s)$-diagram, see Fig. 11.1, (11.15) corresponds to the horizontal lines, whereas (11.16) and (11.18) correspond to lines of slope n.

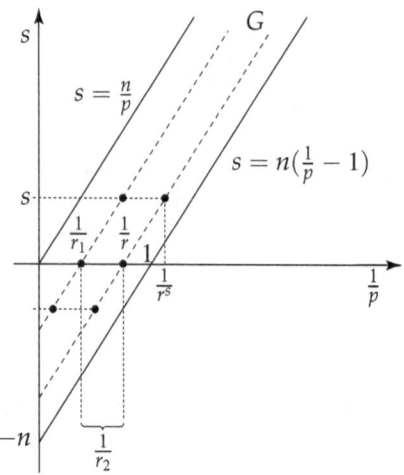

Fig. 11.1

11.5 Hölder inequalities: Preliminaries, the case $s = 0$

We need *pointwise multiplication properties*. Restricted to our later purposes a typical example is given by the question under which conditions

$$H^s_{p_1}(\mathbb{R}^n) H^s_{p_2}(\mathbb{R}^n) \subset H^s_p(\mathbb{R}^n) \tag{11.21}$$

makes sense. Here (11.21) must be understood to mean that there exists a constant $c > 0$ with

$$\|f_1 f_2 \mid H^s_p(\mathbb{R}^n)\| \leq c \|f_1 \mid H^s_{p_1}(\mathbb{R}^n)\| \, \|f_2 \mid H^s_{p_2}(\mathbb{R}^n)\| \tag{11.22}$$

for all $f_j \in H^s_{p_j}(\mathbb{R}^n)$, $j = 1, 2$. Recall that $S(\mathbb{R}^n)$ is dense in all the spaces $B^s_{pq}(\mathbb{R}^n)$ and $F^s_{pq}(\mathbb{R}^n)$ if $\max(p,q) < \infty$, see [Tri83], p. 48. Then one proves first assertions of type (11.22) for $f_j \in S(\mathbb{R}^n)$. The rest is a matter of completion. Using the Fatou property, see [ET96], p. 48, a similar argument can also be used if $p = \infty$ and/or $q = \infty$. We do not stress this point in the sequel. We will be very brief and summarize a few results from [ET96], 2.4, pp. 45–57, which, in turn, are mostly based on [SicT95]. There one finds also the necessary references, additional results and comments. Furthermore a systematic and far-reaching treatment of problems of that type may be found in [RuS96]. If $s = 0$ in (11.21) then one has as a special case the classical Hölder inequality

$$L_{r_1}(\mathbb{R}^n) L_{r_2}(\mathbb{R}^n) \subset L_r(\mathbb{R}^n), \tag{11.23}$$

with $1 \leq r_1 \leq \infty$, $1 \leq r_2 \leq \infty$, $\frac{1}{r_1} + \frac{1}{r_2} = \frac{1}{r} \leq 1$. This corresponds in the $(\frac{1}{p}, s)$-diagram in Fig. 11.1 to the large dots at the level $s = 0$. It will be helpful to look at (11.23), (11.21) and other assertions of that type in a non-symmetric way, interpreting, for instance, $L_{r_2}(\mathbb{R}^n)$ as a multiplier space which transports $L_{r_1}(\mathbb{R}^n)$ into $L_r(\mathbb{R}^n)$. In that case, roughly speaking, it turns out that desired relations of type (11.21) can be obtained by shifting (11.23) along the lines of slope n in the $(\frac{1}{p}, s)$-diagram to the level s, see the large dots in Fig. 11.1 at levels $s \neq 0$. Furthermore, we recall that the sharp embeddings in (11.16), (11.17) and (11.18)–(11.20), characterized by constant differential dimensions, correspond to lines in Fig. 11.1 of slope n. Some notation seem to be quite natural now. Let

$$G = \left\{ (\frac{1}{p}, s) : \quad 0 < p < \infty, \ n(\frac{1}{p} - 1) < s < \frac{n}{p} \right\} \tag{11.24}$$

be the strip indicated in Fig. 11.1. Any line of slope n in this strip is characterized by the point at which it meets the axis $s = 0$: we shall refer to this point, $(\frac{1}{r}, 0)$ say, as the *foot point* of the line. Thus any point on G has coordinates

$$(\frac{1}{r^s}, s) \quad \text{with} \quad \frac{1}{r^s} = \frac{1}{r} + \frac{s}{n} > 0, \quad 1 < r < \infty. \tag{11.25}$$

11.6 Hölder inequalities: the case s > 0
Let $s > 0$, $0 < q_1 \leq \infty$, $0 < q_2 \leq \infty$, $0 < q \leq \infty$,

$$1 < r_1 < \infty, \ 1 < r_2 < \infty, \ \frac{1}{r} = \frac{1}{r_1} + \frac{1}{r_2} < 1. \tag{11.26}$$

(i) *Then*

$$B^s_{r_1^s q_1}(\mathbb{R}^n) B^s_{r_2^s q_2}(\mathbb{R}^n) \subset B^s_{r^s q}(\mathbb{R}^n) \tag{11.27}$$

if, and only if,

$$0 < q_1 \leq r_1, \ 0 < q_2 \leq r_2, \quad \max(q_1, q_2) \leq q \leq \infty. \tag{11.28}$$

(ii) *Also*

$$F^s_{r_1^s q_1}(\mathbb{R}^n) F^s_{r_2^s q_2}(\mathbb{R}^n) \subset F^s_{r^s q}(\mathbb{R}^n) \tag{11.29}$$

if, and only if,

$$\max(q_1, q_2) \leq q \leq \infty. \tag{11.30}$$

Full proofs of these assertions are given in [SicT95] and [RuS96], 4.8, where the latter reference contains a lot of further results of related type. A short proof of the if-part (which covers what we need later on) has also been given in [ET96], 2.4.3. Of interest for us is the following special case of (ii):

(iii) *Let $s \geq 0$, and let r_1, r_2, r and r_1^s, r_2^s, r^s be given by (11.26) and (11.25), then*

$$H^s_{r_1^s}(\mathbb{R}^n) H^s_{r_2^s}(\mathbb{R}^n) \subset H^s_{r^s}(\mathbb{R}^n). \tag{11.31}$$

This follows immediately from (11.29), (11.30) with $q_1 = q_2 = q = 2$ and (10.14), notationally extended to

$$H^s_p(\mathbb{R}^n) = F^s_{p,2}(\mathbb{R}^n), \quad s \in \mathbb{R}, \ 0 < p < \infty, \tag{11.32}$$

(Hardy-Sobolev spaces). (11.27), (11.29) and its special case (11.31) are characterized by the large dots in Fig. 11.1 at the level $s > 0$.

11.7 Hölder inequalities: the case $s < 0$
Recall that $H_p^s(\mathbb{R}^n)$ is now given by (11.32). The extension of (11.31) to $s \in \mathbb{R}$ reads now as follows:

Let $s \in \mathbb{R}$ and let r_1, r_2, r be given by (11.26). Suppose

$$\frac{1}{r_1^s} = \frac{1}{r_1} + \frac{s}{n} > 0.$$

Then

$$H_{r_1^s}^s(\mathbb{R}^n) \, H_{r_2^{|s|}}^{|s|}(\mathbb{R}^n) \subset H_{r^s}^s(\mathbb{R}^n). \tag{11.33}$$

Of course, $r_1^s, r_2^{|s|}, r^s$ have the same meaning as in (11.25). If $s \geq 0$, then (11.33) coincides with (11.31). If $s < 0$ then the situation is characterized by the large dots in Fig. 11.1 at the level $s < 0$. In that case the simple proof, based on (11.31) and duality arguments, may be found in [ET96], p. 56–57.

11.8 Limiting cases
The results formulated in 11.6 and 11.7 apply to the open strip G in (11.24). Later on we need also some limiting assertions. Of peculiar interest is the case $r_1 = \infty$ in (11.26) and (11.27), (11.31). In [ET96], 2.4.4, p. 55, and in [SicT95], Theorem 4.3.1, one finds also in this limiting situation assertions of the same final character as in 11.6. But we restrict ourselves here to a special case which is of interest later on. We complement the notation (11.25) by $\infty^s = \frac{n}{s}$ if $s > 0$. Furthermore we shorten $B_{pp}^s(\mathbb{R}^n)$ by $B_p^s(\mathbb{R}^n)$. We have the following two assertion:

(i) *Let $s > 0$ and $1 < p < \infty$. Then*

$$B_{p^s}^s(\mathbb{R}^n) \, B_{\infty^s,1}^s(\mathbb{R}^n) \subset B_{p^s}^s(\mathbb{R}^n), \tag{11.34}$$

and, hence, $B_{\infty^s,1}^s(\mathbb{R}^n)$ is a space of pointwise multipliers.

(ii) *Let $s < 0$, $1 < p < \infty$, and $\frac{1}{p^s} = \frac{1}{p} + \frac{s}{n} \geq 0$. Then*

$$B_{p^s}^s(\mathbb{R}^n) \, B_{\infty^{|s|},1}^{|s|}(\mathbb{R}^n) \subset B_{p^s}^s(\mathbb{R}^n). \tag{11.35}$$

Part (i) is covered by [SicT95], Theorem 4.3.1, whereas (11.35) follows from (11.34) and similar duality arguments as in [ET96], p. 56–57, based on [Tri83], 2.11.2, p. 178. Further results may be found in [SicT95] and [RuS96].

12 Harmonic and local representations and characterizations

12.1 Introduction
In the preceding Sections 10 and 11 we collected without proofs basic facts and those properties of function spaces which are needed later on with references

mainly to [Tri83], [Tri92], and [ET96]. Beginning with this section the situation is different. We present now new material complementing the quoted literature. We give proofs. On the one hand we prepare the later applications. On the other hand we confess that we hope that some results (especially the subatomic decompositions in Sections 14 and 15) are of self-contained interest and that they serve as the basis for further applications.

12.2 Harmonic representations: preliminaries
Let $f \in C_0^\infty(\mathbb{R}^n) = D(\mathbb{R}^n)$, then

$$u(x,t) = \left(e^{-t|\xi|}\hat{f}\right)^\vee(x) = d_n \int_{\mathbb{R}^n} t \left(|x-y|^2 + t^2\right)^{-\frac{n+1}{2}} f(y)\, dy \qquad (12.1)$$

for some $d_n > 0$ is the harmonic extension of f into

$$\mathbb{R}^{n+1}_+ = \{(x,t): \quad x \in \mathbb{R}^n, \, t > 0\}. \qquad (12.2)$$

In other words, if $\Delta = \sum\limits_{j=1}^{n} \dfrac{\partial^2}{\partial x_j^2}$ applies to the space variables x, then

$$\Delta u(x,t) + \frac{\partial^2 u}{\partial t^2} = 0 \quad \text{in } \mathbb{R}^{n+1}_+. \qquad (12.3)$$

Recall that (12.1) is the well-known *Cauchy-Poisson semi-group*. We refer for details to [Tri78], 2.5.3, pp. 192–196. We have

$$u(x,t) \to f(x) \quad \text{if } t \downarrow 0 \quad \text{and } x \in \mathbb{R}^n \text{ fixed}, \qquad (12.4)$$

and for any $k \in \mathbb{N}_0$,

$$t^k \frac{\partial^k u(x,t)}{\partial t^k} \to 0 \quad \text{if } t \to \infty \quad \text{and } x \in \mathbb{R}^n \text{ fixed}. \qquad (12.5)$$

In particular there are real constants d_l^k with $k \in \mathbb{N}$ and $l = 0, \ldots, k-1$ such that for any $0 < a < b < \infty$ and $x \in \mathbb{R}^n$,

$$\int_a^b t^{k-1} \frac{\partial^k u(x,t)}{\partial t^k}\, dt$$

$$= \tau^{k-1} \frac{\partial^{k-1} u(x,\tau)}{\partial \tau^{k-1}}\bigg|_a^b - (k-1) \int_a^b t^{k-2} \frac{\partial^{k-1} u(x,t)}{\partial t^{k-1}}\, dt$$

$$= \sum_{l=0}^{k-1} d_l^k \, b^l \frac{\partial^l u(x,b)}{\partial t^l} - \sum_{l=0}^{k-1} d_l^k \, a^l \frac{\partial^l u(x,a)}{\partial t^l}. \qquad (12.6)$$

Let $a \downarrow 0$. At least in a formal way, not discussing at this moment the type of convergence, we obtain from (12.4) and (12.6)

$$f(x) = \sum_{\nu=0}^{\infty} \sum_{l=0}^{k-1} c_l^k \left(2^{-(\nu+1)l} \frac{\partial^l u(x, 2^{-\nu-1})}{\partial t^l} - 2^{-\nu l} \frac{\partial^l u(x, 2^{-\nu})}{\partial t^l} \right)$$

$$+ \sum_{l=0}^{k-1} c_l^k \frac{\partial^l u(x, 1)}{\partial t^l}$$

$$= c \sum_{\nu=0}^{\infty} \int_{2^{-\nu-1}}^{2^{-\nu}} t^k \frac{\partial^k u(x, t)}{\partial t^k} \frac{dt}{t} + \sum_{l=0}^{k-1} c_l^k \frac{\partial^l u(x, 1)}{\partial t^l} \qquad (12.7)$$

for suitable real coefficients c and c_l^k which come from the above numbers d_l^k, where $k \in \mathbb{N}$ and $l = 0, \ldots, k-1$. By (12.4) we have $c_0^k = 1$. We call (12.7) a *harmonic representation*, although $u(x, t)$ is harmonic with respect to the $(n+1)$ variables $x \in \mathbb{R}^n$ and $t > 0$ (and not with respect to $x \in \mathbb{R}^n$ alone). It is our aim to give (12.7) a precise meaning not only for $f \in D(\mathbb{R}^n)$ but also for $f \in B_{pq}^s(\mathbb{R}^n)$ or $f \in F_{pq}^s(\mathbb{R}^n)$ as a representation which converges at least in $S'(\mathbb{R}^n)$. If this is so then one can ask whether (12.7) converges even in, say, $B_{pq}^s(\mathbb{R}^n)$ and whether the respective summands can be used to find equivalent quasi-norms similar as in 11.2. It turns out that these two questions have rather different answers [in sharp contrast to, say, 11.2] which originates from the fact that $e^{-|\xi|}$ is not smooth at the origin. Let B be the closed unit ball in $\mathbb{R}^n$ and let $0 < p \leq \infty$; then

$$L_p^B(\mathbb{R}^n) = \left\{ f \in S'(\mathbb{R}^n) : \operatorname{supp} \hat{f} \subset B, \ \|f \mid L_p(\mathbb{R}^n)\| < \infty \right\} \qquad (12.8)$$

are the spaces of entire analytic functions which we studied in detail in [Tri83], 1.4 and 1.5. In particular, $M(\xi) = e^{-|\xi|}$ is called a Fourier multiplier for $L_p^B(\mathbb{R}^n)$ if there is a constant c such that

$$\|(M\hat{f})^\vee \mid L_p(\mathbb{R}^n)\| \leq c \, \|f \mid L_p(\mathbb{R}^n)\| \quad \text{for all } f \in L_p^B(\mathbb{R}^n), \qquad (12.9)$$

see [Tri83], p. 25.

12.3 Proposition *Let $\frac{n}{n+1} < p \leq \infty$; then $M(\xi) = e^{-|\xi|}$ is a Fourier multiplier in $L_p^B(\mathbb{R}^n)$.*

Proof. By (12.1) with $t = 1$ the assertion is obvious if $p \geq 1$. Let $p < 1$. Recall $\langle \xi \rangle = (1 + |\xi|^2)^{\frac{1}{2}}$ and

$$\left(e^{-|\xi|} \right)^\vee (x) = c \, \langle x \rangle^{-n-1}, \quad x \in \mathbb{R}^n, \qquad (12.10)$$

for some $c > 0$, see (12.1) or better [Tri78], 2.5.3, p. 195. By [Tri83], 1.5.2, p. 26, we have to check whether there is a number σ with

$$e^{-|\xi|} \in H_2^\sigma(\mathbb{R}^n) \quad \text{and} \quad \sigma > n\left(\frac{1}{p} - \frac{1}{2}\right). \tag{12.11}$$

By (10.13) and (12.10) this is equivalent to

$$\| \langle \xi \rangle^\sigma \, \langle \xi \rangle^{-n-1} \, | L_2(\mathbb{R}^n) \| < \infty \quad \text{with} \quad \sigma > n\left(\frac{1}{p} - \frac{1}{2}\right). \tag{12.12}$$

But this is the case since $\frac{n}{p} - \frac{n}{2} < \frac{n}{2} + 1$.

12.4 Some technicalities

First we clarify the question of the convergence of (12.7) by reducing the problem to the technique developed in [Tri92]. Let, say, $f \in F_{pq}^s(\mathbb{R}^n)$ for some $s \in \mathbb{R}$, $0 < p < \infty$, $0 < q \le \infty$. Let $k \in \mathbb{N}$ and

$$\psi(\xi) = |\xi|^k e^{-|\xi|} \quad \text{and} \quad \varphi(\xi) = \int_{\frac{1}{2}}^1 \psi(\tau \xi) \frac{d\tau}{\tau}, \quad \xi \in \mathbb{R}^n. \tag{12.13}$$

By (12.1) we have

$$\psi(t D) f(x) = \left(t^k |\xi|^k e^{-t|\xi|} \hat{f} \right)^\vee (x) = (-1)^k t^k \frac{\partial^k u(x,t)}{\partial t^k} \tag{12.14}$$

and

$$\varphi(2^{-\nu} D) f(x) = (-1)^k \int_{2^{-\nu-1}}^{2^{-\nu}} t^k \frac{\partial^k u(x,t)}{\partial t^k} \frac{dt}{t}, \quad \nu \in \mathbb{N}_0. \tag{12.15}$$

If we choose k sufficiently large, that means $k \ge k_0(s,p,q)$, then we can apply [Tri92], 2.6.4, pp. 152–153, and 2.4.1, pp. 100–101, which prove that the sum over $\nu \in \mathbb{N}_0$ in (12.7) converges at least in $S'(\mathbb{R}^n)$. Of course the above arguments apply also to $f \in B_{pq}^s(\mathbb{R}^n)$ for $s \in \mathbb{R}$, $0 < p \le \infty$, $0 < q \le \infty$. Otherwise we return to the meaning of (12.7) in 12.5. In the theorem below we need also some maximal functions related to (12.14) and (12.15). Then the involved $k \in \mathbb{N}$ must be even larger. The phrase «$k \in \mathbb{N}$ sufficiently large» must always be interpreted as: There exists a $k^0(s,p,q)$ in case of the F-spaces and a $k^0(s,p)$ in the case of the B-spaces such that the corresponding assertion holds for any $k \in \mathbb{N}$ with $k \ge k^0$. The parts (ii) and (iii) of the theorem below are counterparts of [Tri92], Theorem 2.6.4, p. 152–153. In contrast to that theorem we have now the restriction $p > \frac{n}{n+1}$ which comes from the first terms on the right-hand sides of (12.18), (12.19), and (12.21), (12.22) below which cause additional problems now. Furthermore, the understanding of (12.20) and (12.23) must also be explained. As far as the related second terms on the right-hand sides in (12.18), (12.19), and (12.21), (12.22) are

concerned we refer to [Tri92], p. 153, and the references given there even under
the assumption $f \in S'(\mathbb{R}^n)$. To avoid any additional technical difficulties caused
by the related first terms we replace now $S'(\mathbb{R}^n)$ by

$$\mathscr{C}^{-\infty}(\mathbb{R}^n) = \bigcup_{\sigma \in \mathbb{R}} \mathscr{C}^\sigma(\mathbb{R}^n), \tag{12.16}$$

see (10.16). As we shall see in Step 1 of the proof of the following theorem $\frac{\partial^l u(x,t)}{\partial t^l}$
makes sense if $f \in \mathscr{C}^{-\infty}(\mathbb{R}^n)$. Then the technique developed in [Tri92], p. 153,
and the references given there can be applied also to these terms. We note that

$$F_{pq}^s(\mathbb{R}^n) \subset \mathscr{C}^{-\infty}(\mathbb{R}^n) \quad \text{and} \quad B_{pq}^s(\mathbb{R}^n) \subset \mathscr{C}^{-\infty}(\mathbb{R}^n) \tag{12.17}$$

for any admitted s, p, q. This follows from (11.17) and (11.14): $\mathscr{C}^{-\infty}(\mathbb{R}^n)$ is
simply the union of all spaces $F_{pq}^s(\mathbb{R}^n)$ and $B_{pq}^s(\mathbb{R}^n)$. In other words, the claimed
characterizations (12.20) and (12.23) follow from the quoted arguments in [Tri92]
if all the other assertions of the following theorem are proved.

12.5 Theorem (i) *Let $s \in \mathbb{R}$, $0 < q \leq \infty$ and $0 < p \leq \infty$ ($p < \infty$ in the case of
the F-spaces). Let $k \in \mathbb{N}$ be sufficiently large in dependence on s, p, q, B-space
or F-space. Let $f \in B_{pq}^s(\mathbb{R}^n)$ or $f \in F_{pq}^s(\mathbb{R}^n)$. Then the right-hand side of (12.7)
converges in $S'(\mathbb{R}^n)$ to f.*

(ii) *Let $s \in \mathbb{R}$, $0 < q \leq \infty$ and $\frac{n}{n+1} < p \leq \infty$. Let $k \in \mathbb{N}$ be sufficiently
large. Then*

$$\|f \mid B_{pq}^s(\mathbb{R}^n)\|_1 = \|u(\cdot, 1) \mid L_p(\mathbb{R}^n)\|$$
$$+ \left(\int_0^1 t^{(k-s)q} \left\| \frac{\partial^k u}{\partial t^k}(\cdot, t) \mid L_p(\mathbb{R}^n) \right\|^q \frac{dt}{t} \right)^{\frac{1}{q}} \tag{12.18}$$

and for any $c > 0$

$$\|f \mid B_{pq}^s(\mathbb{R}^n)\|_2 = \sum_{l=0}^{k-1} \left\| \sup_{|\cdot - y| \leq c} \left| \frac{\partial^l u}{\partial t^l}(y, 1) \right| \mid L_p(\mathbb{R}^n) \right\|$$
$$+ \left(\int_0^1 t^{(k-s)q} \left\| \sup \left| \frac{\partial^k u}{\partial t^k}(y, \tau) \right| \mid L_p(\mathbb{R}^n) \right\|^q \frac{dt}{t} \right)^{\frac{1}{q}} \tag{12.19}$$

(modification if $q = \infty$) are equivalent quasi-norms in $B_{pq}^s(\mathbb{R}^n)$, where sup *is the
supremum taken over $|x - y| \leq ct$ and $t \leq \tau \leq 2t$. Furthermore*

$$B_{pq}^s(\mathbb{R}^n) = \{ f \in \mathscr{C}^{-\infty}(\mathbb{R}^n) : \|f \mid B_{pq}^s(\mathbb{R}^n)\|_r < \infty \}, \quad r = 1, 2. \tag{12.20}$$

(iii) *Let* $s \in \mathbb{R}$, $0 < q \le \infty$, *and* $\frac{n}{n+1} < p < \infty$. *Let* $k \in \mathbb{N}$ *be sufficiently large. Then*

$$\|f \mid F_{pq}^s(\mathbb{R}^n)\|_1 = \|u(\cdot, 1) \mid L_p(\mathbb{R}^n)\|$$

$$+ \left\| \left(\int_0^1 t^{(k-s)q} \left| \frac{\partial^k u}{\partial t^k}(\cdot, t) \right|^q \frac{dt}{t} \right)^{\frac{1}{q}} \mid L_p(\mathbb{R}^n) \right\| \qquad (12.21)$$

and for any $c > 0$

$$\|f \mid F_{pq}^s(\mathbb{R}^n)\|_2 = \sum_{l=0}^{k-1} \left\| \sup_{|\cdot - y| \le c} \left| \frac{\partial^l u}{\partial t^l}(y, 1) \right| \mid L_p(\mathbb{R}^n) \right\|$$

$$+ \left\| \left(\int_0^1 t^{(k-s)q} \sup \left| \frac{\partial^k u}{\partial t^k}(y, \tau) \right|^q \frac{dt}{t} \right)^{\frac{1}{q}} \mid L_p(\mathbb{R}^n) \right\| \qquad (12.22)$$

(modification if $q = \infty$*) are equivalent quasi-norms in* $F_{pq}^s(\mathbb{R}^n)$ *where* sup *is the supremum taken over* $|x - y| \le ct$ *and* $t \le \tau \le 2t$*. Furthermore*

$$F_{pq}^s(\mathbb{R}^n) = \{f \in \mathscr{C}^{-\infty}(\mathbb{R}^n) : \|f \mid F_{pq}^s(\mathbb{R}^n)\|_r < \infty\}, \quad r = 1, 2. \qquad (12.23)$$

Proof.
Step 1. To prove (i) we begin with some preliminaries. Let $\varrho \in S(\mathbb{R}^n)$ with

$$\varrho(x) = 1 \quad \text{if} \quad |x| \le 1 \quad \text{and} \quad \varrho(x) = 0 \quad \text{if} \quad |x| \ge \frac{3}{2}, \qquad (12.24)$$

as in (10.3). Let $f \in F_{pq}^s(\mathbb{R}^n)$ or $f \in B_{pq}^s(\mathbb{R}^n)$. Put

$$f = f_1 + f_2 = (\varrho \hat{f})^\vee + ((1 - \varrho)\hat{f})^\vee. \qquad (12.25)$$

Let $\varepsilon > 0$. By the lifting assertions described in (10.12) and the known Fourier multiplier properties for the spaces under consideration, see [Tri83], 2.3.7, p. 57, it follows that $(e^{-\varepsilon|\xi|} \hat{f_2})^\vee$ belongs to $F_{pq}^\sigma(\mathbb{R}^n)$ or $B_{pq}^\sigma(\mathbb{R}^n)$ for all $\sigma > 0$. By (11.16), (11.17) and (10.16) this function is smooth and bounded in $\mathbb{R}^n$. Let $u_2(x, t)$ be given by (12.1) with f_2 in place of f, then $u_2(x, t)$ is a classical harmonic function in $\mathbb{R}_+^n$ which is bounded in every half-space

$$\mathbb{R}_\delta^n = \{(x, t) : x \in \mathbb{R}^n, t > \delta\} \quad \text{with} \quad \delta > 0. \qquad (12.26)$$

Next we apply the Plancherel-Polya-Nikolskij inequality, see [Tri83], 1.3.2, p. 17, and 1.4.1, p. 22, to f_1 and obtain that f_1 is a smooth bounded function in $\mathbb{R}^n$. Hence, $u_1(x, t)$ given by (12.1) with f_1 in place of f, has the same properties as

$u_2(x,t)$. Consequently, also $u(x,t)$ is a classical harmonic function in $\mathbb{R}^n_+$ which is bounded in any half-space $\mathbb{R}^n_\delta$ given by (12.26). That means that all the terms in (12.7) are bounded smooth functions in $\mathbb{R}^n$.

Step 2. We prove a special case of (i). Let

$$\varphi_0(\xi) = \left(\sum_{l=0}^{k-1} (-1)^l \, c_l^k \, |\xi|^l \right) e^{-|\xi|}, \quad \xi \in \mathbb{R}^n, \qquad (12.27)$$

where c_l^k are the same constants as in (12.7), in particular $c_0^k = 1$. By (12.1), the corresponding reformulation of the right-hand side of (12.7), 12.4 and Step 1 it follows

$$(\varphi_0(t\cdot)\hat{f})^{\vee} = \varphi_0(tD)f \to g \quad \text{in} \quad S'(\mathbb{R}^n) \quad \text{if} \quad t \to 0. \qquad (12.28)$$

Strictly speaking we must replace t by $2^{-\nu}$ and let $\nu \to \infty$. We have to prove $g = f$. We assume $f \in L_p^B(\mathbb{R}^n)$, see (12.8). Preparing also the proof of parts (ii) and (iii) and modifying Proposition 12.3 we first remark that $|\xi|$ is a Fourier multiplier in $L_p^B(\mathbb{R}^n)$ with $p > \frac{n}{n+1}$. This follows from [Tri92], Remark 1 on p. 107, since

$$1 > \sigma_p = n \left(\frac{1}{p} - 1 \right)_+ . \qquad (12.29)$$

Let M be the related multiplier constant. Expanding $e^{-t|\xi|}$ in its Taylor series we find

$$\left\| \left((e^{-t|\xi|} - 1)\hat{f} \right)^{\vee} | L_p(\mathbb{R}^n) \right\| \leq c\, t\, e^{tM} \|f \,|\, L_p(\mathbb{R}^n)\|, \qquad (12.30)$$

where c does not depend on $0 < t \leq 1$. Again by the Plancherel-Polya-Nikolskij inequality in [Tri83], 1.3.2, p. 17, and 1.4.1, p. 22, we have also (12.30) with ∞ on the left-hand side instead of p. It follows

$$u(x,t) \to f \quad \text{in} \quad S'(\mathbb{R}^n) \quad \text{if} \quad t \to 0, \qquad (12.31)$$

what corresponds to the first summand in (12.27). As for the other term in (12.27) with $l = 1, \ldots, k - 1$ we have obvious counterparts of (12.31). In particular the related expressions tend in $S'(\mathbb{R}^n)$ to 0 if $t \to 0$. This proves (12.28) with $g = f$ in this special case.

Step 3. Now we are in the position to finish the proof of (i). By the embedding assertion (11.17) it is sufficient to prove $f = g$ in (12.28) for $f \in \mathscr{C}^s(\mathbb{R}^n) = B_{\infty,\infty}^s(\mathbb{R}^n)$. We rely on the duality

$$(B_{1,1}^{-s}(\mathbb{R}^n))' = \mathscr{C}^s(\mathbb{R}^n), \quad s \in \mathbb{R}^n, \qquad (12.32)$$

see [Tri83], 2.11.2, p. 178. Let

$$\chi \in S(\mathbb{R}^n) \quad \text{with} \quad supp\, \hat{\chi} \quad \text{compact}. \qquad (12.33)$$

By the preceding step we have

$$\varphi_0(tD)\chi \to \chi \quad \text{in} \quad L_1(\mathbb{R}^n) \quad \text{if} \quad t \to 0 \tag{12.34}$$

and, by the assumption for the support of $\hat{\chi}$ also

$$\varphi_0(tD)\chi \to \chi \quad \text{in} \quad B_{1,1}^{-s}(\mathbb{R}^n) \quad \text{if} \quad t \to 0. \tag{12.35}$$

Now let $f \in \mathscr{C}^s(\mathbb{R}^n)$. Then (12.28), (12.32) and (12.35) prove

$$g(\chi) = \lim_{t \to 0}(\varphi_0(tD)f)(\chi) = \lim_{t \to 0} f(\varphi_0(tD)\chi) = f(\chi) \tag{12.36}$$

and hence $f = g$.

Step 4. We prove (iii). The proof of (ii) is the same. First we remark that the second terms of the right-hand sides of (12.21) and (12.22) are covered by the technique developed in [Tri92], where we proved corresponding results in 2.6.4, p. 152–153, but with

$$\|\varrho(D)f \,|\, L_p(\mathbb{R}^n)\|$$

in place of the first terms in (12.21) and (12.22), where ϱ is given by (12.24). As far as the insertion of the supremum in the second term in (12.22) is concerned we refer also to the technique developed in [Tri92], in particular characterizations via maximal functions in [Tri92], Theorem 2.4.2(iii), pp. 111–114, and pp. 115–116. This applies to all p with $0 < p < \infty$. In other words we have to prove that $\|\varrho(D)f \,|\, L_p(\mathbb{R}^n)\|$ can be replaced by the first terms of the right-hand sides of (12.21) and (12.22), where now the restriction $p > \frac{n}{n+1}$ comes in. Let $f \in F_{pq}^s(\mathbb{R}^n)$ be decomposed as in (12.25). As mentioned there we have

$$\begin{aligned} &\|u(\cdot,1) \,|\, L_p(\mathbb{R}^n)\| \\ &\leq c \left\| \left(e^{-|\xi|}\varrho(\xi)\hat{f}\right)^{\vee} |\, L_p(\mathbb{R}^n) \right\| + c \,\|f \,|\, F_{pq}^s(\mathbb{R}^n)\|. \end{aligned} \tag{12.37}$$

Since now $p > \frac{n}{n+1}$ is assumed we can apply Proposition 12.3 and obtain

$$\begin{aligned} \|u(\cdot,1) \,|\, L_p(\mathbb{R}^n)\| &\leq c \,\|\varrho(D)f \,|\, L_p(\mathbb{R}^n)\| + c \,\|f \,|\, F_{pq}^s(\mathbb{R}^n)\| \\ &\leq c' \,\|f \,|\, F_{pq}^s(\mathbb{R}^n)\|. \end{aligned} \tag{12.38}$$

By the arguments in Step 2 one can replace u in (12.38) by $\frac{\partial^l u}{\partial t^l}$. The insertion of the supremum in the first term on the right-hand side of (12.22) is again a matter of the technique of maximal functions covered by the above references, see also [Tri83], 1.4.1, p. 22. Hence the quasi-norm in (12.21) and (12.22) can be estimated from above by $c \,\|f \,|\, F_{pq}^s(\mathbb{R}^n)\|$. As for the converse assertion it remains to prove that $\|\varrho(D)f\,|\,L_p(\mathbb{R}^n)\|$ can be estimated from above by the right-hand side of (12.21). By (12.27), (12.14), (12.15), and (12.7) or direct reasoning we have

$$1 = \varphi_0(\xi) + c \sum_{\nu=0}^{\infty} \int_{2^{-\nu-1}}^{2^{-\nu}} t^k |\xi|^k e^{-t|\xi|} \frac{dt}{t} \tag{12.39}$$

and after multiplication with $\varrho(\xi)$ a corresponding decomposition of $\varrho(D)f$. The terms originating from the sum over ν fit in our previous scheme and can be estimated from above by the second terms in (12.21). Furthermore as mentioned in Step 2 we have

$$\left\| \left(|\xi|^l e^{-|\xi|} \varrho(\xi)\hat{f} \right)^{\vee} | L_p(\mathbb{R}^n) \right\|$$
$$\leq c \left\| \left(e^{-|\xi|} \varrho(\xi)\hat{f} \right)^{\vee} | L_p(\mathbb{R}^n) \right\| \tag{12.40}$$

since $|\xi|^l$ are Fourier multipliers in $L_p^B(\mathbb{R}^n)$ with $p > \frac{n}{n+1}$. Hence it remains to estimate the right-hand side of (12.40). But the last term in

$$\left\| \left(e^{-|\xi|} \varrho(\xi)\hat{f} \right)^{\vee} | L_p(\mathbb{R}^n) \right\| \leq c \left\| \left(e^{-|\xi|} \hat{f} \right)^{\vee} | L_p(\mathbb{R}^n) \right\|$$
$$+ c \left\| \left((1 - \varrho(\xi)) e^{-|\xi|} \hat{f} \right)^{\vee} | L_p(\mathbb{R}^n) \right\| \tag{12.41}$$

has been treated in the first step, (12.24) and (12.25), and can be estimated from above by the second terms on the right-hand side of (12.21). However the first term on the right-hand side of (12.41) is just the wanted term $\|u(\cdot,1)\,|\,L_p(\mathbb{R}^n)\|$ and, hence, we are through. Finally by our remarks at the end of 12.4 this covers also the characterization (12.23).

12.6 Necessity of $p > \frac{n}{n+1}$

In connection with the parts (ii) and (iii) of the above theorem the question arises whether these assertions can be extended to values $0 < p \leq \frac{n}{n+1}$. We prove that this is not the case. Let $p \leq \frac{n}{n+1}$ and assume that (12.21) is an equivalent quasi-norm in $F_{pq}^s(\mathbb{R}^n)$. In particular then we have

$$\left\| \left(e^{-|\xi|} \hat{f} \right)^{\vee} | L_p(\mathbb{R}^n) \right\| \leq c \|f \,|\, F_{pq}^s(\mathbb{R}^n)\|. \tag{12.42}$$

Using again the splitting (12.25) it follows by the arguments given there that we have also

$$\left\| \left(\varrho(\xi) e^{-|\xi|} \hat{f} \right)^{\vee} | L_p(\mathbb{R}^n) \right\| \leq c \|f \,|\, F_{pq}^s(\mathbb{R}^n)\| \tag{12.43}$$

and

$$\left\| \left(e^{-|\xi|} \hat{f} \right)^{\vee} | F_{pq}^s(\mathbb{R}^n) \right\| \leq c \|f \,|\, F_{pq}^s(\mathbb{R}^n)\|. \tag{12.44}$$

Hence $e^{-|\xi|}$ must be a Fourier multiplier in $F_{pq}^s(\mathbb{R}^n)$. Then $e^{-|\xi|}$ is also a Fourier multiplier in $B_{pq}^s(\mathbb{R}^n)$. This follows by real interpolation, see [Tri83], 2.4.2, p. 64. Hence by (12.10) and [Tri83], 2.6.3, p. 121,

$$\langle x \rangle^{-n-1} = c \, (e^{-|\xi|})\widehat{}\,(x) \in B_{p,\infty}^{n(\frac{1}{p}-1)}(\mathbb{R}^n). \tag{12.45}$$

In particular by (11.6) we have

$$\int_{\mathbb{R}^n} k_0(y) \, \langle x - y \rangle^{-n-1} \, dy \in L_p(\mathbb{R}^n) \tag{12.46}$$

and hence

$$\langle x \rangle^{-n-1} \in L_p(\mathbb{R}^n), \quad p \leq \frac{n}{n+1}. \tag{12.47}$$

But this is wrong. In other words, there is no chance to extend the parts (ii) and (iii) of Theorem 12.5 to values $p \leq \frac{n}{n+1}$.

12.7 Remark Theorem 12.5 is the basis for our later considerations about sub-atomic (or quarkonial) decompositions of (vector-valued) function spaces, where the fact that $u(x,t)$ is harmonic in $\mathbb{R}_+^{n+1}$ is decisive. We reduced the second terms on the right-hand sides of (12.18), (12.19) and (12.21), (12.22) to what has been said in [Tri92], 2.4.1 etc. about the possibilities to replace the original functions φ_j in (10.3)–(10.5) by more general functions. We refer in this connection also to the recent papers [BPT96a,b]. But in sharp contrast to our situation here, both in [Tri92] (and the underlying papers) and in the papers just mentioned it is al-ways assumed that the starting term related to $\varphi = \varphi_0$ in (10.6), (10.7) is (at least sufficiently) smooth such that the theory applies to $0 < p \leq \infty$. It should be mentioned that for the homogeneous counterparts of $F_{pq}^s(\mathbb{R}^n)$ and $B_{pq}^s(\mathbb{R}^n)$ (see [Tri83], Ch. 5, for definitions) there are no restrictions for p, also not in connection with harmonic characterizations, see [BPT96a,b]. One can replace in Theorem 12.5 harmonic representations and characterizations by *thermic repre-sentations* and *characterizations*, see [Tri92], 2.6.4, p. 152, and [BPT96a,b]. Then there are no restrictions for p. But it is at least not immediately clear, whether the thermic version of Theorem 12.5 is useful for our later purposes.

12.8 Local representations: preliminaries
A first discussion about known results connected with local means was given in 11.2. Now we ask for the counterparts of (12.7), (12.39), and Theorem 12.5 in terms of local means. Let $k_M(t,f)(x)$ and the kernels k_M be given by (11.3) and (11.1), (11.2). The counterpart of (12.7) is given by

$$f = k_0(1,f)(x) + \sum_{\nu=0}^{\infty} k_N(2^{-\nu},f)\,(x), \tag{12.48}$$

convergence being at least in $S'(\mathbb{R}^n)$, with (11.5). In particular, (11.6), (11.7), and (11.8), (11.9) are equivalent quasi-norms in $F_{pq}^s(\mathbb{R}^n)$ and $B_{pq}^s(\mathbb{R}^n)$, respectively. By

(11.3) and in analogy to (12.39), the above identity is, at least formally, equivalent to

$$1 = \widehat{k_0}(\xi) + \sum_{\nu=0}^{\infty} \widehat{k_N}(2^{-\nu}\xi), \quad \xi \in \mathbb{R}^n. \tag{12.49}$$

To construct functions with the required properties we follow 11.2 and assume that $k_0 \in S(\mathbb{R}^n)$ has a compact support and $\widehat{k_0}(0) = 1$. Recall that $\widehat{k_0}(\xi)$ is an entire analytic function, and let

$$\widehat{k_0}(\xi) = 1 + \sum_{|\gamma| \ge 1} c_\gamma \xi^\gamma, \quad \xi \in \mathbb{R}^n, \tag{12.50}$$

be its Taylor expansion. Assume that $c = c_\gamma \ne 0$ for $\gamma = (1, 0, \ldots, 0)$. Then there are two numbers λ_1 and λ_2 with

$$\lambda_1 + \lambda_2 = 1, \quad \lambda_1 + 2\lambda_2 = 0. \tag{12.51}$$

Replacing $k_0(x)$ by

$$k_0'(x) = \lambda_1 k_0(x) + \lambda_2 2^{-n} k_0\left(\frac{x}{2}\right),$$

then k_0' has the same properties as k_0, but $\widehat{k_0'}(\xi)$ has no term with ξ_1 (and also not with $\xi_2, \ldots, \xi_n$). Iterating this procedure it follows that we may assume $c_\gamma = 0$ with $1 \le |\gamma| \le 2N$ in (12.50) and hence

$$k_N(x) = 2^n k_0(2x) - k_0(x), \quad x \in \mathbb{R}^n, \tag{12.52}$$

satisfies

$$\widehat{k_N}(\xi) = \widehat{k_0}\left(\frac{\xi}{2}\right) - \widehat{k_0}(\xi) = O(|\xi|^{2N}), \quad \xi \in \mathbb{R}^n. \tag{12.53}$$

In (11.2) one does not really need the special construction given there. It is sufficient to know that $\widehat{k_N}(\xi)$ decays strongly enough at the origin. This is ensured by (12.53) and hence we have (11.6)–(11.9) with (11.5) as respective quasi-norms for this choice of k_0 and k_N. We have (12.49) since

$$\widehat{k_0}(\xi) + \sum_{\nu=0}^{\infty} \widehat{k_N}(2^{-\nu}\xi) = \lim_{\nu \to \infty} \widehat{k_0}(2^{-\nu}\xi) = 1 \tag{12.54}$$

pointwise. Since (11.3) can be reformulated as

$$k_N(2^{-\nu}, f)^\frown(\xi) = \widehat{k_N}(-2^{-\nu}\xi)\hat{f}(\xi), \quad \nu \in \mathbb{N}_0, \tag{12.55}$$

(12.48) follows from (12.54). We refer also to [Tri92], 3.3.1, 3.3.2, pp. 173–177, for a similar construction in a somewhat different context. Now we are in the position to formulate the counterpart of Theorem 12.5 including maximal versions of the equivalent quasi-norms in (11.6)–(11.9), which will be of some service for us later on. Recall that we introduced $\mathscr{C}^{-\infty}(\mathbb{R}^n)$ in (12.16).

12.9 Theorem (i) *Let* $s \in \mathbb{R}$, $0 < q \le \infty$ *and* $0 < p \le \infty$ ($p < \infty$ *in case of F-spaces). Let* k_0 *and* k_N *be the functions above with* $N \in \mathbb{N}$ *in* (12.53) *sufficiently large in dependence on* s, p, q. *Let* $f \in B_{pq}^s(\mathbb{R}^n)$ *or* $f \in F_{pq}^s(\mathbb{R}^n)$. *Then the right-hand side of* (12.48) *converges in* $S'(\mathbb{R}^n)$ *to* f.

(ii) *Let* $s \in \mathbb{R}$, $0 < q \le \infty$ *and* $0 < p \le \infty$. *Let again* $N \in \mathbb{N}$ *be sufficiently large. Then* (11.8), (11.9), *and for any* $c > 0$

$$
\|f \mid B_{pq}^s(\mathbb{R}^n)\|_* = \left\| \sup_{|\cdot - y| \le c} |k_0(1,f)(y)| \mid L_p(\mathbb{R}^n) \right\|
$$
$$
+ \left(\sum_{\nu=0}^{\infty} 2^{\nu s q} \left\| \sup_{|\cdot - y| \le c2^{-\nu}} |k_N(2^{-\nu}, f)(y)| \mid L_p(\mathbb{R}^n) \right\|^q \right)^{\frac{1}{q}}
\tag{12.56}
$$

(modification if $q = \infty$) *are equivalent quasi-norms in* $B_{pq}^s(\mathbb{R}^n)$. *Furthermore,*

$$
B_{pq}^s(\mathbb{R}^n) = \{ f \in \mathscr{C}^{-\infty}(\mathbb{R}^n) : \|f \mid B_{pq}^s(\mathbb{R}^n)\|_* < \infty \}.
\tag{12.57}
$$

(iii) *Let* $s \in \mathbb{R}$, $0 < q \le \infty$ *and* $0 < p < \infty$. *Let again* $N \in \mathbb{N}$ *be sufficiently large. Then* (11.6), (11.7), *and for any* $c > 0$

$$
\|f \mid F_{pq}^s(\mathbb{R}^n)\|_* = \left\| \sup_{|\cdot - y| \le c} |k_0(1,f)(y)| \mid L_p(\mathbb{R}^n) \right\|
$$
$$
+ \left\| \left(\sum_{\nu=0}^{\infty} 2^{\nu s q} \sup_{|\cdot - y| \le c2^{-\nu}} |k_N(2^{-\nu}, f)(y)|^q \right)^{\frac{1}{q}} \mid L_p(\mathbb{R}^n) \right\|
\tag{12.58}
$$

(modification if $q = \infty$) *are equivalent quasi-norms in* $F_{pq}^s(\mathbb{R}^n)$. *Furthermore,*

$$
F_{pq}^s(\mathbb{R}^n) = \{ f \in \mathscr{C}^{-\infty}(\mathbb{R}^n) : \|f \mid F_{pq}^s(\mathbb{R}^n)\|_* < \infty \}.
\tag{12.59}
$$

Proof. In contrast to the proof of Theorem 12.5 there are now no complications caused by the first terms on the right-hand sides of (12.18), (12.19) etc. or by the bad behaviour of φ_0 in (12.39), (12.27). We have the same situation here as for the second terms on the right-hand sides of (12.18), (12.19) etc. In other words, both parts (ii) and (iii) of the theorem are covered by the technique developed in [Tri92], see in particular [Tri92], 2.4.1, 2.4.2, 2.4.6, pp. 100–114 and 122–124, as far as the F-spaces are concerned and their easier B-counterparts in [Tri92], 2.5.1, 2.5.3, pp. 131–135 and 138–139. In particular our a priori assumption $f \in \mathscr{C}^{-\infty}(\mathbb{R}^n)$ in (12.57) and (12.59) ensures that (14) on p. 111 in [Tri92] is applicable, where it can be seen that the maximal function used there can be replaced by the maximal functions in (12.56) and (12.58). As for part (i) of the theorem we are in the same situation as in Step 3 of the proof of Theorem 12.5: the convergence in $S'(\mathbb{R}^n)$ of the right-hand side of (12.48) to some $g \in S'(\mathbb{R}^n)$ is ensured by the above quoted references in [Tri92]. To prove $f = g$ we argue as in the Steps 2 and 3 of the proof of Theorem 12.5. Let $\psi \in S(\mathbb{R}^n)$ with a compact support and $\psi(x) = 1$ in $B = \{y : |y| \leq 1\}$. Recall that $L_p^B(\mathbb{R}^n)$ with $0 < p \leq \infty$ is given by (12.8). By the multiplier theorem in [Tri83], p. 26, the counterpart of (12.30) is given by

$$
\begin{aligned}
\|k_0(tD)f - f \,|\, L_p(\mathbb{R}^n)\| \\
= \left\| \left(\left(\widehat{k_0}\,(t\xi)\,\psi(\xi) - \psi(\xi) \right) \hat{f} \right)^{\vee} \,|\, L_p(\mathbb{R}^n) \right\| \\
\leq c \sum_{|\alpha| \leq m} \left\| D^\alpha \left(\widehat{k_0}(t\xi) - 1 \right) \psi(\xi) \,|\, L_2(\mathbb{R}^n) \right\| \; \|f \,|\, L_p(\mathbb{R}^n)\|
\end{aligned}
\tag{12.60}
$$

for $m \in \mathbb{N}$, $m > \sigma_p + \frac{n}{2}$ and $f \in L_p^B(\mathbb{R}^n)$. By (12.50) (or (12.54)) the first factor on the right-hand side of (12.60) tends to zero if $t \to 0$. Hence we obtain the counterpart of (12.34)

$$
k_0(tD)f \to f \quad \text{in} \quad L_1(\mathbb{R}^n) \quad \text{if} \quad t \to 0.
\tag{12.61}
$$

Now the desired assertion $f = g$ follows by the arguments in the third step of the proof of Theorem 12.5.

12.10 Remark Characterizations of $B_{pq}^s(\mathbb{R}^n)$ and $F_{pq}^s(\mathbb{R}^n)$ in terms of local means (with and without suprema) may also be found in the papers [BPT96a,b] mentioned above.

13 Atomic decompositions

13.1 Introduction

Atomic and sometimes also subatomic representations and decompositions play a decisive role in our later considerations. This justifies the development here of the corresponding theory in detail and with full proofs. Atomic decompositions of function spaces are nowadays well established. They have a history of some twenty years. In [Tri92], 1.9, we gave a historically-minded report, which will not be repeated here, see also [Tri92], 3.2, for further references. We only mention that the (smooth) atoms in $B_{pq}^s(\mathbb{R}^n)$ and $F_{pq}^s(\mathbb{R}^n)$ as they are used nowadays go essentially back to M. Frazier and B. Jawerth, see [FrJ85], [FrJ90], and [FJW91]. As far as significant contributions of other mathematicians are concerned we refer to the literature given in [Tri92], 1.9, especially to the papers by Ju. V. Netrusov mentioned there. We present new proofs of this theory, using the harmonic and local representations obtained in the preceding Section 12. Based on harmonic representations we derive in Section 14 subatomic (or quarkonial) decompositions, and extend in Section 15 the technique of atomic and subatomic decompositions to

Banach-space-valued function spaces of the above type. These results are published here for the first time. We are exclusively interested in smooth atoms (and quarks) in $\mathbb{R}^n$. There is a growing interest in atoms in smooth and non-smooth domains in $\mathbb{R}^n$ (and even on fractals) and in reducing the necessary smoothness of the atoms as far as possible. Again we do not go into detail and refer to [TrW96a] and [ET96], 2.5, pp. 57–65, and the literature mentioned there.

13.2 Heuristics and notation

Since we do not assume that the reader is familiar with atomic decompositions in function spaces it seems to be appropriate to begin with some heuristical considerations making clear what is going on. Let, say, $f \in B_{pq}^s(\mathbb{R}^n)$ and let $k_M(t,f)$ be the local means described in 11.2. According to Theorem 12.9(i) and (12.48)

$$f = k_0(1,f)(x) + \sum_{\nu=1}^{\infty} k_N(2^{-\nu+1}, f)(x), \tag{13.1}$$

convergence being in $S'(\mathbb{R}^n)$, may be assumed in addition. By (11.3) it is suggested to decompose $k_N(2^{-\nu}, f)(x)$ with respect to a lattice in $\mathbb{R}^n$ with side-length $2^{-\nu}$. But first we introduce some useful notation.

Recall that $\mathbb{Z}^n$ stands for the lattice of all points in $\mathbb{R}^n$ with integer-valued components. Furthermore, $Q_{\nu m}$ denotes a cube in $\mathbb{R}^n$ with sides parallel to the axes of coordinates, centred at $2^{-\nu}m$, and with side length $2^{-\nu}$, where $m \in \mathbb{Z}^n$ and $\nu \in \mathbb{N}_0$. Let Q be a cube in $\mathbb{R}^n$ and $r > 0$ then rQ is the cube in $\mathbb{R}^n$ concentric with Q and with side length r times the side length of Q. Finally, let

$$\psi \in S(\mathbb{R}^n), \qquad supp\,\psi \quad \text{compact}, \tag{13.2}$$

and

$$\sum_{m \in \mathbb{Z}^n} \psi(x - m) = 1 \quad \text{for} \quad x \in \mathbb{R}^n. \tag{13.3}$$

Then $\psi_{\nu m}(x) = \psi(2^\nu x - m)$ is a resolution of unity adapted to $2^{-\nu}\mathbb{Z}^n$ and there is a constant $d > 0$ such that

$$supp\,\psi_{\nu m} \subset d\,Q_{\nu m}, \quad \nu \in \mathbb{N}_0, \ m \in \mathbb{Z}^n. \tag{13.4}$$

We multiply the ν-th term in (13.1) with that resolution of unity and obtain

$$f = \sum_{m \in \mathbb{Z}^n} \psi_{0m}(x)\, k_0(1,f)(x)$$
$$+ \sum_{\nu=1}^{\infty} \sum_{m \in \mathbb{Z}^n} \psi_{\nu m}(x)\, k_N(2^{-\nu+1}, f)(x). \tag{13.5}$$

We need some normalizations of the so-obtained building blocks and also some restrictions for s and p. Let

$$s > \sigma_p = n(\frac{1}{p} - 1)_+, \ 0 < q \leq \infty, \quad \text{and} \quad K \in \mathbb{N} \quad \text{with} \quad K > s. \quad (13.6)$$

Let

$$\lambda_{\nu m} = 2^{\nu(s-\frac{n}{p})} \sum_{|\gamma| \leq K} \sup_{y \in dQ_{\nu m}} \left| (D^\gamma k_N)(2^{-\nu+1}, f)(y) \right| \quad (13.7)$$

if $\nu \in \mathbb{N}_0$ and $m \in \mathbb{Z}^n$ (with k_0 and 1 in place of k_N and $2^{-\nu+1}$ if $\nu = 0$.) After this normalization one arrives at the typical *optimal atomic decomposition* of f,

$$f = \sum_{\nu=0}^\infty \sum_{m \in \mathbb{Z}^n} \lambda_{\nu m} a_{\nu m}(x), \quad \text{convergence being in } S'(\mathbb{R}^n), \quad (13.8)$$

where $\lambda_{\nu m}$ is given by (13.7) and

$$a_{\nu m}(x) = \lambda_{\nu m}^{-1} \psi_{\nu m}(x) k_N(2^{-\nu+1}, f)(x), \quad \nu \in \mathbb{N}_0, \ m \in \mathbb{Z}^n, \quad (13.9)$$

(above modifications if $\nu = 0$). Of course if $\lambda_{\nu m} = 0$ for some ν and m, then the corresponding term in (13.8) is simply omitted. To explain the normalizing factors in (13.7) and (13.8) we collect now the typical properties of the atom $a_{\nu m}(x)$. First we have

$$\text{supp}\, a_{\nu m} \subset dQ_{\nu m}, \quad \nu \in \mathbb{N}_0, \ m \in \mathbb{Z}^n. \quad (13.10)$$

Furthermore, by (11.3), (13.7) and (13.9) it follows

$$|D^\gamma a_{\nu m}(x)| \leq c\, 2^{-\nu(s-\frac{n}{p})+\nu|\gamma|}, \quad |\gamma| \leq K, \quad (13.11)$$

where c is independent of ν, m, and x. To understand the reason for the factor $2^{-\nu(s-\frac{n}{p})}$ we put first $s = 0$ and obtain by (13.9) and (13.11) that $\|a_{\nu m}\,|\,L_p(\mathbb{R}^n)\|$ are uniformly bounded. But this normalization property holds under the above circumstances (13.6) for all spaces $B_{pq}^s(\mathbb{R}^n)$ and $F_{pq}^s(\mathbb{R}^n)$ with $s - \frac{n}{p}$ in place of $-\frac{n}{p}$, see [ET96], 2.3.2, pp. 35–36. Here one needs (13.6). In other words we have always

$$\|a_{\nu m}\,|\,B_{pq}^s(\mathbb{R}^n)\| = O(1) \quad (13.12)$$

independently of ν and m. Finally let

$$\lambda = \{\lambda_{\nu m} : \nu \in \mathbb{N}_0, m \in \mathbb{Z}^n\}$$

and

$$\|\lambda\,|\,b_{pq}\| = \left(\sum_{\nu=0}^\infty \left(\sum_m |\lambda_{\nu m}|^p \right)^{\frac{q}{p}} \right)^{\frac{1}{q}}. \quad (13.13)$$

We wish to apply Theorem 12.9(ii). If we replace k_0 and k_N in (12.56) by $D^\gamma k_0$ and $D^\gamma k_N$, respectively, for some multi-index γ, then the corresponding quasi-norm in (12.56) can be estimated from above by $c \, \|f \mid B^s_{pq}(\mathbb{R}^n)\|$. This assertion is covered by the technique developed in [Tri92], see the detailed references given in the proof of Theorem 12.9. But now we have

$$\|\lambda \mid b_{pq}\| \sim \|f \mid B^s_{pq}(\mathbb{R}^n)\|, \tag{13.14}$$

where «$\sim$» indicates equivalence constants which are independent of f. In other words, an atomic decomposition of $f \in B^s_{pq}(\mathbb{R}^n)$ under the restrictions (13.6) is a representation of type (13.8) where $a_{\nu m}(x)$ are normalized building blocks, called *atoms*, satisfying (13.10)–(13.12) and the coefficients $\{\lambda_{\nu m}\}$ generate an equivalent quasi-norm according to (13.14).

13.3 Definition (i) *Let $K \in \mathbb{N}_0$ and $d > 1$. A K times differentiable complex-valued function $a(x)$ in $\mathbb{R}^n$ (continuous if $K = 0$) is called an 1-atom (more precisely 1_K-atom) if*

$$\operatorname{supp} a \subset dQ_{0m} \quad \text{for some} \quad m \in \mathbb{Z}^n \tag{13.15}$$

and

$$|D^\alpha a(x)| \leq 1 \quad \text{for} \quad |\alpha| \leq K. \tag{13.16}$$

(ii) *Let $s \in \mathbb{R}$, $0 < p \leq \infty$, $K \in \mathbb{N}_0$, $L + 1 \in \mathbb{N}_0$, and $d > 1$. A K times differentiable complex-valued function $a(x)$ in $\mathbb{R}^n$ (continuous if $K = 0$) is called an (s,p)-atom (more precisely $(s,p)_{K,L}$-atom) if for some $\nu \in \mathbb{N}_0$*

$$\operatorname{supp} a \subset d\, Q_{\nu m} \quad \text{for some} \quad m \in \mathbb{Z}^n, \tag{13.17}$$

$$|D^\alpha a(x)| \leq 2^{-\nu(s - \frac{n}{p}) + |\alpha|\nu} \quad \text{for} \quad |\alpha| \leq K \tag{13.18}$$

and

$$\int_{\mathbb{R}^n} x^\beta \, a(x)\, dx = 0 \quad \text{if} \quad |\beta| \leq L. \tag{13.19}$$

13.4 Comments The cubes $Q_{\nu m}$ have been introduced in 13.2. In particular, Q_{0m} in (13.15) are cubes with side-length 1. The value of the number d in (13.15) and (13.17) is unimportant. It simply makes clear that at the level ν some controlled overlapping of the supports of $a_{\nu m}(x)$ must be allowed. The moment conditions (13.19) can be reformulated as

$$(D^\beta \hat{a})(0) = 0 \quad \text{if} \quad |\beta| \leq L, \tag{13.20}$$

which shows that a sufficiently strong decay of $\hat{a}(\xi)$ at the origin is required. If $L = -1$ then (13.19) simply means that there are no moment conditions. In our heuristical considerations in 13.2 we provided an understanding (so we hope) both of (13.17) and (13.18), see (13.10) and (13.11), respectively, but not of (13.19) or (13.20) so far. In the construction (13.9) the function $k_N(2^{-\nu+1}, f)(x)$ has the desired moment conditions, see (11.3) with (11.2) or (12.53). But the multiplication with $\psi_{\nu m}(x)$ in (13.9) destroys this property. It turns out that for large values of s as in (13.6) moment conditions are not necessary (but desirable). But for small values of s moment conditions will be indispensable and they must be restored later on. Furthermore if K and L are sufficiently large (in dependence on s, p, q, B or F spaces) it comes out that the above atoms are normalized building blocks in $B^s_{pq}(\mathbb{R}^n)$ and $F^s_{pq}(\mathbb{R}^n)$ as in (13.12). Similar as in 13.2 we write in the sequel $a_{\nu m}(x)$ instead of $a(x)$ if this atom is located at $Q_{\nu m}$ according to (13.15) and (13.17), that means

$$supp\, a_{\nu m} \subset d\, Q_{\nu m}, \quad \nu \in \mathbb{N}_0, \; m \in \mathbb{Z}^n, \, d > 1. \tag{13.21}$$

As explained in 13.2 an atomic decomposition is an interplay between (sufficiently smooth) building blocks $a_{\nu m}(x)$ and some sequence spaces according to (13.8) and (13.14). Besides b_{pq} given by (13.13) we need its counterpart f_{pq}. For that purpose we denote by $\chi^{(p)}_{\nu m}$ the p-normalized characteristic function of the cube $Q_{\nu m}$, that means

$$\chi^{(p)}_{\nu m}(x) = 2^{\frac{\nu n}{p}} \quad \text{if} \quad x \in Q_{\nu m} \quad \text{and} \quad \chi^{(p)}_{\nu m}(x) = 0 \quad \text{if } x \notin Q_{\nu m}, \tag{13.22}$$

where $\nu \in \mathbb{N}_0$, $m \in \mathbb{Z}^n$, and $0 < p \leq \infty$. Of course,

$$\|\chi^{(p)}_{\nu m} \,|\, L_p(\mathbb{R}^n)\| = 1. \tag{13.23}$$

13.5 Definition *Let* $0 < p \leq \infty$, $0 < q \leq \infty$, *and*

$$\lambda = \{\lambda_{\nu m} \in \mathbb{C} : \nu \in \mathbb{N}_0, \; m \in \mathbb{Z}^n\}.$$

Then

$$b_{pq} = \left\{ \lambda : \|\lambda\,|\, b_{pq}\| = \left(\sum_{\nu=0}^{\infty} \left(\sum_{m \in \mathbb{Z}^n} |\lambda_{\nu m}|^p \right)^{\frac{q}{p}} \right)^{\frac{1}{q}} < \infty \right\} \tag{13.24}$$

and

$$f_{pq} = \left\{ \lambda : \|\lambda\,|\, f_{pq}\| = \left\| \left(\sum_{\nu=0}^{\infty} \sum_{m \in \mathbb{Z}^n} |\lambda_{\nu m}\, \chi^{(p)}_{\nu m}(\cdot)|^q \right)^{\frac{1}{q}} |\, L_p(\mathbb{R}^n) \right\| < \infty \right\} \tag{13.25}$$

(with the usual modification if $p = \infty$ *or/and* $q = \infty$*).*

13.6 Proposition *Let* $0 < p \leq \infty$ *and* $0 < q \leq \infty$. *Then* b_{pq} *and* f_{pq} *are quasi-Banach spaces. Furthermore*

$$b_{p,\min(p,q)} \subset f_{pq} \subset b_{p,\max(p,q)}, \qquad (13.26)$$

and, in particular, $b_{pp} = f_{pp}$.

Proof. The proof that b_{pq} and f_{pq} are quasi-Banach spaces is standard. By (13.23) we have

$$\| \lambda \mid b_{pq} \| = \left(\sum_{\nu=0}^{\infty} \left\| \sum_{m \in \mathbb{Z}^n} \lambda_{\nu m} \, \chi_{\nu m}^{(p)}(\cdot) \mid L_p(\mathbb{R}^n) \right\|^q \right)^{\frac{1}{q}}. \qquad (13.27)$$

Compared with $\| \lambda \mid f_{pq} \|$ the two quasi-norms $L_p(\mathbb{R}^n)$ and l_q in b_{pq} and f_{pq} interchange. But then (13.26) follows from the triangle inequality for Banach spaces by standard arguments.

13.7 Remark By the theorem below, (13.26) is the counterpart of (11.15). We complement the notations introduced in 13.2 and used below by

$$\sigma_p = n\left(\frac{1}{p} - 1\right)_+ \quad \text{and} \quad \sigma_{pq} = n\left(\frac{1}{\min(p,q)} - 1\right)_+ \qquad (13.28)$$

where $0 < p \leq \infty$ and $0 < q \leq \infty$, see (11.4). As usual for any $\varkappa \in \mathbb{R}$ we put $\varkappa_+ = \max(\varkappa, 0)$ and $[\varkappa]$ stands for the largest integer smaller than or equal to $\varkappa$.

13.8 Theorem (i) *Let* $0 < p \leq \infty$, $0 < q \leq \infty$, *and* $s \in \mathbb{R}$. *Let* $K \in \mathbb{N}_0$ *and* $L + 1 \in \mathbb{N}_0$ *with*

$$K \geq (1 + [s])_+ \quad \text{and} \quad L \geq \max(-1, [\sigma_p - s]) \qquad (13.29)$$

be fixed. Then $f \in S'(\mathbb{R}^n)$ *belongs to* $B_{pq}^s(\mathbb{R}^n)$ *if, and only if, it can be represented as*

$$f = \sum_{\nu=0}^{\infty} \sum_{m \in \mathbb{Z}^n} \lambda_{\nu m} \, a_{\nu m}(x), \quad \text{convergence being in } S'(\mathbb{R}^n), \qquad (13.30)$$

where the $a_{\nu m}(x)$ *are* 1_K*-atoms* $(\nu = 0)$ *or* $(s, p)_{K,L}$*-atoms* $(\nu \in \mathbb{N})$ *according to Definition 13.3, with* (13.21) *and* $\lambda \in b_{pq}$. *Furthermore*

$$\inf \| \lambda \mid b_{pq} \| \qquad (13.31)$$

where the infimum is taken over all admissible representations (13.30), *is an equivalent quasi-norm in* $B_{pq}^s(\mathbb{R}^n)$.

   (ii) *Let* $0 < p < \infty$, $0 < q \leq \infty$, *and* $s \in \mathbb{R}$. *Let* $K \in \mathbb{N}_0$ *and* $L + 1 \in \mathbb{N}_0$ *with*

$$K \geq (1 + [s])_+ \quad \text{and} \quad L \geq \max(-1, [\sigma_{pq} - s]) \qquad (13.32)$$

be fixed. Then $f \in S'(\mathbb{R}^n)$ *belongs to* $F_{pq}^s(\mathbb{R}^n)$ *if, and only if, it can be represented by* (13.30), *where the atoms* $a_{\nu m}(x)$ *have the same meaning as in part* (i) *(now perhaps with a different value of L) and* $\lambda \in f_{pq}$. *Furthermore*

$$\inf \| \lambda \, | \, f_{pq} \| \tag{13.33}$$

where the infimum is taken over all admissible representations (13.30), *is an equivalent quasi-norm in* $F_{pq}^s(\mathbb{R}^n)$.

Proof.

Step 1. We begin with a few remarks about the structure of the proof. As we shall see, what we did so far in 13.2 will prove the existence of an *optimal atomic decomposition* (13.30) for $f \in B_{pq}^s(\mathbb{R}^n)$, complemented by (13.14) under the restrictions (13.6) and with $L = -1$ in the above meaning. As will be shown in the Steps 3 and 4 below this is the crucial assertion to prove the «only-if-parts» of the theorem. In the sequel we use the notation *optimal atomic decomposition* in the understanding of the equivalence (13.14) with its obvious $F_{pq}^s(\mathbb{R}^n)$-counterpart for all $s \in \mathbb{R}$, $0 < p \leq \infty$ ($p < \infty$ in the F-case), $0 < q \leq \infty$. In other words we have mainly to concentrate on the problem whether (13.30) converges in $S'(\mathbb{R}^n)$ and represents under the formulated circumstances an element of $B_{pq}^s(\mathbb{R}^n)$ or $F_{pq}^s(\mathbb{R}^n)$, and that the corresponding quasi-norms can be estimated from above by $c \, \| \lambda \, | \, b_{pq} \|$ or $c \, \| \lambda \, | \, f_{pq} \|$, respectively. This will be done in Step 2 below. Since we have only a controlled overlapping of the supports of the atoms $a_{\nu m}(x)$ for fixed ν, convergence in $S'(\mathbb{R}^n)$ means

$$\lim_{\mu \to \infty} \sum_{\nu=0}^{\mu} \left(\sum_m \cdots \right),$$

where the inner sum causes no problems. It comes out that the convergence in $S'(\mathbb{R}^n)$ of the right-hand side of (13.30) in this sense is ensured by the required properties of the involved atoms and $\lambda \in b_{pq}$ or $\lambda \in f_{pq}$. *In particular, convergence in* $S'(\mathbb{R}^n)$ *in* (13.30) *is not an additional assumption but a result.* But we shift this question to Corollary 13.9 below.

Step 2. We prove the if-part of (ii) in the just explained understanding. The proof of the if-part of (i) is the same but technically simpler. We rely on the equivalent quasi-norm in $F_{pq}^s(\mathbb{R}^n)$ given by (11.6) and the underlying local means according to (11.1)–(11.3), where N in (11.5) may be chosen arbitrarily large. Let $a_{\nu m}(x)$ with $\nu \in \mathbb{N}_0$ and $m \in \mathbb{Z}^n$ be an $(s, p)_{K,L}$-atom according to Definition 13.3 with (13.21), where K and L are fixed integers satisfying (13.32) (in case of $\nu = 0$ that means $a_{\nu m}(x)$ is a 1_K-atom). We have for $j \in \mathbb{N}$,

$$2^{js} k_N(2^{-j}, a_{\nu m})(x) = 2^{js} \int_{\mathbb{R}^n} \Delta^N k^0(y) \, a_{\nu m}(x + 2^{-j} y) \, dy. \tag{13.34}$$

We have to distinguish between $j \geq \nu$ and $j < \nu$. The exceptional values $\nu = 0$ and/or $j = 0$ corresponding to 1_K-atoms and the first summand in (11.6), respectively, can be incorporated in the following considerations after the necessary modifications. We shall not stress this point and assume $\nu \in \mathbb{N}$ and $j \in \mathbb{N}$. Let $j \geq \nu$. We put

$$a_{\nu m}(x) = 2^{-\nu(s-\frac{n}{p})} a^{\nu m}(2^{\nu} x - m) \tag{13.35}$$

and observe that $a^{\nu m}$ is a 1_K-atom with respect to the unit cube at the origin. Let $K = 2M$ with $M \in \mathbb{N}_0$ for simplicity. By the arguments below the necessary modifications are clear if K is an odd integer. We insert (13.35) in (13.34), choose $N > M$, and obtain by partial integration

$$2^{js} k_N(2^{-j}, a_{\nu m})(x) = 2^{s(j-\nu)+\nu\frac{n}{p}} \int_{\mathbb{R}^n} \Delta^N k^0(y) a^{\nu m}(2^{\nu} x - m + 2^{\nu-j} y) \, dy$$

$$= 2^{-(K-s)(j-\nu)} 2^{\nu\frac{n}{p}} \int_{\mathbb{R}^n} \Delta^{N-M} k^0(y)(\Delta^M a^{\nu m})(2^{\nu} x - m + 2^{\nu-j} y) \, dy. \tag{13.36}$$

Since both k^0 and $\Delta^M a^{\nu m}$ have supports in a ball centred at the origin of some radius $c > 0$ and $j \geq \nu$ it follows that

$$2^{js} \left| k_N(2^{-j}, a_{\nu m})(x) \right| \leq c \, 2^{-(K-s)(j-\nu)} \tilde{\chi}_{\nu m}^{(p)}(x), \quad j \geq \nu, \tag{13.37}$$

where $\tilde{\chi}_{\nu m}^{(p)}(x)$ is the p-normalized characteristic function defined in (13.22), (13.23) with $cQ_{\nu m}$ instead of $Q_{\nu m}$. Let $j < \nu$ and put again $k_N(y) = \Delta^N k^0(y)$. Then the integration in

$$2^{js} k_N(2^{-j}, a_{\nu m})(x) = 2^{j(s+n)} \int k_N(2^j y) a_{\nu m}(x+y) \, dy \tag{13.38}$$

can be restricted to $\{y : |y| \leq c2^{-j}\}$ for some $c > 0$. Furthermore with L given by (13.32) we expand $k_N(2^j y)$ up to the order L with respect to the off-point $2^{-\nu} m - x$ and obtain

$$k_N(2^j y) = \sum_{|\beta| \leq L} c_\beta(x) \, (y - 2^{-\nu} m + x)^\beta + 2^{j(L+1)} O\left(|x + y - 2^{-\nu} m|^{L+1} \right). \tag{13.39}$$

We insert (13.39) in (13.38). By (13.19) the terms with $|\beta| \leq L$ vanish. Since

$$|a_{\nu m}(x+y)| \leq 2^{-\nu(s-\frac{n}{p})} \tilde{\chi}_{\nu m}(x+y), \tag{13.40}$$

where $\tilde{\chi}_{\nu m}(x)$ is the characteristic function of $dQ_{\nu m}$ according to (13.21), we obtain

$$2^{js} \left| k_N(2^{-j}, a_{\nu m})(x) \right|$$

$$\leq c \, 2^{j(s+n)} \, 2^{-\nu(s-\frac{n}{p})} \, 2^{(j-\nu)(L+1)} \int_{|y| \leq c2^{-j}} \tilde{\chi}_{\nu m}(x+y) \, dy. \tag{13.41}$$

Recall that $j < \nu$. The last integral is at most $c2^{-\nu n}$ and it is zero if x is outside a cube $c2^{\nu-j} Q_{\nu m}$ (centred at $2^\nu m$ and with side-length $c2^{-j}$). Hence

$$\int_{|y| \leq c2^{-j}} \tilde{\chi}_{\nu m}(x+y)\, dy \leq c\, 2^{-\nu n}\, \chi(c2^{\nu-j} Q_{\nu m})(x), \tag{13.42}$$

where $\chi(c2^{\nu-j} Q_{\nu m})(x)$ is the characteristic function of the indicated cube. Recall that the Hardy-Littlewood maximal function of an integrable function $g(x)$ is given by

$$(Mg)(x) = \sup |Q|^{-1} \int_Q |g(y)|\, dy \tag{13.43}$$

where the supremum is taken over all cubes Q with $x \in Q$. Let $g = \chi_{\nu m}$ be the characteristic function of $Q_{\nu m}$, $x \in c2^{\nu-j} Q_{\nu m}$ and $Q = c2^{\nu-j} Q_{\nu m}$, then

$$(M\chi_{\nu m})(x) \geq |Q|^{-1} 2^{-\nu n} \geq c\, 2^{-(\nu-j)n}, \quad x \in c\, 2^{\nu-j} Q_{\nu m}. \tag{13.44}$$

Let $0 < w < \min(1, p, q)$. By (13.44), inserted in (13.42), we have

$$\int_{|y| \leq c2^{-j}} \tilde{\chi}_{\nu m}(x+y)\, dy \leq c\, 2^{-\nu n}\, 2^{(\nu-j)\frac{n}{w}} (M\chi_{\nu m})^{\frac{1}{w}}(x), \quad x \in \mathbb{R}^n. \tag{13.45}$$

Replacing $\chi_{\nu m}$ in (13.45) by $\chi_{\nu m}^{(p)}$ according to (13.22) and inserting the estimate (13.45) in (13.41) we obtain

$$2^{js} \left| k_N(2^{-j}, a_{\nu m})(x) \right|$$
$$\leq c\, 2^{-(\nu-j)(s+L+1+n-\frac{n}{w})} \left(M\, \chi_{\nu m}^{(p)\, w} \right)^{\frac{1}{w}}(x), \quad x \in \mathbb{R}^n. \tag{13.46}$$

By (13.32) and (13.28) the number w can be chosen such that $s+L+1+n-\frac{n}{w} > 0$. Hence for some $\varkappa > 0$ we arrive at

$$2^{js} \left| k_N(2^{-j}, a_{\nu m})(x) \right| \leq c\, 2^{-(\nu-j)\varkappa} \left(M\, \chi_{\nu m}^{(p)\, w} \right)^{\frac{1}{w}}(x), \quad j < \nu. \tag{13.47}$$

Combining (13.37) and (13.47) we obtain for $q \leq 1$

$$\left| 2^{js}\, k_N \left(2^{-j}, \sum_{\nu,m} \lambda_{\nu m}\, a_{\nu m} \right) \right|^q$$
$$\leq c \sum_{\nu \leq j} \sum_m |\lambda_{\nu m}|^q\, 2^{-\varrho(j-\nu)q}\, \tilde{\chi}_{\nu m}^{(p)\, q}(x) \tag{13.48}$$
$$+ c \sum_{j < \nu} \sum_m |\lambda_{\nu m}|^q\, 2^{-(\nu-j)\varkappa q} \left(M\, \chi_{\nu m}^{(p)\, w} \right)^{\frac{q}{w}}(x)$$

for some $\varrho > 0$ and $\varkappa > 0$. We sum over j, take the $\frac{1}{q}$-power and afterwards the $L_p(\mathbb{R}^n)$-quasi-norm and arrive at

$$\left\| \left(\sum_{j=1}^{\infty} 2^{jsq} \left| k_N(2^{-j}, \sum_{\nu,m} \lambda_{\nu m} a_{\nu m})(\cdot) \right|^q \right)^{\frac{1}{q}} \mid L_p(\mathbb{R}^n) \right\|$$

$$\leq c \left\| \left(\sum_{\nu,m} |\lambda_{\nu m}|^q \, \tilde{\chi}_{\nu m}^{(p)\,q}(\cdot) \right)^{\frac{1}{q}} \mid L_p(\mathbb{R}^n) \right\| \tag{13.49}$$

$$+ c \left\| \left(\sum_{\nu,m} |\lambda_{\nu m}|^q \left(M \chi_{\nu m}^{(p)\,w} \right)^{\frac{q}{w}}(\cdot) \right)^{\frac{1}{q}} \mid L_p(\mathbb{R}^n) \right\|.$$

The modification of (13.48) if $1 < q \leq \infty$ is clear. Hence (13.49) holds for any $0 < q \leq \infty$. The first summand on the right-hand side is just what we want, see (13.33), since $\tilde{\chi}_{\nu m}^{(p)}$ can be replaced by $\chi_{\nu m}^{(p)}$. With $g_{\nu m}(x) = \lambda_{\nu m} \, \chi_{\nu m}^{(p)}(x)$ the second summand on the right-hand side can be written as

$$c \left\| \left(\sum_{\nu,m} (M g_{\nu m}^w)(\cdot)^{\frac{q}{w}} \right)^{\frac{w}{q}} \mid L_{\frac{p}{w}}(\mathbb{R}^n) \right\|^{\frac{1}{w}}. \tag{13.50}$$

Since $1 < \frac{q}{w} \leq \infty$ and $1 < \frac{p}{w} < \infty$ we can apply the vector-valued maximal inequality of Fefferman and Stein, see [Tri92], 2.2.2, p. 89, and the references given there to the original papers by Hardy, Littlewood, Wiener, Fefferman and Stein. A short proof may be found in [Tor86], pp. 303–305. Application of this theorem gives again the wanted estimate, see (13.25). As we said, the terms with $\nu = 0$ and/or $j = 0$ are also covered by this technique. We shift some technicalities mentioned at the end of Step 1 concerning the convergence of (13.30) in $S'(\mathbb{R}^n)$ and the question whether the resulting $f \in S'(\mathbb{R}^n)$ belongs to $F_{pq}^s(\mathbb{R}^n)$ to the proof of Corollary 13.9 below. Then we obtain by (11.6) and (13.25)

$$\|f \mid F_{pq}^s(\mathbb{R}^n)\| \leq c \, \|\lambda \mid f_{pq}\|, \tag{13.51}$$

which completes the proof of the if-part of (ii). The proof of the if-part of (i) is similar, but simpler. One needs only the scalar Hardy-Littlewood maximal theorem which holds also for $p = \infty$.

Step 3. It remains to prove that any $f \in B_{pq}^s(\mathbb{R}^n)$ or $f \in F_{pq}^s(\mathbb{R}^n)$ can be decomposed by (13.30) and that (13.31) or (13.33) are equivalent quasi-norms in $B_{pq}^s(\mathbb{R}^n)$ or $F_{pq}^s(\mathbb{R}^n)$, respectively. Let $L = -1$, $K \in \mathbb{N}$ with $K > s$, and

$$s > \sigma_p, \quad 0 < p \leq \infty, \quad 0 < q \leq \infty, \tag{13.52}$$

then the desired assertion in case of the spaces $B_{pq}^s(\mathbb{R}^n)$ is covered by our arguments in 13.2, see in particular (13.6) and (13.14). Although the very special construction in 13.2 caused no problems about the convergence of (13.8), (13.30) in $S'(\mathbb{R}^n)$, the restrictions $L = -1$, $K > s$ and (13.52) have their origin in (13.29) and the discussion of convergence in (13.30) in a more general situation, see also the proof of the Corollary 13.9 below. First we extend the considerations given in 13.2 to the spaces $F_{pq}^s(\mathbb{R}^n)$. Again let $L = -1$, $K \in \mathbb{N}$ with $K > s$ and

$$s > \sigma_{pq}, \quad 0 < p < \infty, \quad 0 < q \leq \infty. \tag{13.53}$$

Let $f \in F_{pq}^s(\mathbb{R}^n)$. We follow the arguments given in 13.2. We have (13.1)–(13.5), (13.53) in place of (13.6), (13.7)–(13.11). By the same reasoning as after (13.13) one can replace k_0 and k_N in (12.58) by $D^\gamma k_0$ and $D^\gamma k_N$, at least as an estimate from above by

$$c \, \|f \mid F_{pq}^s(\mathbb{R}^n)\|.$$

Having this in mind it follows easily from (13.25) that

$$\|\lambda \mid f_{pq}\| \sim \|f \mid F_{pq}^s(\mathbb{R}^n)\|. \tag{13.54}$$

This completes the proof also for the F-spaces under the above restrictions.

Step 4. Finally we extend the proof of the existence of optimal atomic decompositions to arbitrary values of $s \in \mathbb{R}$ and L restricted by (13.29) and (13.32). Let, say, $f \in F_{pq}^s(\mathbb{R}^n)$ with $s \in \mathbb{R}$, $0 < p < \infty$, $0 < q \leq \infty$. Replacing $\langle \xi \rangle^\sigma$ in (10.11) and (10.12) by $1 + |\xi|^{2M}$ with $M \in \mathbb{N}$ and its inverse we have respective counterparts of (10.12). Hence f can be represented as

$$f = g + (-\Delta)^M g, \quad \|g \mid F_{pq}^{s+2M}(\mathbb{R}^n)\| \sim \|f \mid F_{pq}^s(\mathbb{R}^n)\|. \tag{13.55}$$

We apply this argument to g with $s + 2M$ in place of s. Iteration yields

$$f = f_1 + (-\Delta)^M f_2, \quad f_1 \in F_{pq}^{s+2kM}(\mathbb{R}^n), \quad f_2 \in F_{pq}^{s+2M}(\mathbb{R}^n) \tag{13.56}$$

with

$$\|f_1 \mid F_{pq}^{s+2kM}(\mathbb{R}^n)\| + \|f_2 \mid F_{pq}^{s+2M}(\mathbb{R}^n)\| \sim \|f \mid F_{pq}^s(\mathbb{R}^n)\|, \tag{13.57}$$

where $k \in \mathbb{N}$ can be chosen arbitrarily large. By (11.17) and (10.16) we have

$$\|f_1 \mid \mathscr{C}^\sigma(\mathbb{R}^n)\| \leq c \|f_1 \mid F_{pq}^{s+2kM}(\mathbb{R}^n)\| \leq c' \|f \mid F_{pq}^s(\mathbb{R}^n)\|, \tag{13.58}$$

where $\sigma \leq s + 2kM - \frac{n}{p}$ can also be chosen arbitrarily large. Let $K > s$ be given. We choose $\sigma > K$ and decompose

$$f_1(x) = \sum_{m \in \mathbb{Z}^n} \lambda_{0m} \, a_{0m}(x) \tag{13.59}$$

with

$$\lambda_{0m} = c_2 \sum_{|\alpha| \le K} \sup_{|y-m| \le c_1} |D^\alpha f_1(y)|, \quad m \in \mathbb{Z}^n, \tag{13.60}$$

and

$$a_{0m}(x) = \lambda_{0m}^{-1} \psi(x-m) f_1(x), \tag{13.61}$$

where $c_1 > 0$ and $c_2 > 0$ are sufficiently large and ψ has the properties (13.2), (13.3). Then a_{0m} are 1_K-atoms according to Definition 13.3(i) with the support property (13.21). Furthermore by (13.58) and [Tri92], p. 125, formula (7), we have

$$\left(\sum_{m \in \mathbb{Z}^n} |\lambda_{0m}|^p \right)^{\frac{1}{p}} \le c_1 \sum_{|\alpha| \le K} \left\| \sup_{|\cdot - y| \le c_2} |D^\alpha f_1(y)| \mid L_p(\mathbb{R}^n) \right\| \tag{13.62}$$

$$\le c_3 \|f_1 \mid F_{pq}^{s+2kM}(\mathbb{R}^n)\| \le c_4 \|f \mid F_{pq}^s(\mathbb{R}^n)\|,$$

where $k \in \mathbb{N}$ is chosen sufficiently large. As for f_2 in (13.56) we assume that M is large. Then we can apply Step 3 to f_2. Assume that we have an optimal decomposition (13.30) with f_2 and, say, $a_{\nu m}^{(2)}(x)$, in place of f and $a_{\nu m}(x)$, respectively, where $a_{\nu m}^{(2)}(x)$ are $(s+2M,p)_{K+2M,-1}$-atoms according to Definition 13.3. Then it can be seen easily that $(-\Delta)^M a_{\nu m}^{(2)}$ are $(s,p)_{K,2M-1}$-atoms. Furthermore, by (13.54) with f_2 and $s+2M$ in place of f and s, and (13.57) we obtain the desired optimal atomic decomposition with $L \le 2M - 1$.

13.9 Corollary (i) *Let $0 < p \le \infty$, $0 < q \le \infty$, and $s \in \mathbb{R}$. Let $K \in \mathbb{N}_0$ and $L + 1 \in \mathbb{N}_0$ with*

$$K \ge (1 + [s])_+ \quad and \quad L \ge \max(-1, [\sigma_p - s]). \tag{13.63}$$

Then

$$\sum_{\nu=0}^{\infty} \sum_{m \in \mathbb{Z}^n} \lambda_{\nu m} a_{\nu m}(x) \tag{13.64}$$

converges in $S'(\mathbb{R}^n)$ where $a_{\nu m}$ are 1_K-atoms ($\nu = 0$) or $(s,p)_{K,L}$-atoms ($\nu \in \mathbb{N}$) according to Definition 13.3, with (13.21) and $\lambda \in b_{pq}$.

(ii) *Let $0 < q \le \infty$, $s \in \mathbb{R}$, and $0 < p \le \infty$ ($p < \infty$ in the F-case). Let $\varepsilon > 0$ and $f \in B_{pq}^s(\mathbb{R}^n)$ or $f \in F_{pq}^s(\mathbb{R}^n)$. Let $K \in \mathbb{N}_0$ and $L + 1 \in \mathbb{N}_0$ be fixed satisfying (13.29) or (13.32), respectively. If $\nu_0 \in \mathbb{N}_0$ is chosen sufficiently large (in dependence on ε) then there are optimal atomic decompositions*

$$f = \sum_{\nu=\nu_0}^{\infty} \sum_{m \in \mathbb{Z}^n} \lambda_{\nu m} a_{\nu m}(x), \quad convergence \ being \ in \quad S'(\mathbb{R}^n), \tag{13.65}$$

where $a_{\nu m}$ are $(s,p)_{K,-1}$-atoms ($\nu = \nu_0$) or $(s,p)_{K,L}$-atoms ($\nu > \nu_0$) according to Definition 13.3, with (13.21) and $\lambda \in b_{pq}$ or $\lambda \in f_{pq}$ (Definition 13.5 appropriately modified),

$$\|\lambda \mid b_{pq}\| \sim \|f \mid B^s_{pq}(\mathbb{R}^n)\| \quad or \quad \|\lambda \mid f_{pq}\| \sim \|f \mid F^s_{pq}(\mathbb{R}^n)\|, \tag{13.66}$$

respectively, and where the summation over m in $\sum\limits_{m\in\mathbb{Z}^n}^{\nu}$ is restricted to those cubes $Q_{\nu m}$ with

$$Q_{\nu m} \subset (\operatorname{supp} f)_{\varepsilon} = \{y \in \mathbb{R}^n : \operatorname{dist}(y, \operatorname{supp} f) < \varepsilon\}. \tag{13.67}$$

Proof.
Step 1. We prove (i) and assume first $1 \le p \le \infty$. Let $\varphi \in S(\mathbb{R}^n)$. By (13.17)–(13.19) and the Taylor expansion of φ up to the order L with respect to the off-points $2^{-\nu}m$ we obtain for fixed ν

$$\int_{\mathbb{R}^n} \sum_m \lambda_{\nu m} a_{\nu m}(y)\varphi(y)\,dy$$

$$= \int_{\mathbb{R}^n} \sum_m \lambda_{\nu m} 2^{-\nu(L+1)} a_{\nu m}(y) \left(\varphi(y) - \sum_{|\beta|\le L} c^{\nu,m}_{\beta}(y-2^{-\nu}m)^{\beta} \right) 2^{\nu(L+1)}\,dy. \tag{13.68}$$

The last factor in the integral can be uniformly estimated from above by

$$c \langle y\rangle^{-M} \sup_{x\in\mathbb{R}^n} \langle x\rangle^M \sum_{|\gamma|\le L+1} |D^{\gamma}\varphi(x)|, \tag{13.69}$$

where $M > 0$ is at our disposal. By (13.63) we have $L + 1 > -s$ and hence by (13.18)

$$2^{-\nu(L+1)} |a_{\nu m}(y)| \le c\, 2^{\frac{\nu n}{p}} 2^{-\nu\varkappa} \quad in \quad c\, Q_{\nu m} \tag{13.70}$$

for some $\varkappa > 0$. Since we have in addition the factor $\langle y\rangle^{-M}$ it follows by Hölder's inequality that

$$\left| \int_{\mathbb{R}^n} \sum_m \lambda_{\nu m} a_{\nu m}(y)\, \varphi(y)\, dy \right|$$

$$\le c\, 2^{-\nu\varkappa} \left(\sum_m |\lambda_{\nu m}|^p \right)^{\frac{1}{p}} \sup_{x\in\mathbb{R}^n} \langle x\rangle^M \sum_{|\alpha|\le L+1} |D^{\gamma}\varphi(x)|. \tag{13.71}$$

Since $\varkappa > 0$ and $\lambda \in b_{pq} \subset b_{p\infty}$ the convergence of (13.64) in $S'(\mathbb{R}^n)$ is now clear (including an estimate as in (13.71)). Now let $0 < p < 1$. Then we have $L + 1 > -s + \frac{n}{p} - n$ and by (13.18) we obtain (13.70) with νn in place of $\frac{\nu n}{p}$.

We arrive at (13.71) with $\sum |\lambda_{\nu m}|$ in place of $(\sum |\lambda_{\nu m}|^p)^{\frac{1}{p}}$. Now the desired convergence in $S'(\mathbb{R}^n)$ follows from $\lambda \in b_{pq} \subset b_{1,\infty}$.

Step 2. We prove (ii). A dilation argument $x \to cx$ shows that we may replace $\sum\limits_{\nu=0}^{\infty} \sum\limits_m$ in (13.30) by $\sum\limits_{\nu=\nu_0}^{\infty} \sum\limits_m$. Hence the main assertion of part (ii) is the support property (13.66), (13.67). Let $s > \sigma_p$ or $s > \sigma_{pq}$, respectively, and $L = -1$. That means, no moment conditions are required. Then the multiplication of the representation (13.30), where $\sum\limits_{\nu=0}^{\infty}$ is replaced by $\sum\limits_{\nu=\nu_0}^{\infty}$, with a suitable cut-off function proves (13.65)–(13.67). To extend this assertion to arbitrary, say, $f \in F_{pq}^s(\mathbb{R}^n)$ and admitted L we complement (13.55) and (13.56) by a closer look about the needed supports of the involved distributions. We multiply (13.55) with a suitable cut-off function φ which is identically 1 near *supp* f and has a support in $(supp\, f)_\varepsilon$. We obtain

$$f = \varphi g + (-\Delta)^M \varphi g + \sum c_{\alpha\beta} D^\beta \varphi D^\alpha g \tag{13.72}$$

where the last sum, taken over $|\alpha| + |\beta| = 2M$, $|\alpha| \le 2M - 1$, has similar support properties and belongs to $F_{pq}^{s+1}(\mathbb{R}^n)$. Its quasi-norm can be estimated from above by $c\,\|f\,|\,F_{pq}^s(\mathbb{R}^n)\|$. We apply the same argument to this summand now with $s + 1$ in place of s. Iteration yields

$$f = g_1 + (-\Delta)^M g_2, \quad g_j \in F_{pq}^{s+2M}(\mathbb{R}^n), \tag{13.73}$$

and

$$supp\, g_j \subset (supp\, f)_\varepsilon, \quad j = 1, 2. \tag{13.74}$$

Now we apply this argument to g_1 and obtain again by iteration the decomposition (13.56), (13.57) with the additional property

$$supp\, f_j \subset (supp\, f)_\varepsilon, \quad j = 1, 2. \tag{13.75}$$

If k is chosen large enough we can use the above argument for the case $L = -1$. The rest is now the same (with ν_0 in place of 0) as in Step 4 of the proof of Theorem 13.8.

13.10 Harmonic atoms

We based the proof of the existence of optimal atomic decompositions in Theorem 13.8 on representation formulas connected with local means, see (13.1) and (13.5). The crucial step was done in 13.2. Afterwards we extended these optimal atomic decompositions in the Steps 3 and 4 of the proof of Theorem 13.8 by some modifications and a lifting argument to all spaces in question. Comparing the Theorems 12.9 and 12.5 it is quite clear that the considerations in 13.2 have a harmonic counterpart at least if $p > \frac{n}{n+1}$. We deal with this possibility with some care because it is the basis for our later subatomic (or quarkonial) decompositions. In other words, 13.10 is the harmonic counterpart of 13.2. Let, say, $f \in F_{pq}^s(\mathbb{R}^n)$. According now to Theorem 12.5(i) and (12.7),

$$f = \sum_{r=0}^{k-1} c_r^k \frac{\partial^r u(x,1)}{\partial t^r} + c \sum_{\nu=0}^{\infty} \int_{2^{-\nu-1}}^{2^{-\nu}} t^k \frac{\partial^k u(x,t)}{\partial t^k} \frac{dt}{t} \tag{13.76}$$

is a harmonic representation. It is the counterpart of (13.1). Although the respective counterparts of (13.5)–(13.11) seem to be more or less obvious, we need some extra considerations to prepare our later applications. Let $\mu \in \mathbb{N}$ be fixed. Let

$$\nu \in \mathbb{N} \quad \text{with} \quad \nu \geq \mu, \quad m \in \mathbb{Z}^n, \quad \text{and} \quad l = 0, 1, \ldots, 2^\mu - 1. \tag{13.77}$$

Then $B_{\nu,m,l}$ denotes the ball in $\mathbb{R}_+^{n+1}$, centred at $(2^{-\nu}m, 2^{-\nu+\mu}+l2^{-\nu})$ with radius $2^{-\nu+\mu-2}$, see Fig. 13.1. Let again $Q_{\nu m}$ be the $\mathbb{R}^n$-cubes introduced in 13.2.

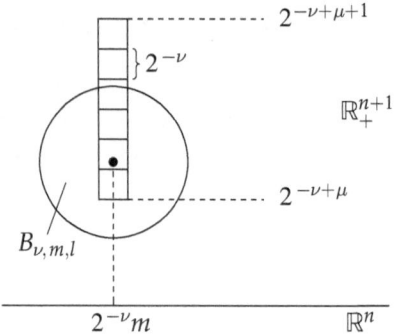

Fig. 13.1

We subdivide the tall rectangles

$$Q_{\nu m} \times (2^{-\nu+\mu}, 2^{-\nu+\mu+1}) \tag{13.78}$$

in $\mathbb{R}_+^{n+1}$ into 2^μ cubes of side-length $2^{-\nu}$ in $\mathbb{R}_+^{n+1}$ in the indicated way. The above balls $B_{\nu,m,l}$ are obviously related to these cubes. We always assume $p > \frac{n}{n+1}$ and that k in (13.76) is chosen large enough such that Theorem 12.5(iii) can be applied. The counterpart of (13.7) is given by

$$\lambda_{\nu m} = 2^{\nu(s-\frac{n}{p})} 2^{-\nu k} \sup \left| \frac{\partial^k u(y,t)}{\partial t^k} \right|, \quad \nu > \mu, \quad m \in \mathbb{Z}^n, \tag{13.79}$$

where the supremum is taken over

$$|2^{-\nu}m - y| \leq c\,2^{-\nu+\mu-1} \quad \text{and} \quad c^{-1}2^{-\nu+\mu} \leq t \leq c\,2^{-\nu+\mu+1}$$

for some sufficiently large $c \geq 1$. In case of $\nu = \mu$ we put

$$\lambda_{\mu m} = \sum_{\nu=0}^{k-1} \sup \left| \frac{\partial^r u(y,t)}{\partial t^r} \right| |c_r^k|, \quad m \in \mathbb{Z}^n, \tag{13.80}$$

where the supremum is taken over $|2^{-\mu}m - y| \leq c$ and $\frac{1}{2c} \leq t \leq \frac{3}{2}c$ with $c > 1$ sufficiently large, and c_r^k are the coefficients in (13.76). The first terms on the right-hand side of (12.22) can be modified by (13.80). Then it follows easily from Theorem 12.5 that

$$\|\lambda \mid f_{pq}\| \sim \|f \mid F_{pq}^s(\mathbb{R}^n)\|, \quad \lambda = \{\lambda_{\nu m}, \ \nu \geq \mu, \ m \in \mathbb{Z}^n\}, \tag{13.81}$$

see (13.25), obviously modified by $\nu \geq \mu$ in place of $\nu \in \mathbb{N}_0$. Let again ψ be given by (13.2)–(13.4). The counterparts of the atoms $a_{\nu m}$ in (13.9) are now given by

$$a_{\nu m}(x) = \sum_{l=0}^{2^\mu - 1} a_{\nu m,l}(x), \quad \nu > \mu, \quad m \in \mathbb{Z}^n, \tag{13.82}$$

with

$$a_{\nu m,l}(x) = c \, \lambda_{\nu m}^{-1} \psi(2^\nu x - m) \int_{2^{-\nu+\mu}+l2-\nu}^{2^{-\nu+\mu}+(l+1)2^{-\nu}} t^k \frac{\partial^k u(x,t)}{\partial t^k} \frac{dt}{t} \tag{13.83}$$

and

$$a_{\mu m}(x) = \sum_{r=0}^{k-1} a_{\mu m,r}(x), \quad m \in \mathbb{Z}^n, \tag{13.84}$$

with

$$a_{\mu m,r}(x) = \lambda_{\mu m}^{-1} \psi(2^\mu x - m) \frac{\partial^r u(x,1)}{\partial t^r} c_r^k. \tag{13.85}$$

By (13.76), (13.79), (13.80), (13.82)–(13.85) it follows that

$$f = \sum_{\nu=\mu}^{\infty} \sum_{m \in \mathbb{Z}^n} \lambda_{\nu m} a_{\nu m}(x), \quad \text{convergence being in} \quad S'(\mathbb{R}^n), \tag{13.86}$$

is the optimal decomposition in *harmonic atoms* we are looking for. We check the necessary properties of $a_{\nu m}(x)$ and prepare our later applications. By construction we have

$$\operatorname{supp} a_{\nu m} \subset d \, Q_{\nu m}, \quad \nu \geq \mu, \quad m \in \mathbb{Z}^n, \tag{13.87}$$

see (13.4) and (13.10). To prove the counterpart of (13.11) we first recall the well-known formula for the solution of the Dirichlet boundary value problem for the

Laplacian in a ball. Let $W(X_1, \ldots, X_N)$ be harmonic in $K_R = \{X \in \mathbb{R}^N : |X| \le R\}$. Then

$$W(X) = \frac{R^2 - |X|^2}{R|\omega_N|} \int_{|Y|=R} \frac{W(Y)}{|X - Y|^N} \, ds_Y, \quad |X| < R, \tag{13.88}$$

where $|\omega_N|$ is the volume of the unit sphere in $\mathbb{R}^N$. Hence

$$|D^\alpha W(X)| \le c_{\alpha,\varkappa} R^{-|\alpha|} \sup_{|Y|=R} |W(Y)| \quad \text{if} \quad |X| \le \varkappa R \tag{13.89}$$

for some $0 < \varkappa < 1$, where $c_{\alpha,\varkappa}$ depends on α and $\varkappa$ but not on R. We apply the estimate (13.89) to $W(X) = \frac{\partial^k u(x,t)}{\partial t^k}$, the ball $B_{\nu,m,l}$ in place of K_R, $R = 2^{-\nu+\mu-2}$ and $\varkappa \sim 2^{-\mu}$. By (13.79) and (13.83) we obtain

$$|D^\gamma a_{\nu m}(x)| \le c \, 2^{-\nu(s-\frac{n}{p})+\nu|\gamma|}, \quad \nu \ge \mu, \tag{13.90}$$

where c depends on γ and μ, but not on ν and m. We incorporated the case $\nu = \mu$ which causes no problem. Hence by Definition 13.3, up to unimportant constants, $a_{\nu m}$ are $(s,p)_{K,-1}$-atoms. We call them *harmonic atoms* and by (13.81) we have an optimal decomposition in harmonic atoms given by (13.86). We formalize this result and extend it to the B-case. We use the word *harmonic atom*, somewhat vaguely, if it is the product of some standard functions and a function which originates from f and is harmonic in $\mathbb{R}^{n+1}_+$. Recall that $\mathscr{C}^{-\infty}(\mathbb{R}^n)$ had been defined in (12.16), (12.17). The numbers σ_p and σ_{pq} have the same meaning as in (13.28). Finally, b_{pq} and f_{pq} are the obviously modified ($\nu \ge \mu$ in place of $\nu \in \mathbb{N}_0$) sequence spaces given by Definition 13.5.

13.11 Theorem *Let $\mu \in \mathbb{N}$ and let $a_{\nu m}(x)$ be the harmonic atoms introduced in (13.84), (13.85) if $\nu = \mu$ and in (13.82), (13.83) if $\nu > \mu$, where $m \in \mathbb{Z}^n$. Here $k \in \mathbb{N}$ and c_r^k are chosen such that (13.76) is a harmonic representation of a given $f \in \mathscr{C}^{-\infty}(\mathbb{R}^n)$ according to Theorem 12.5 (i).*

(i) Let $\frac{n}{n+1} < p \le \infty$, $0 < q \le \infty$, $s > \sigma_p$ and $K \in \mathbb{N}$ with $K > s$. Let $f \in B_{pq}^s(\mathbb{R}^n)$. Then

$$f = \sum_{\nu=\mu}^{\infty} \sum_{m \in \mathbb{Z}^n} \lambda_{\nu m} a_{\nu m}(x), \quad \text{convergence being in } S'(\mathbb{R}^n), \tag{13.91}$$

where $a_{\nu m}(x)$ are the above harmonic $(s,p)_{K,-1}$-atoms, and the coefficients $\lambda_{\nu m}$ are given by (13.79) and (13.80), is an optimal atomic decomposition according to Theorem 13.8 (i) with

$$\|\lambda \,|\, b_{pq}\| \sim \|f \,|\, B_{pq}^s(\mathbb{R}^n)\| \tag{13.92}$$

(in the understanding of (13.14) and with the modification indicated in front of the theorem).

(ii) *Let $\frac{n}{n+1} < p < \infty$, $0 < q \leq \infty$, $s > \sigma_{pq}$ and $K \in \mathbb{N}$ with $K > s$. Let $f \in F_{pq}^s(\mathbb{R}^n)$. Then, in modification of part (i), (13.91) is an optimal atomic decomposition according to Theorem 13.8 (ii) with*

$$\|\lambda \mid f_{pq}\| \sim \|f \mid F_{pq}^s(\mathbb{R}^n)\|. \tag{13.93}$$

Proof. Part (ii) is covered by the considerations in 13.10, see in particular (13.81). The restriction $s > \sigma_{pq}$ is caused by Theorem 13.8 and Corollary 13.9. Obvious modifications prove also part (i). Of course the constant c in (13.90) can be shifted to the coefficients $\lambda_{\nu m}$ in order to be in perfect agreement with what is called an atom according to Definition 13.3.

13.12 The case $0 < p \leq \frac{n}{n+1}$

In connection with the above theorem two remarks seem to be in order. First the role of the number μ in 13.10 and 13.11 remained unclear so far. But in our later considerations in connection with subatomic (or quarkonial) decompositions it will be vital that μ can be chosen arbitrarily large. Secondly for the same reason we are interested to have also decompositions by harmonic atoms if $0 < p \leq \frac{n}{n+1}$. For that purpose we must modify the above considerations somewhat. Let, say,

$$f \in F_{pq}^s(\mathbb{R}^n), \quad 0 < p < \infty, \quad 0 < q \leq \infty, \quad s > \sigma_{pq}. \tag{13.94}$$

Let ϱ be the same function as in (12.24) and let as in (12.25)

$$f = f_1 + f_2 = (\varrho \hat{f})^\vee + ((1 - \varrho) \hat{f})^\vee. \tag{13.95}$$

By Step 4 of the proof of Theorem 12.5 applied to f_2 it follows that (12.21) and (12.22) are equivalent quasi-norms without any restriction for p, that means $0 < p < \infty$. Of course,

$$f \quad \text{and} \quad u(x, t) = \left(e^{-t|\xi|} \hat{f} \right)^\vee (x)$$

must be replaced by

$$f_2 \quad \text{and} \quad u_2(x, t) = \left(e^{-t|\xi|} \hat{f_2} \right)^\vee (x).$$

Let $a_{\nu m}^2(x)$ and $\lambda_{\nu m}^2$ be related harmonic $(s, p)_{K, -1}$-atoms and coefficients according to Theorem 13.11(ii). Then we have by the arguments in 13.10

$$f_2 = \sum_{\nu = \mu}^{\infty} \sum_{m \in \mathbb{Z}^n} \lambda_{\nu m}^2 a_{\nu m}^2(x) \tag{13.96}$$

and

$$\|\lambda^2 \,|\, f_{pq}\| \sim \|f_2 \,|\, F^s_{pq}(\mathbb{R}^n)\| \le c \,\|f \,|\, F^s_{pq}(\mathbb{R}^n)\|. \tag{13.97}$$

Since $f_1 \in \mathscr{C}^\sigma(\mathbb{R}^n) = B^\sigma_{\infty,\infty}(\mathbb{R}^n)$ for any $\sigma \in \mathbb{R}$ we have again by the above theorem

$$f_1 = \sum_{\nu=\mu}^\infty \sum_{m\in\mathbb{Z}^n} \lambda^1_{\nu m} a^1_{\nu m}(x), \tag{13.98}$$

where $a^1_{\nu m}(x)$ are the related harmonic $(\sigma,\infty)_{K,-1}$-atoms and

$$\sup_{\nu,m} |\lambda^1_{\nu m}| = \|\lambda^1 \,|\, b_{\infty\infty}\| \sim \|f^1 \,|\, \mathscr{C}^\sigma(\mathbb{R}^n)\| \le c \,\|f \,|\, F^s_{pq}(\mathbb{R}^n)\|. \tag{13.99}$$

By (13.95)–(13.99) we have a modified atomic decomposition. Of peculiar interest for us is the case in which f has compact support. Recall that $(supp\,f)_\varepsilon$ stands for the ε-neighbourhood of $supp\,f$, see (13.67), $\varepsilon > 0$. Let $\varphi \in S(\mathbb{R}^n)$ with

$$\varphi(x) = 1 \quad \text{if} \quad x \in (supp\,f)_\varepsilon \quad \text{and} \quad supp\,\varphi \subset (supp\,f)_{2\varepsilon}. \tag{13.100}$$

Then by (13.95), (13.96), and (13.98) we have

$$f = \sum_{\nu=\mu}^\infty \sum_{m\in\mathbb{Z}^n} \left(\lambda^1_{\nu m} \varphi\, a^1_{\nu m}(x) + \lambda^2_{\nu m} \varphi\, a^2_{\nu m}(x) \right). \tag{13.101}$$

By our agreement in front of Theorem 13.11 both $\varphi\, a^1_{\nu m}$ and $\varphi\, a^2_{\nu m}$ are called harmonic atoms. Of course, the summation in (13.101) can be restricted to those atoms for which $(supp\,f)_{2\varepsilon}$ and $d\,Q_{\nu m}$ have a non-empty intersection, see (13.87). For fixed ν there are at most $c\,2^{\nu n}$ terms with this property. Let $\varkappa > \frac{n}{p}$ and $\sigma - \varkappa > s - \frac{n}{p}$ then by (13.90) it follows that $\varphi\, a^3_{\nu m}$ with $a^3_{\nu m} = 2^{\nu\varkappa} a^1_{\nu m}$ are harmonic $(s,p)_{K,-1}$-atoms. On the other hand, put $\lambda^3_{\nu m} = 2^{-\nu\varkappa} \lambda^1_{\nu m}$, we find by (13.26) and (13.99) for some u,

$$\|\lambda^3 \,|\, f_{pq}\| \le \left(\sum_{\nu=\mu}^\infty \left(\sum_m{}^\nu 2^{-\nu\varkappa p} |\lambda^1_{\nu m}|^p \right)^{\frac{u}{p}} \right)^{\frac{1}{u}} \tag{13.102}$$

$$\le c\,\|\lambda^1 \,|\, b_{\infty\infty}\| \le c'\,\|f \,|\, F^s_{pq}(\mathbb{R}^n)\|,$$

where $\sum_m^\nu$ is the sum over those $c\,2^{\nu n}$ elements $m \in \mathbb{Z}^n$ which are of interest in (13.101). But now we have a similar situation as in Theorem 13.11. We formalize this result and extend it to the B-case.

13.13 Corollary *Let* $\mu \in \mathbb{N}$ *and let* $a_{\nu m}^2(x)$ *and* $a_{\nu m}^3(x)$ *be the above harmonic atoms with respect to the compactly supported* f. *Let* φ *be given by* (13.100).

(i) *Let* $0 < p \leq \infty$, $0 < q \leq \infty$, $s > \sigma_p$ *and* $K \in \mathbb{N}$ *with* $K > s$. *Let* $f \in B_{pq}^s(\mathbb{R}^n)$. *Then*

$$f = \sum_{\nu=\mu}^{\infty} \sum_{m \in \mathbb{Z}^n} \left(\lambda_{\nu m}^2 \, \varphi \, a_{\nu m}^2(x) + \lambda_{\nu m}^3 \, \varphi \, a_{\nu m}^3(x) \right), \qquad (13.103)$$

convergence being in $S'(\mathbb{R}^n)$, *is an optimal atomic decomposition where the harmonic* $(s, p)_{K, -1}$-*atoms* $\varphi \, a_{\nu m}^2$ *and* $\varphi \, a_{\nu m}^3$ *have the above meaning, and*

$$\|\lambda^2 \,|\, b_{pq}\| + \|\lambda^3 \,|\, b_{pq}\| \sim \|f \,|\, B_{pq}^s(\mathbb{R}^n)\|. \qquad (13.104)$$

(ii) *Let* $0 < p < \infty$, $0 < q \leq \infty$, $s > \sigma_{pq}$ *and* $K \in \mathbb{N}$ *with* $K > s$. *Let* $f \in F_{pq}^s(\mathbb{R}^n)$. *Then* (13.103) *is an optimal atomic decomposition with*

$$\|\lambda^2 \,|\, f_{pq}\| + \|\lambda^3 \,|\, f_{pq}\| \sim \|f \,|\, F_{pq}^s(\mathbb{R}^n)\|. \qquad (13.105)$$

Proof. Part (ii) is covered by 13.12 and Theorem 13.8, where the latter ensures the equivalence assertion in (13.105). The proof of part (i) is the same.

14 Subatomic decompositions

14.1 Building blocks: molecules, atoms, splines, wavelets, and quarks
Let f be an element of $B_{pq}^s(\mathbb{R}^n)$ or of $F_{pq}^s(\mathbb{R}^n)$. By Theorem 13.8 one can decompose f in smooth building blocks, in that case atoms, and coefficients belonging to some sequence spaces. This possibility proved to be a very efficient tool in many applications, for instance, in connection with mapping properties of pseudodifferential operators, say, of order zero, see [Torr91] or [Tri92], Ch. 6, and the references given there. Roughly speaking, a smooth zero-order pseudodifferential operator maps an atom of type (13.17)–(13.19) in a molecule, again related to that cube $Q_{\nu m}$. *Molecules* are generalized atoms. Instead of (13.17) it is only assumed that $a(x)$ decays outside of $dQ_{\nu m}$ sufficiently strong, say, like $dist(x, Q_{\nu m})^{-\varkappa}$ with $\varkappa > 0$ large enough. Then (13.18) must also be changed appropriately, whereas (13.19) remains unchanged. This technique has been developed especially by the American school, we refer to [FrJ90], [FJW91] and [Torr91]. We do not go into detail. Our aim here is just opposite: We do not wish to generalize the atoms but to specialize them. The motivation may be taken from the constructions of optimal atomic decompositions in the Theorems 13.8 and 13.11. In the first case, the optimal atomic decomposition via local means was given by (13.1), (13.5) and (13.9). In the second case, based on harmonic extensions

$$u(x, t) = \left(e^{-t|\xi|} \hat{f} \right)^{\vee}(x),$$

the optimal atomic decomposition was obtained by (13.76), (13.82)–(13.86). In both constructions, besides f, only very few special functions are involved: the resolution of unity (13.2), (13.3), the kernel functions k_0, k_N in connections with the local means, and, on the Fourier side, $e^{-t|\xi|}$ and some of its derivatives, as far as the approach via harmonic functions is concerned. These basic functions are dilated by $x \to 2^\nu x$ and translated by $x \to x + m$, where $\nu \in \mathbb{N}_0$ and $m \in \mathbb{Z}^n$. Motivated by these observations one can ask whether, say, $f \in F_{pq}^s(\mathbb{R}^n)$ can be represented as

$$f = \sum_{\gamma \in \Gamma} \sum_{\nu=0}^{\infty} \sum_{m \in \mathbb{Z}^n} \lambda_{\nu m}^\gamma \, \Phi_{\nu m}^\gamma(x), \qquad (14.1)$$

convergence being in $S'(\mathbb{R}^n)$. Here Γ stands for a finite or countably infinite index-set, whereas the sums over $\nu \in \mathbb{N}_0$ (or, slightly more general, $\nu \geq \mu$ for a given $\mu \in \mathbb{N}_0$) and $m \in \mathbb{Z}^n$ fit in our previous schemes of atomic decompositions in the Theorems 13.8 and 13.11 and the Corollaries 13.9 and 13.13. By Definition 13.3 atoms are characterized in qualitative terms. In contrast, in (14.1) we are asking for representations via functions of the type

$$\Phi_{\nu m}^\gamma(x) = 2^{-\nu(s-\frac{n}{p})} \, \Phi^\gamma(2^\nu x - m) \qquad (14.2)$$

where $\Phi^\gamma(x)$ are distinguished (mother, father, or basic) atoms. As in 13.2 and 13.3 let $Q = Q_{00}$ be the cube of side-length 1 and centred at the origin. By (14.1), (14.2), Definition 13.3 and the above-quoted theorems and corollaries it is reasonable to assume that each function $\Phi^\gamma(x)$ is a 1_K-atom according to Definition 13.3, complemented by some moment conditions of type (13.19) in dependence on $\gamma \in \Gamma$:

$$supp \, \Phi^\gamma \subset d \, Q \quad \text{for some} \quad d > 1, \quad \gamma \in \Gamma, \qquad (14.3)$$

$$|D^\alpha \, \Phi^\gamma(x)| \leq 1 \quad \text{for} \quad |\alpha| \leq K \in \mathbb{N}_0, \quad \gamma \in \Gamma, \qquad (14.4)$$

and

$$\int_{\mathbb{R}^n} x^\beta \, \Phi^\gamma(x) \, dx = 0 \quad \text{if} \quad |\beta| \leq L(\gamma), \quad \gamma \in \Gamma, \qquad (14.5)$$

where $L(\gamma) + 1 \in \mathbb{N}_0$, see 13.4 for explanations. The question is whether functions Φ^γ with the desired properties exist such that, say, any $f \in F_{pq}^s(\mathbb{R}^n)$ can be represented by (14.1) and, in case of a finite index-set Γ,

$$\sum_\gamma \|\lambda^\gamma \,|\, f_{pq}\| \sim \|f \,|\, F_{pq}^s(\mathbb{R}^n)\|, \qquad (14.6)$$

where

$$\lambda^\gamma = \{\lambda_{\nu m}^\gamma \in \mathbb{C} : \nu \in \mathbb{N}_0, \, m \in \mathbb{Z}^n\},$$

and f_{pq} is the sequence space introduced in Definition 13.5. Hence in our previous notation, (14.1), (14.6) is an optimal atomic decomposition via distinguished atoms generated in the described way, see 13.2, Theorem 13.11 and Corollary 13.13. Of course there is an obvious counterpart for the B-spaces with b_{pq} in (14.6) in place of f_{pq}. Apart from *card* $\Gamma < \infty$, the case where $\Gamma = \mathbb{N}_0^n$ consists of all multi-indices in $\mathbb{R}^n$ will be of interest for us later on. Then

$$\sup_{\gamma \in \mathbb{N}_0^n} 2^{\varkappa |\gamma|} \|\lambda^\gamma \,|\, f_{pq}\| \sim \|f \,|\, F_{pq}^s(\mathbb{R}^n)\| \tag{14.7}$$

for some $\varkappa > 0$ may serve as a desirable substitute of (14.6). However the first attempt is to keep *card* $\Gamma < \infty$ as small as possible. Then we arrive at the problem of *spline representations, spline bases* and their *wavelet* counterparts. Spline representations and spline bases in connection with function spaces have been considered for some twenty years with a culmination around 1980 by the work of Ciesielski, Figiel and many other mathematicians. We gave in [Tri83], 2.12.3, pp. 184–187, a brief description of some results available at that time, and the related references, see also [Mey92], 3.1, pp. 66–71, for further information. In the last decade splines are complemented and superceded by wavelets. In both versions, splines and wavelets, one looks for explicit constructions of finitely many basic splines or wavelets (with compact or non-compact supports) which satisfy the above requirements. We do not go into detail and refer to [Mey92] and [Dau92]. Wavelet constructions in connection with the spaces $B_{pq}^s(\mathbb{R}^n)$ and $F_{pq}^s(\mathbb{R}^n)$ may be found in [FJW91], [FrJ90], Theorem 4.2 and Corollary 4.3, [Sic90], and [RuS96], 2.3.3. In any case some efforts are needed to construct wavelets having the desired properties. It turns out that representations of type (14.1) are needed later on to calculate the entropy numbers of compact embeddings between function spaces, say, on fractals. To circumvent the indicated problems with the construction of splines and wavelets we shall look for sub-decompositions of optimal decompositions of $f \in F_{pq}^s(\mathbb{R}^n)$ or $f \in B_{pq}^s(\mathbb{R}^n)$ in harmonic atoms obtained in Theorem 13.11 and Corollary 13.13. These harmonic atoms are given, for example, by (13.82), (13.83). Recall that

$$u(x, t) = \left(e^{-t|\xi|} \hat{f} \right)^{\vee}(x) \quad \text{and also} \quad t^k \frac{\partial^k u(x, t)}{\partial t^k}$$

are harmonic in $\mathbb{R}_+^{n+1}$. Hence they can be *expanded*, at least locally, *in a Taylor series*. If one wishes to get rid of the restrictions $s > \sigma_{pq}$ or $s > \sigma_p$ in Theorem 13.11 and Corollary 13.13 then some lifting of type (13.56) is needed. Clipping together these arguments it sounds at least plausible that the functions

$$\Phi^{\gamma, L}(x) = (-\Delta)^{\frac{L+1}{2}} x^\gamma \psi(x), \quad \gamma \in \mathbb{N}_0^n, \quad \frac{L+1}{2} \in \mathbb{N}_0, \tag{14.8}$$

may serve as elementary building blocks in (14.1). Here ψ is given by (13.2)–(13.4) and Δ stands for the Laplacian. Since functions of type (14.8) decompose (harmonic) atoms we call them *quarks*. In other words, our intention in this section is to obtain *subatomic* or *quarkonial* or, for short, *qu*-decompositions of type (14.1), (14.7). Compared with wavelet decompositions, advantages and disadvantages of *qu*-decompositions are quite clear. First, in (14.1) with $\Gamma = \mathbb{N}_0^n$ we have now infinitely many sums over $(\nu, m) \in \mathbb{N}_0 \times \mathbb{Z}^n$. But this is well compensated by (14.7) with $\varkappa > 0$. It even turns out that $\varkappa$ must be chosen sufficiently large and can be chosen arbitrarily large (at the expense of the equivalence constants). Secondly, sometimes splines or wavelets are (Schauder) bases in the related spaces, see the literature mentioned above. Presumably nothing of this can be expected for *qu*-decompositions. Wavelet or spline decompositions can be constructed in such a way that the corresponding coefficients $\lambda_{\nu m}^\gamma$ depend linearly on f, sometimes as

$$\lambda_{\nu m}^\gamma = f(\Psi_{\nu m}^\gamma), \tag{14.9}$$

where $\Psi_{\nu m}^\gamma$ are also (dilated and translated) wavelets. We rely here on 13.11 and 13.13 where the corresponding coefficients are not of type (14.9) and hence also the resulting coefficients in the related *qu*-decomposition have not this property. But using the technique developed in connection with Theorem 4.2 in [FrJ90] it should be possible to reformulate Theorem 13.11 and Corollary 13.13 in such a way that one has (14.9) for suitable functions $\Psi_{\nu m}^\gamma$. Such a property is afterwards passed on to the related *qu*-decompositions. We do not stress this point in the sequel although it might well be of some interest in connection with numerical applications (if there are any). However in 14.15 and 14.16 we outline an alternative way to get subatomic decompositions. In that case *the coefficients* $\lambda_{\nu m}^\gamma$ *depend linearly on* f *according to* (14.9). More interesting for us is the somewhat surprising possibility to extend the quarkonial (and, afterwards, as a consequence, the atomic) decompositions to vector-valued spaces $B_{pq}^s(\mathbb{R}^n, E)$ and $F_{pq}^s(\mathbb{R}^n, E)$, where E is a complex Banach space. Let $\Phi_{\nu m}^{\gamma, L}(x)$ be given by (14.2) with $\Phi^{\gamma, L}(x)$, introduced in (14.8), in place of $\Phi^\gamma(x)$. Then the related building blocks, the E-valued quarks, are given by

$$\Phi_{\nu m}^{\gamma, L}(x)\, e_{\nu m}^\gamma, \quad e_{\nu m}^\gamma \in E, \quad \|e_{\nu m}^\gamma \,|\, E\| \le 1. \tag{14.10}$$

We proceed as follows. In Section 14 we deal with *qu*-decompositions in the scalar case. In Section 15 we give a brief introduction in the theory of the E-valued spaces of type B_{pq}^s and F_{pq}^s, where the quarkonial and atomic decompositions of these spaces may be considered as the main results.

14.2 Definition *Let* $Q_{\nu m}$ *with* $\nu \in \mathbb{N}_0$ *and* $m \in \mathbb{Z}^n$ *be the cubes introduced in* 13.2. *Let, in particular,* $Q = Q_{00}$ *be the cube with side-length 1 centred at the origin. Let*

$$\psi \in S(\mathbb{R}^n) \quad with \quad supp\,\psi \subset dQ \quad for\ some \quad d > 1 \tag{14.11}$$

and

$$\sum_{m \in \mathbb{Z}^n} \psi(x - m) = 1 \quad for\ x \in \mathbb{R}^n. \tag{14.12}$$

Let $s \in \mathbb{R}$, $0 < p \leq \infty$, $\frac{L+1}{2} \in \mathbb{N}_0$, $\gamma \in \mathbb{N}_0^n$ (the set of all multi-indices in $\mathbb{R}^n$) and let $\psi^\gamma(x) = x^\gamma \psi(x)$. Then

$$(\gamma qu)_{\nu m}^L(x) = 2^{-\nu(s-\frac{n}{p})} \left((-\Delta)^{\frac{L+1}{2}} \psi^\gamma\right)(2^\nu x - m) \tag{14.13}$$

is called an $(s,p)_L - \gamma$-quark, related to $Q_{\nu m}$. Let $(\gamma qu)_{\nu m} = (\gamma qu)_{\nu m}^L$ if $L = -1$.

14.3 Remark Recall that $\Delta = \sum_{j=1}^n \frac{\partial^2}{\partial x_j^2}$ stands for the Laplacian in $\mathbb{R}^n$ and $x^\beta = x_1^{\beta_1} \cdots x_n^{\beta_n}$ where $x = (x_1, \ldots, x_n) \in \mathbb{R}^n$ and $\beta = (\beta_1, \ldots, \beta_n) \in \mathbb{N}_0^n$ is a multi-index. Up to normalizing constants, $(s,p)_L - \gamma$-quarks are $(s,p)_{K,L}$-atoms for any given $K \in \mathbb{N}_0$. This follows immediately from Definition 13.3, (13.21), and

$$\int_{\mathbb{R}^n} x^\beta (\gamma qu)_{\nu m}^L(x)\, dx = 0 \quad \text{if} \quad |\beta| \leq L. \tag{14.14}$$

The normalizing constants by which the $(s,p)_L - \gamma$-quark must be divided to become an $(s,p)_{K,L}$-atom can be estimated from above by $c\, 2^{\varkappa|\gamma|}$ where $c > 0$ and $\varkappa > 0$ are independent of γ (but may depend on L and K). In the theorem below we use the sequence spaces b_{pq} and f_{pq} introduced in Definition 13.5 with respect to the sequences

$$\lambda^\gamma = \{\lambda_{\nu m}^\gamma \in \mathbb{C} : \nu \in \mathbb{N}_0,\, m \in \mathbb{Z}^n\}$$

where now $\gamma \in \mathbb{N}_0^n$ is a multi-index. Of course, σ_p and σ_{pq} have the same meaning as in (13.28).

14.4 Theorem (i) Let $\frac{n}{n+1} < p \leq \infty$, $0 < q \leq \infty$ and $s \in \mathbb{R}$. Let $M \in \mathbb{N}$ with $M > \sigma_p$ and $M > s$; and L with $\frac{L+1}{2} \in \mathbb{N}_0$ and $L \geq \max(-1, [\sigma_p - s])$ be fixed. Let $(\gamma qu)_{\nu m}$ be $(M,p)_{-1} - \gamma$-quarks and let $(\gamma qu)_{\nu m}^L$ be $(s,p)_L - \gamma$-quarks. There is a number $\varkappa > 0$ with the following property. Let $\mu > \varkappa$. Then $f \in S'(\mathbb{R}^n)$ belongs to $B_{pq}^s(\mathbb{R}^n)$ if, and only if, it can be represented as

$$f = \sum_{\gamma \in \mathbb{N}_0^n} \sum_{\nu=0}^\infty \sum_{m \in \mathbb{Z}^n} \left(\varrho_{\nu m}^\gamma (\gamma qu)_{\nu m}(x) + \lambda_{\nu m}^\gamma (\gamma qu)_{\nu m}^L(x)\right), \tag{14.15}$$

convergence being in $S'(\mathbb{R}^n)$ (first m, then ν, then γ), and

$$\sup_\gamma 2^{\mu|\gamma|} \left(\|\varrho^\gamma \,|\, b_{pq}\| + \|\lambda^\gamma \,|\, b_{pq}\|\right) < \infty. \tag{14.16}$$

Furthermore the infimum in (14.16) over all admissible representations (14.15) is an equivalent quasi-norm in $B_{pq}^s(\mathbb{R}^n)$.

(ii) Let $\frac{n}{n+1} < p < \infty$, $0 < q \leq \infty$ and $s \in \mathbb{R}$. Let $M \in \mathbb{N}$ with $M > \sigma_{pq}$ and $M > s$; and L with $\frac{L+1}{2} \in \mathbb{N}_0$ and $L \geq \max(-1, [\sigma_{pq} - s])$ be fixed. The quarks *and* $\varkappa$ *have the same meaning as in part* (i) *(now, perhaps, with a different value of* L*). Let* $\mu > \varkappa$. *Then* $f \in S'(\mathbb{R}^n)$ *belongs to* $F_{pq}^s(\mathbb{R}^n)$ *if, and only if, it can be represented by* (14.15) *and*

$$\sup_{\gamma} 2^{\mu|\gamma|} \left(\|\varrho^\gamma \,|\, f_{pq}\| + \|\lambda^\gamma \,|\, f_{pq}\| \right) < \infty. \tag{14.17}$$

Furthermore the infimum in (14.17) *over all admissible representations* (14.15) *is an equivalent quasi-norm in* $F_{pq}^s(\mathbb{R}^n)$.

Proof.
Step 1. According to Definition 13.3 and the remarks in 14.3, for any fixed $K \in \mathbb{N}_0$ the above quarks $(\gamma qu)_{\nu m}$ and $(\gamma qu)_{\nu m}^L$ are $(M, p)_{K, -1}$-atoms and $(s, p)_{K, L}$-atoms, respectively, multiplied with normalizing constants which can be estimated from above by $c\, 2^{\varkappa|\gamma|}$, see (14.14) as far as the required moment conditions are concerned. Under the assumption (14.17) it follows from Corollary 13.9 and Theorem 13.8 that for any fixed $\gamma \in \mathbb{N}_0^n$,

$$f^\gamma = \sum_{\nu=0}^{\infty} \sum_{m \in \mathbb{Z}^n} \left(\varrho_{\nu m}^\gamma (\gamma qu)_{\nu m}(x) + \lambda_{\nu m}^\gamma (\gamma qu)_{\nu m}^L(x) \right) \tag{14.18}$$

converges in $S'(\mathbb{R}^n)$, $f^\gamma \in F_{pq}^s(\mathbb{R}^n)$, and

$$\|f^\gamma \,|\, F_{pq}^s(\mathbb{R}^n)\| \le c\, 2^{\varkappa|\gamma|} \left(\|\varrho^\gamma \,|\, f_{pq}\| + \|\lambda^\gamma \,|\, f_{pq}\| \right) \tag{14.19}$$

where c and $\varkappa$ are independent of γ (the latter assertion follows from Step 2 of the proof of Theorem 13.8). Since $\mu > \varkappa$ in (14.17) may be assumed we obtain that

$$f = \sum_{\gamma \in \mathbb{N}_0^n} f^\gamma \tag{14.20}$$

converges in $F_{pq}^s(\mathbb{R}^n)$ and that for some $c > 0$

$$\|f \,|\, F_{pq}^s(\mathbb{R}^n)\| \le c \sup_{\gamma} 2^{\mu|\gamma|} \left(\|\varrho^\gamma \,|\, f_{pq}\| + \|\lambda^\gamma \,|\, f_{pq}\| \right). \tag{14.21}$$

It should be noted that we did not use the restriction $p > \frac{n}{n+1}$. *In other words the above reasoning applies to* $0 < p < \infty$, $0 < q \le \infty$, *and* $s \in \mathbb{R}$. Of course, the same can be done for $B_{pq}^s(\mathbb{R}^n)$ in place of $F_{pq}^s(\mathbb{R}^n)$, based on (14.16), *for all* $0 < p \le \infty$, $0 < q \le \infty$, $s \in \mathbb{R}$.

Step 2. We prove the «only if»-assertion of part (ii) for large s. The proof of the corresponding assertion of part (i) is the same. Let $f \in F_{pq}^s(\mathbb{R}^n)$ with $\frac{n}{n+1} < p < \infty$, $0 < q \le \infty$ and $s > \sigma_{pq}$. We apply Theorem 13.11(ii). Let f be represented by the optimal decomposition in the harmonic atoms $a_{\nu m}(x)$ with $\nu \ge \mu$ and $m \in \mathbb{Z}^n$, given by (13.91), and (13.82)–(13.85), respectively. Let $B_{\nu, m, l}$ be the balls of radius $2^{-\nu+\mu-2}$ and centred at $(2^{-\nu}m, 2^{-\nu+\mu} + l\,2^{-\nu})$ as described in 13.10 and Fig. 13.1. Since $\frac{\partial^k u(x,t)}{\partial t^k}$ in (13.83) is harmonic in $\mathbb{R}^{n+1}_+$ it is also analytic and we expand this function at the off-point $(2^{-\nu}m, 2^{-\nu+\mu} + l\,2^{-\nu})$ in its Taylor series. We use the following assertion about harmonic functions which will be proved in 14.5 below:

Let $N \in \mathbb{N}$ be given. There are two numbers $c > 0$ and $0 < \tau < 1$ which depend on N such that

$$|D^\alpha w(0)| \leq c\,\alpha!\,\tau^{-|\alpha|}\,\sup_{|y|=1}\,|w(y)| \tag{14.22}$$

for any $\alpha \in \mathbb{N}_0^N$ and any complex-valued function $w(y)$ which is harmonic in a neighbourhood of the unit ball $\{y \in \mathbb{R}^N : |y| \leq 1\}$ in $\mathbb{R}^N$.

We apply (14.22) with $N = n+1$ and $y = (x,t)$ to $w(x,t) = \frac{\partial^k u(x,t)}{\partial t^k}$ with respect to the above ball $B_{\nu,m,l}$ and obtain by a dilation and translation argument

$$|D^\alpha w(2^{-\nu}m, 2^{-\nu+\mu} + l\,2^{-\nu})|$$
$$\leq c\,\alpha!\,\varrho^{|\alpha|}\,2^{\nu|\alpha|-\mu|\alpha|}\,\sup_{(x,t)\in B_{\nu,m,l}}\,|w(x,t)| \tag{14.23}$$

for some $\varrho > 1$ and a constant $c > 0$ which is independent of ν, μ and α. Put $\alpha = (\gamma, \beta)$ with $\gamma \in \mathbb{N}_0^n$ and $\beta \in \mathbb{N}_0$. Of course, γ refers to $x \in \mathbb{R}^n$ and β to $t \in \mathbb{R}_+$. We expand $w(x,t)$ at $(2^{-\nu}m, 2^{-\nu+\mu} + l\,2^{-\nu})$ and insert the result in (13.83). We obtain

$$a_{\nu m,l}(x) = c\,\lambda_{\nu m}^{-1}\,\psi(2^\nu x - m)\int_{2^{-\nu+\mu}+l2^{-\nu}}^{2^{-\nu+\mu}+(l+1)2^{-\nu}} t^k$$

$$\times \sum_{\gamma\in\mathbb{N}_0^n,\beta\in\mathbb{N}_0} \frac{1}{\gamma!\,\beta!}\,D^{(\gamma,\beta)}\,w(2^{-\nu}m, 2^{-\nu+\mu} + l2^{-\nu})\,(x - 2^{-\nu}m)^\gamma \tag{14.24}$$

$$\times\,(t - 2^{-\nu+\mu} - l2^{-\nu})^\beta\frac{dt}{t} = \sum_{\gamma\in\mathbb{N}_0^n} c_\gamma\,\psi(2^\nu x - m)\,(2^\nu x - m)^\gamma.$$

By (13.79) and (14.23) the coefficients c_γ can be estimated by

$$|c_\gamma| \leq c\,2^{-\nu(s-\frac{n}{p})-\nu|\gamma|}\,2^{\mu k}\,\varrho^{|\gamma|}\,2^{\nu|\gamma|-\mu|\gamma|}\sum_{\beta=0}^{\infty} \varrho^\beta\,2^{\nu\beta-\mu\beta}\,2^{-\nu\beta} \tag{14.25}$$

where again c is independent of μ. We choose μ such that $\varrho\,2^{-\mu} < 1$. Then the sum over β in (14.25) converges and we obtain by (14.24), (14.25), and (14.13) with $L = -1$

$$a_{\nu m,l}(x) = \sum_{\gamma\in\mathbb{N}_0^n} \eta_{\nu m,l}^\gamma\,(\gamma q u)_{\nu m}(x) \tag{14.26}$$

with
$$|\eta^{\gamma}_{\nu m,l}| \leq c\, 2^{\mu k}\,(\varrho\, 2^{-\mu})^{|\gamma|}, \tag{14.27}$$

where $c > 0$ and $\varrho > 1$ are independent of μ and γ. In particular, to compensate ϱ in the desired way one can replace μ by $M\mu$ with $M \in \mathbb{N}$. Then (14.26), (14.27) and summation over l in (13.82) yield

$$a_{\nu m}(x) = \sum_{\gamma \in \mathbb{N}_0^n} \eta^{\gamma}_{\nu m}\,(\gamma q u)_{\nu m}(x), \quad \nu > \mu, \quad m \in \mathbb{Z}^n, \tag{14.28}$$

with
$$|\eta^{\gamma}_{\nu m}| \leq c\, 2^{\mu \delta}\, 2^{-\mu|\gamma|} \tag{14.29}$$

for some $c > 0$ and $\delta > 0$, which are independent of μ and γ (and ν, m). By (13.91) and the notation introduced in (14.15) we have

$$\lambda_{\nu m}\, a_{\nu m}(x) = \sum_{\gamma \in \mathbb{N}_0^n} \lambda^{\gamma}_{\nu m}\,(\gamma q u)_{\nu m}(x) \tag{14.30}$$

with
$$\lambda^{\gamma}_{\nu m} = \lambda_{\nu m}\, \eta^{\gamma}_{\nu m}, \quad \nu > \mu, \quad m \in \mathbb{Z}^n, \quad \gamma \in \mathbb{N}_0^n. \tag{14.31}$$

By (13.93) and the notation introduced in front of the above theorem we arrive at

$$2^{\mu|\gamma|}\, \|\lambda^{\gamma}\,|\,f_{pq}\| \leq c\, 2^{\mu\delta}\, \|f\,|\,F^s_{pq}(\mathbb{R}^n)\|, \quad \gamma \in \mathbb{N}_0^n, \tag{14.32}$$

where $c > 0$ and $\delta > 0$ are independent of μ and γ. With the necessary modifications the above arguments apply also to the harmonic atoms $a_{\mu m}(x)$ given by (13.84), (13.85). Now (13.91), (14.30) and (14.32) complete the proof in that case. Of course it is immaterial whether the summation over ν in f_{pq} starts with μ or with 0 (one has simply to add some terms with $\lambda^{\gamma}_{\nu m} = 0$ if $\nu < \mu$). Furthermore in that case the terms with $\varrho^{\gamma}_{\nu m}$ in (14.15) are not needed.

Step 3. We prove the «only if»-assertion of part (ii) for $\frac{n}{n+1} < p < \infty,\, 0 < q \leq \infty$, $s \in \mathbb{R}$, which covers again the corresponding assertion in part (i). By (13.56) we decompose $f \in F^s_{pq}(\mathbb{R}^n)$ as

$$f = f_1 + (-\Delta)^{\frac{L+1}{2}} f_2, \quad f_1 \in F^M_{pq}(\mathbb{R}^n), \quad f_2 \in F^{s+L+1}_{pq}(\mathbb{R}^n), \tag{14.33}$$

with the counterpart of (13.57). Of course we may assume that M and L are so large that Step 2 can be applied to f_1 and f_2. Hence the ϱ-terms in (14.15) and (14.16) result from the corresponding decomposition of f_1. Let the corresponding qu-decomposition of f_2 be given in terms of λ-coefficients and with respect of the $(s + L + 1, p)_{-1} - \gamma$-quarks, denoted temporarily by $(\gamma q u)^{-1}_{\nu m}(x)$. By (14.13) we have

$$(-\Delta)^{\frac{L+1}{2}}\,(\gamma q u)^{-1}_{\nu m}(x) = 2^{-\nu(s+L+1-\frac{n}{p})}\left((-\Delta)^{\frac{L+1}{2}}\,\psi^{\gamma}\right)(2^{\nu}x - m)\, 2^{\nu(L+1)} \tag{14.34}$$

which are the desired $(s,p)_L - \gamma$-quarks. The counterpart of (13.57) and (14.32) prove finally

$$\sup_{\gamma} 2^{\mu|\gamma|}\,(\|\varrho^{\gamma}\,|\,f_{pq}\| + \|\lambda^{\gamma}\,|\,f_{pq}\|) \leq c\, 2^{\mu\delta}\, \|f\,|\,F^s_{pq}(\mathbb{R}^n)\| \tag{14.35}$$

for some $c > 0$ and $\delta > 0$ which are independent of μ.

14.5 A property of harmonic functions

We prove (14.22). In (13.88) with $R = 1$ we replace $X = (X_1, \ldots, X_N) \in \mathbb{R}^N$ by $Z = (Z_1, \ldots, Z_N) \in \mathbb{C}^N$. For Y with $|Y| = 1$ we expand

$$|Z - Y|^{-N} = \left[\sum_{j=1}^{N} (Z_j - Y_j)^2 \right]^{-\frac{N}{2}}$$

near the origin in $\mathbb{C}^N$ in its Taylor series. The coefficients can be estimated uniformly with respect to Y. By standard arguments, for instance the N-dimensional Cauchy formula as in (14.62), (14.63) below, one obtains (14.22) with W in place of w.

14.6 Comment As far as the dependence on μ in the above theorem is concerned we proved a bit more than stated: Under the assumptions of part (ii) there are two constants $c > 0$ and $\delta > 0$ such that for all $\mu > \kappa$ and all $f \in F_{pq}^s(\mathbb{R}^n)$

$$\inf_{\gamma \in \mathbb{N}_0^n} \sup 2^{\mu|\gamma|} \left(\|\varrho^\gamma \,|\, f_{pq}\| + \|\lambda^\gamma \,|\, f_{pq}\| \right) \leq c \, 2^{\mu\delta} \, \|f \,|\, F_{pq}^s(\mathbb{R}^n)\| \qquad (14.36)$$

where the infimum is taken over all admissible *qu*-representations (14.15). Of course, under the conditions of part (i) there is a counterpart of (14.36) with b_{pq} and $B_{pq}^s(\mathbb{R}^n)$ in place of f_{pq} and $F_{pq}^s(\mathbb{R}^n)$.

14.7 Corollary (i) *Let* $\frac{n}{n+1} < p \leq \infty$, $0 < q \leq \infty$, *and* $s > \sigma_p$. *Let* $\mu > \varkappa$ *as in Theorem* 14.4. *Then* $f \in S'(\mathbb{R}^n)$ *belongs to* $B_{pq}^s(\mathbb{R}^n)$ *if, and only if, it can be represented as*

$$f = \sum_{\gamma \in \mathbb{N}_0^n} \sum_{\nu=0}^{\infty} \sum_{m \in \mathbb{Z}^n} \lambda_{\nu m}^\gamma \, (\gamma qu)_{\nu m}(x), \qquad (14.37)$$

convergence being in $S'(\mathbb{R}^n)$ *(first* m, *then* ν, *then* γ*), where* $(\gamma qu)_{\nu m}$ *are* $(s, p)_{-1} - \gamma$*-quarks, and*

$$\sup_\gamma 2^{\mu|\gamma|} \|\lambda^\gamma \,|\, b_{pq}\| < \infty. \qquad (14.38)$$

Furthermore, the infimum in (14.38) *over all admissible representations* (14.37) *is an equivalent quasi-norm in* $B_{pq}^s(\mathbb{R}^n)$.

(ii) *Let* $\frac{n}{n+1} < p < \infty$, $0 < q \leq \infty$ *and* $s > \sigma_{pq}$. *Let* $\mu > \varkappa$. *Then* $f \in S'(\mathbb{R}^n)$ *belongs to* $F_{pq}^s(\mathbb{R}^n)$ *if, and only if, it can be represented by* (14.37) *where the quarks have the same meaning as in part (i) and*

$$\sup_\gamma 2^{\mu|\gamma|} \|\lambda^\gamma \,|\, f_{pq}\| < \infty. \qquad (14.39)$$

Furthermore, the infimum in (14.39) *over all admissible representations* (14.37) *is an equivalent quasi-norm in* $F_{pq}^s(\mathbb{R}^n)$.

Proof. The assertion of the corollary is completely covered by the Steps 1 and 2 of the proof of Theorem 14.4.

14.8 Remark The number $\varkappa > 0$ plays the same role as in Theorem 14.4. It must be sufficiently large. Furthermore, there is an obvious counterpart of 14.6 and (14.36) where now only the λ-terms are needed.

14.9 An example As explained in (10.16)–(10.18),

$$\mathscr{C}^s(\mathbb{R}^n) = B^s_{\infty,\infty}(\mathbb{R}^n)$$

with $s > 0$ are the classical Hölder-Zygmund spaces. Let ψ be the function introduced in (14.11), (14.12). Let $\mu > 0$ be given and sufficiently large. Then $f \in L_\infty(\mathbb{R}^n)$ belongs to $\mathscr{C}^s(\mathbb{R}^n)$ if, and only if, it can be represented as

$$f = \sum_{\gamma \in \mathbb{N}_0^n} \sum_{\nu=0}^{\infty} \sum_{m \in \mathbb{Z}^n} 2^{-\nu s - \mu |\gamma|} \lambda_{\gamma \nu m} (2^\nu x - m)^\gamma \psi(2^\nu x - m) \qquad (14.40)$$

with

$$\{\lambda_{\gamma \nu m} : \quad \gamma \in \mathbb{N}_0^n, \ \nu \in \mathbb{N}_0, \ m \in \mathbb{Z}^n\} \in l_\infty, \qquad (14.41)$$

convergence being in, say, $L_\infty(\mathbb{R}^n)$. This is a special case of Corollary 14.7(i) with $p = q = \infty$. Furthermore,

$$\|f \mid \mathscr{C}^s(\mathbb{R}^n)\| \sim \inf \|\{\lambda_{\gamma \nu m}\} \mid l_\infty\| \qquad (14.42)$$

where the infimum is taken over all representations (14.40) for fixed μ. In other words, the right-hand side of (14.42) is an equivalent norm in $\mathscr{C}^s(\mathbb{R}^n)$. A word about the somewhat sinister role played by μ seems to be in order. By (14.11) we have

$$|(2^\nu x - m)^\gamma \psi(2^\nu x - m)| \leq c\, 2^{\varkappa |\gamma|} \qquad (14.43)$$

for some $c > 0$ and $\varkappa > 0$ which are independent of γ. By Theorem 13.8 for fixed $\gamma \in \mathbb{N}_0^n$ the resulting elements in (14.40) belong to $\mathscr{C}^s(\mathbb{R}^n)$. If $\mu > \varkappa$ then it follows also $f \in \mathscr{C}^s(\mathbb{R}^n)$. The coefficients $\lambda_{\gamma \nu m} = \lambda_{\gamma \nu m}(\mu)$ in (14.40) and also the equivalence constants in (14.42) depend on μ. By (14.29) and (14.31) we have

$$\lambda_{\gamma \nu m}(\mu) = O(2^{\mu \delta}) \quad \text{for some} \quad \delta > 0, \qquad (14.44)$$

and an obvious counterpart of (14.32) according to Remark 14.8. This clarifies the interplay between the strong decay of $2^{-\mu |\gamma|} \lambda_{\gamma \nu m}$ if $|\gamma| \to \infty$ and the possibly large coefficients $\lambda_{\gamma \nu m}$ if $|\gamma|$ is small and the corresponding bad equivalence constants in (14.42). Of course these remarks apply to all spaces $B^s_{pq}(\mathbb{R}^n)$ and $F^s_{pq}(\mathbb{R}^n)$ under consideration so far.

14.10 The case $0 < p \leq \frac{n}{n+1}$

Let f be given by (14.15) with (14.16) or (14.17), then f belongs to $B_{pq}^s(\mathbb{R}^n)$ or $F_{pq}^s(\mathbb{R}^n)$, respectively, without any restriction for p. This was proved in Step 1 of the proof of Theorem 14.4. The restriction $p > \frac{n}{n+1}$ was needed in Step 2 of this proof because we used Theorem 13.11. Under the additional assumption that f has a compact support we proved in Corollary 13.13 that there are optimal decompositions by somewhat modified harmonic atoms. Now we can apply Step 2 of the proof of Theorem 14.4 to this modified situation. We give a brief description. Let $s > \sigma_p$ in the B-case and $s > \sigma_{pq}$ in the F-case. Let $f \in B_{pq}^s(\mathbb{R}^n)$ or $f \in F_{pq}^s(\mathbb{R}^n)$ with $supp\, f$ compact. Let now $0 < p \leq \infty$ (with $p < \infty$ in the F-case) and $0 < q \leq \infty$. Let φ be the same function as in (13.100). Then we have (13.103) and (13.104) or (13.105), respectively. The harmonic $(s, p)_{K,-1}$-atoms $a_{\nu m}^2(x)$ and $a_{\nu m}^3(x)$ can be treated as in Step 2 of the proof of Theorem 14.4, where the two resulting qu-decompositions can be added. Multiplying the result with φ we obtain

$$f = \sum_{\gamma \in \mathbb{N}_0^n} \sum_{\nu=0}^{\infty} \sum_{m \in \mathbb{Z}^n} \lambda_{\nu m}^{\gamma} \, \varphi(x) \, (\gamma qu)_{\nu m}(x), \qquad (14.45)$$

where $(\gamma qu)_{\nu m}$ are $(s, p)_{-1} - \gamma$-quarks according to Definition 14.2 and

$$\sup_{\gamma} 2^{\mu|\gamma|} \, \|\lambda^{\gamma} \,|\, b_{pq}\| \, < \infty, \qquad (14.46)$$

respectively

$$\sup_{\gamma} 2^{\mu|\gamma|} \, \|\lambda^{\gamma} \,|\, f_{pq}\| \, < \infty. \qquad (14.47)$$

Let now $0 < p \leq \infty$ ($p < \infty$ in the F-case), $0 < q \leq \infty$ and $s \in \mathbb{R}$. Then we can apply Step 3 of the proof of Theorem 14.4, based on (13.73), (13.74). But now we do not have the elegant formula (14.34) since the factor φ comes in. So we stick at the left-hand side of (14.34) but with the factor φ in addition. We have also the counterpart of (14.35). Finally we remark that Step 1 of the proof of Theorem 14.4 can also be applied to this slightly modified situation. We summarize these results.

14.11 Corollary (i) *Let* $0 < p \leq \infty$, $0 < q \leq \infty$, *and* $s \in \mathbb{R}$. *Let* $M, L, \varkappa$ *as in Theorem 14.4 (i). Let* $\mu > \varkappa$. *Then* $f \in S'(\mathbb{R}^n)$ *with* $supp\, f$ *compact belongs to* $B_{pq}^s(\mathbb{R}^n)$ *if, and only if, it can be represented as*

$$f = \sum_{\gamma \in \mathbb{N}_0^n} \sum_{\nu=0}^{\infty} \sum_{m \in \mathbb{Z}^n} \Big(\varrho_{\nu m}^{\gamma} \, \varphi(x) \, (\gamma qu)_{\nu m}(x)$$
$$+ \, \lambda_{\nu m}^{\gamma} \, (-\Delta)^{\frac{L+1}{2}} \big[\varphi \, (\gamma qu)_{\nu m}^*\big] (x) \Big), \qquad (14.48)$$

where φ *is given by* (13.100), $(\gamma qu)_{\nu m}$ *are* $(M, p)_{-1} - \gamma$-*quarks*, $(\gamma qu)_{\nu m}^*$ *are* $(s + L + 1, p)_{-1} - \gamma$-*quarks, with* (14.16). *Again the infimum in* (14.16) *over* *all admissible representations* (14.48) *is an equivalent quasi-norm in* $B_{pq}^s(\mathbb{R}^n)$. *If, in addition,* $s > \sigma_p$, *then* (14.48), (14.16) *can be replaced by* (14.45), (14.46) *including the corresponding equivalence assertion.*

(ii) *Let* $0 < p < \infty$, $0 < q \le \infty$ *and* $s \in \mathbb{R}$. *Let* $M, L, \varkappa$ *as in Theorem 14.4* (ii). *Let* $\mu > \varkappa$. *Then* $f \in S'(\mathbb{R}^n)$ *with* supp f *compact belongs to* $F_{pq}^s(\mathbb{R}^n)$ *if, and only if, it can be represented by* (14.48) *with* (14.17), *including the equivalence assertion. If, in addition,* $s > \sigma_{pq}$, *then* (14.48), (14.17) *can be replaced by* (14.45), (14.47) *including the corresponding equivalence assertion.*

Proof. This corollary complements Theorem 14.4 and Corollary 14.7. Since

$$(-\Delta)^{\frac{L+1}{2}} \left[\varphi(\gamma q u)_{\nu m}^* \right](x) \quad \text{are } (s, p)_{K,L}\text{-atoms,}$$

up to constants, one can apply Step 1 of the proof of Theorem 14.4. The necessary remarks about the construction of optimal decompositions have been made in 14.10.

14.12 Support property

The above corollary is also a modification of Corollary 13.9(ii). In both cases we used (13.73), (13.74). One can replace $\sum\limits_{\nu=0}^{\infty}$ in (14.48) by $\sum\limits_{\nu=\nu_0}^{\infty}$ to adapt (14.48) to the support assertion in Corollary 13.9(ii).

14.13 Taylor expansions for tempered distributions

The γ-quarks introduced in Definition 14.2 and its use in Theorem 14.4 and in the Corollaries 14.7 and 14.11 may also be interpreted as a substitute of the Taylor expansions of analytic functions. Assume for simplicity that $f \in B_{pq}^s(\mathbb{R}^n)$ meets the requirements of Corollary 14.7 such that we have the decomposition (14.37). By (14.13) with $L = -1$, the sum over γ in (14.37) is the usual Taylor expansion, which has here its origin in the Taylor expansions of harmonic functions. Also the sum over m in (14.37) for fixed ν is quite natural and resembles the «global» Taylor expansion of an analytic function in a sequence of overlapping circles or balls and summing up the results by using (14.12). Compared with these globalized (and uglified) Taylor expansions of analytic functions, the summation over ν in (14.37) is new. It is caused by the fact that f, in general, is not determined by local germs, in contrast to analytic functions. It might be of some interest that the decompositions in Theorem 14.4 and in the Corollaries 14.7 and 14.11 can be extended to any $f \in S'(\mathbb{R}^n)$ with the outcome that

in essence Taylor expansions for analytic functions in $\mathbb{C}^n$ *and for* $f \in S'(\mathbb{R}^n)$ *are very similar.*

We outline the procedure and put for that purpose $\langle x \rangle^\sigma = (1 + |x|^2)^{\frac{\sigma}{2}}$ if $x \in \mathbb{R}^n$ and $\sigma \in \mathbb{R}$. For any $f \in S'(\mathbb{R}^n)$ there are two real numbers σ and s such that

$$\langle x \rangle^{-\sigma} f \in \mathscr{C}^s(\mathbb{R}^n) \tag{14.49}$$

where again $\mathscr{C}^s(\mathbb{R}^n) = B_{\infty,\infty}^s(\mathbb{R}^n)$ are the Hölder-Zygmund spaces. To prove this assertion we first recall that for suitable numbers $\varkappa > 0$, $c > 0$, $c' > 0$, and $N \in \mathbb{N}$,

$$|f(\varphi)| \leq c \sup \langle x \rangle^{\varkappa} |D^{\alpha} \varphi(x)| \leq c' \sup |D^{\alpha} (\langle x \rangle^{\varkappa} \varphi(x))| \qquad (14.50)$$

for all $\varphi \in S(\mathbb{R}^n)$, where the supremum is taken over all $x \in \mathbb{R}^n$ and $|\alpha| \leq N$. The rest is now a matter of duality and embedding. As for the related weighted spaces we refer to [SchT87], 5.1 and [ET96], Ch. 4, where the latter is based on [HaT94a,b]. There one finds the necessary details. The technique developed in 12.5, 13.10, and 13.11 can be extended to these weighted spaces. Assume $s > 0$ in (14.49) for sake of simplicity. Then we can apply (14.40). Hence, f can be extended in $S'(\mathbb{R}^n)$ by

$$f = \sum_{\gamma \in \mathbb{N}_0^n} \sum_{\nu=0}^{\infty} \sum_{m \in \mathbb{Z}^n} 2^{-\nu s - \mu|\gamma|} \lambda_{\gamma\nu m} \langle 2^{-\nu} m \rangle^{\sigma} (2^{\nu} x - m)^{\gamma} \psi(2^{\nu} x - m) \qquad (14.51)$$

with (14.41). One can replace $\mathscr{C}^s(\mathbb{R}^n)$ in (14.49) by $B_{pq}^s(\mathbb{R}^n)$ or $F_{pq}^s(\mathbb{R}^n)$ with $s \in \mathbb{R}$, $1 \leq p \leq \infty$ ($p < \infty$ in the F-case) and $1 \leq q \leq \infty$. If $s > 0$ then one can use Corollary 14.7, otherwise one has to rely on Theorem 14.4. In any case if $f \in S'(\mathbb{R}^n)$ is given one can ask for an optimal choice of σ and of an appropriate space in place of $\mathscr{C}^s(\mathbb{R}^n)$ in (14.49) and to expand f afterwards in the indicated way. Maybe the Hilbert scale $H^s(\mathbb{R}^n) = B_{2,2}^s(\mathbb{R}^n)$ might be of peculiar interest.

14.14 Taylor expansions in weighted spaces
As indicated above there is a theory of weighted spaces of type

$$B_{pq}^s(\mathbb{R}^n, w(\cdot)) \quad \text{and} \quad F_{pq}^s(\mathbb{R}^n, w(\cdot)),$$

where one has to replace $L_p(\mathbb{R}^n)$ in Definition 10.3 by its weighted counterpart $L_p(\mathbb{R}^n, w(\cdot))$. This can be done in the framework of $S'(\mathbb{R}^n)$ if the weight $w(x)$ is of polynomial growth, see [HaT94a,b] and [ET96], Ch. 4. On the other hand, if $w(x)$ is, for example, of growth $e^{\pm|x|^{\beta}}$ with $0 < \beta < 1$, then one needs ultra-distributions, see [SchT87], 5.1. In all these cases one has isomorphic maps $f \mapsto w(\cdot)f$ of the weighted spaces onto the corresponding unweighted spaces. In that way one can extend the theory developed in this section, especially Theorem 14.4 and the Corollaries 14.7 and 14.11, to these weighted spaces as it has been done in (14.49) in a special case. It is somewhat surprising that this may also work in the case of exponential weights of type $w(x) = e^{\pm|x|}$ in $\mathbb{R}^n$ where Fourier-analytical definitions of type 10.3 break down. We refer to [Schott96a,b].

14.15 Alternative proof of subatomic decompositions
We outline an alternative proof of subatomic decompositions for the spaces $B_{pq}^s(\mathbb{R}^n)$. Comparing Theorem 14.4 and Corollary 14.7 we may assume that s is large. Let

$$0 < p \leq \infty, \quad 0 < q \leq \infty, \quad s > \sigma_p = n(\frac{1}{p} - 1)_+. \qquad (14.52)$$

We sketch briefly a new proof of (14.37), (14.38) avoiding harmonic functions.

Let φ_k be given by (10.3)–(10.5), and let $f \in B^s_{pq}(\mathbb{R}^n)$. Then we have

$$\hat{f}(\xi) = \sum_{j=0}^{\infty} \varphi_j(\xi)\hat{f}(\xi), \quad \xi \in \mathbb{R}^n. \tag{14.53}$$

Let Q_j be a cube in $\mathbb{R}^n$ centred at the origin and with side-length, say, $2\pi 2^j$. In particular we have *supp* $\varphi_j \subset Q_j$. We interpret $\varphi_j \hat{f}$ as a periodic distribution and expand it in Q_j by

$$\varphi_j \hat{f}(\xi) = \sum_{k \in \mathbb{Z}^n} b_{jk} \exp(-i2^{-j}k\xi), \quad \xi \in Q_j, \tag{14.54}$$

with

$$b_{jk} = c\, 2^{-jn} \int_{Q_j} \exp(i2^{-j}k\xi)\,(\varphi_j\hat{f})(\xi)\,d\xi = c'\, 2^{-jn}\,(\varphi_j\hat{f})^{\vee}(2^{-j}k). \tag{14.55}$$

By [Tri83], 1.3.4 and 1.4.1, pp. 20, 22, we may assume

$$\sum_k \left| (\varphi_j\hat{f})^{\vee}(2^{-j}k) \right|^p \sim 2^{jn} \int_{\mathbb{R}^n} |(\varphi_j\hat{f})^{\vee}(\xi)|^p\,d\xi \tag{14.56}$$

(modification if $p = \infty$): If necessary we replace $\varphi(x)$ by $\varphi(cx)$ with some $c > 0$. Hence, by (14.55)

$$\left\| (\varphi_j\hat{f})^{\vee} \,|\, L_p(\mathbb{R}^n) \right\| \sim 2^{nj(1-\frac{1}{p})} \left(\sum_{k \in \mathbb{Z}^n} |b_{jk}|^p \right)^{\frac{1}{p}}, \quad j \in \mathbb{N}_0. \tag{14.57}$$

Let $\varrho \in S(\mathbb{R}^n)$, $\varrho_j(\xi) = \varrho(2^{-j}\xi)$, $\varrho_j(\xi) = 1$ if $\xi \in$ *supp* φ_j and *supp* $\varrho_j \subset Q_j$. We multiply (14.54) with ϱ_j and extend it from Q_j to $\mathbb{R}^n$. Then we have

$$\begin{aligned}
\left(\varphi_j\hat{f} \right)^{\vee}(x) &= \sum_{k \in \mathbb{Z}^n} b_{jk}\,\check{\varrho}_j(x - 2^{-j}k) \\
&= 2^{jn} \sum_k b_{jk}\,\check{\varrho}(2^j x - k) \\
&= \sum_k d_{jk} 2^{-j(s-\frac{n}{p})}\,\check{\varrho}(2^j x - k), \quad x \in \mathbb{R}^n,
\end{aligned} \tag{14.58}$$

with

$$\begin{aligned}
\left(\sum_k |d_{jk}|^p \right)^{\frac{1}{p}} &= 2^{js+jn(1-\frac{1}{p})} \left(\sum_k |b_{jk}|^p \right)^{\frac{1}{p}} \\
&\sim 2^{js} \left\| (\varphi_j\hat{f})^{\vee} \,|\, L_p(\mathbb{R}^n) \right\|
\end{aligned} \tag{14.59}$$

where we used (14.57). Hence, we obtain

$$\left(\sum_{j=0}^{\infty}\left(\sum_{k}|d_{jk}|^p\right)^{\frac{q}{p}}\right)^{\frac{1}{q}} \sim \|f\,|\,B_{pq}^s(\mathbb{R}^n)\|. \qquad (14.60)$$

The entire analytic function $\check{\varrho}(x) \in S(\mathbb{R}^n)$ can be extended from $\mathbb{R}^n$ to $\mathbb{C}^n$. By the Paley-Wiener-Schwartz theorem we have for any $\varkappa > 0$ and an appropriate $c_\varkappa$

$$|\check{\varrho}(x+iy)| \le c_\varkappa\, e^{c|y|}\, \langle x\rangle^{-\varkappa}, \qquad (14.61)$$

$x \in \mathbb{R}^n$, and $y \in \mathbb{R}^n$, where $\langle x\rangle = (1+|x|^2)^{\frac{1}{2}}$, see e. g. [Tri83], 1.2.1, p. 13. Iterative application of Cauchy's representation theorem in the complex plane yields

$$\check{\varrho}(z_1,\ldots,z_n)$$
$$= (2\pi i)^{-n} \int_{|\zeta_1-z_1|=1} \cdots \int_{|\zeta_n-z_n|=1} \frac{\check{\varrho}(\zeta_1,\ldots,\zeta_n)}{(\zeta_1-z_1)\cdots(\zeta_n-z_n)}\, d\zeta_1\cdots d\zeta_n, \qquad (14.62)$$

where $z_k \in \mathbb{C}$. By (14.61) we obtain in particular

$$|D^\alpha \check{\varrho}(x)| \le c'_\varkappa \alpha!\langle x\rangle^{-\varkappa}, \quad x \in \mathbb{R}^n, \qquad (14.63)$$

where $c'_\varkappa$ does not depend on $x \in \mathbb{R}^n$ and the multi-index α. Let ψ be the same function as in Definition 14.2 and let as there $\psi^\gamma(x) = x^\gamma \psi(x)$ where $\gamma \in \mathbb{N}_0^n$. We expand $\check{\varrho}(2^j x - k)$ in (14.58) at $2^{-j-\mu}l$ where $l \in \mathbb{Z}^n$ and $\mu \in \mathbb{N}$ fixed. Then we obtain

$$\psi(2^{j+\mu}x - l)\,\check{\varrho}(2^j x - k)$$
$$= \sum_\gamma \frac{2^{j|\gamma|}}{\gamma!}\,(D^\gamma \check{\varrho})(2^{-\mu}l - k)\,(x - 2^{-j-\mu}l)^\gamma\,\psi(2^{j+\mu}x - l) \qquad (14.64)$$
$$= \sum_\gamma \frac{D^\gamma \check{\varrho}(2^{-\mu}l - k)}{\gamma!}\,\psi^\gamma(2^{j+\mu}x - l)\,2^{-\mu|\gamma|}.$$

We insert this equality in (14.58) and have

$$(\varphi_j \hat{f})^\vee(x) = 2^{-j(s-\frac{n}{p})} \sum_{k\in\mathbb{Z}^n} d_{jk} \sum_{l\in\mathbb{Z}^n} \psi(2^{j+\mu}x - l)\,\check{\varrho}(2^j x - k)$$
$$= 2^{-j(s-\frac{n}{p})} \sum_{\gamma\in\mathbb{N}_0^n}\sum_{l\in\mathbb{Z}^n} \psi^\gamma(2^{j+\mu}x - l) \sum_{k\in\mathbb{Z}^n} \frac{D^\gamma\check{\varrho}(2^{-\mu}l - k)}{\gamma!}\,d_{jk}\,2^{-\mu|\gamma|}$$
$$= \sum_{\gamma\in\mathbb{N}_0^n}\sum_{l\in\mathbb{Z}^n} \lambda_{j+\mu,l}^\gamma\,(\gamma qu)_{j+\mu,l}(x)\,2^{\mu(s-\frac{n}{p})} \qquad (14.65)$$

where

$$(\gamma q u)_{jl}(x) = 2^{-j(s-\frac{n}{p})}\, \psi^\gamma(2^j x - l)$$

are the $(s,p)_{-1} - \gamma$-quarks according to Definition 14.2 and $\lambda^\gamma_{j+\mu,l}$ is the sum over k in the last but one line. We may replace $j + \mu$ by j. By (14.63) and (14.59), (14.60) it follows that

$$\left(\sum_{j=0}^{\infty} \left(\sum_l |\lambda^\gamma_{jl}|^p \right)^{\frac{q}{p}} \right)^{\frac{1}{q}} \leq c_\mu\, 2^{-\mu|\gamma|}\, \|f\,|\,B^s_{pq}(\mathbb{R}^n)\|, \qquad (14.66)$$

where the constant c_μ in (14.66) depends on μ, but not on γ. This is the counterpart of (14.38). Summation of (14.65) over j results in (14.37). *In other words, we proved Corollary* 14.7 *(i) now under the assumption* (14.52).

14.16 Comment Extension of 14.15 to arbitrary $f \in B^s_{pq}(\mathbb{R}^n)$ with $0 < p \leq \infty$, $0 < q \leq \infty$ and $s \in \mathbb{R}$ causes no trouble. Also an extension of this method to $F^s_{pq}(\mathbb{R}^n)$ is possible. This has been proved recently by W. Farkas, [Far97], not only for isotropic but also for anisotropic spaces (at least for large values of s in the anisotropic case). Furthermore, the above method has the following advantages:

(i) *The coefficients* λ^γ_{jl} *in (14.37) depend linearly on* f . This follows from (14.65), (14.58) and (14.55). This linear dependence can be written in the form (14.9) where Ψ^γ_{jl} can be calculated from φ_j in (14.55) and $\breve{\varrho}$ in (14.65) (and may depend also on s, p, q). Furthermore, (14.38) with these coefficients is an equivalent quasi-norm in $B^s_{pq}(\mathbb{R}^n)$. This follows from (14.66) and our previous arguments in Step 1 of the proof of Theorem 14.4.

(ii) It seems to be possible to extend this technique not only to anisotropic spaces, but also to spaces with dominating mixed derivatives, periodic spaces etc., at least in case of spaces with sufficiently large smoothness.

15 A digression: Vector-valued function spaces

15.1 Introduction
We introduced in Definition 10.3 the $\mathbb{C}$-valued spaces $B^s_{pq}(\mathbb{R}^n)$ and $F^s_{pq}(\mathbb{R}^n)$. The aim of Section 15 is to replace $\mathbb{C}$ by an arbitrary complex Banach space E. There arise several questions. First of all one has to ask for the technical background to extend 10.2 and 10.3 from $\mathbb{C}$-valued to E-valued function spaces. But both E-valued L_p-spaces and E-valued distributions, including Fourier transforms, have been studied in detail in literature. In 15.2 we collect very briefly the facts needed and give some references. The next major question is about special cases as described in 10.5, including equivalent norms and characterizations by means of differences and derivatives of E-valued functions, Littlewood-Paley properties

etc. It comes out that assertions mentioned in 10.5 can be extended form the $\mathbb{C}$-valued to the E-valued case if the way how they are obtained is characterized by integral representations and the technique of maximal inequalities, now combined with the Hahn-Banach theorem (for the latter reason E cannot be replaced by a quasi-Banach space). This applies, for instance, to the E-valued extensions of (10.16)–(10.19), the E-valued Hölder-Zygmund and (classical) Besov spaces. The resulting E-valued spaces share some distinguished properties with their scalar ancestors, for instance, Fourier multiplier assertions of Michlin-Hörmander type. On the other hand, there is no problem to introduce independently E-valued (fractional) Sobolev spaces by the obvious counterparts of (10.10) and (10.13). Asking for E-valued counterparts of (10.14) (Littlewood-Paley assertions) or (10.15), the situation changes drastically. By rule of thumb all scalar-valued assertions of the indicated type can be extended to their E-valued counterparts if E, maybe after re-norming, is a Hilbert space. Otherwise the situation is rather delicate but nowadays well understood. For example, the E-valued counterpart of (10.15) holds if, and only if, E is a so-called *UMD space*, which is a severe restriction compared with the collection of all Banach spaces. We return to questions of this type at the end of Section 15, collecting a few facts and references. But our main aim here is completely different. We introduce the spaces $B_{pq}^s(\mathbb{R}^n, E)$ and $F_{pq}^s(\mathbb{R}^n, E)$ for the same parameters as in the scalar case and for arbitrary complex Banach spaces E imitating 10.3. This makes sense and one has a lot of equivalent quasi-norms, covering, for example, the E-valued counterparts of (10.16), (10.18). But more important, harmonic representations according to Theorem 12.5 and, at the end, qu-decompositions and atomic decompositions as treated in the two preceding sections have their natural counterparts. Let $(\gamma qu)_{\nu m}^L(x)$ be the scalar $(s, p)_L - \gamma$-quarks introduced in Definition 14.2 then one has qu-decompositions as in Theorem 14.4 and in the Corollaries 14.7 and 14.11 now with the elementary building blocks

$$(\gamma qu)_{\nu m}^L(x)\, e_{\nu m}^{\gamma, L}, \quad e_{\nu m}^{\gamma, L} \in E, \quad \|e_{\nu m}^{\gamma, L} \mid E\| \leq 1. \tag{15.1}$$

Besides the total decomposition in $\mathbb{R}^n$ by quarks one has in addition a decoupling of E and $\mathbb{R}^n$, which in applications might be very useful. Afterwards, one can replace the quarks in (15.1) by more general atoms, or by molecules. After the decoupling this is a scalar matter. The results in this section will not be needed in what follows in this book, but they may be useful in such areas as (operator-valued) linear and quasi-linear parabolic and evolutionary equations. We will be very brief, describing the set-up and indicating the necessary changes compared with the scalar case. On the one hand this section is outside of the main stream of this book. But on the other hand it might well be understood as some motivation for using *these* E-valued function spaces in relevant applications.

15.2 Notation and basic facts
Let E be a complex Banach space. We assume that the reader is familiar with the *Bochner integral* for E-valued Lebesgue measurable functions $f(x)$ in $\mathbb{R}^n$.

The basic references are [DuS58] and [DiU77]. The latter book is devoted to the delicate interplay between vector measures and functions on the one hand and the geometry of the underlying Banach spaces E on the other hand. A very readable survey of all the aspects needed here has been given by H.-J. Schmeisser in [Schm87].

The collection of all strongly Lebesgue-measurable functions $f : \mathbb{R}^n \to E$ with

$$\|f \mid L_p(\mathbb{R}^n, E)\| = \left(\int_{\mathbb{R}^n} \|f(x) \mid E\|^p \, dx \right)^{\frac{1}{p}} < \infty, \quad 0 < p \leq \infty, \qquad (15.2)$$

(usual modification if $p = \infty$) is a quasi-Banach space (Banach space $p \geq 1$). We use the well-known properties of the Bochner integral. For instance, if $T \in L(E_0, E_1)$ is a linear and bounded operator from the Banach space E_0 in the Banach space E_1 and if f is an E_0-integrable function, then

$$T \left(\int_{\mathbb{R}^n} f(x) \, dx \right) = \int_{\mathbb{R}^n} (Tf)(x) \, dx. \qquad (15.3)$$

This applies in particular to linear and continuous functionals $e' \in E'$ of E.

Next we recall some basic notation for E-valued tempered distributions. The basic reference here is [Schw57/8] dealing with distributions in topological vector spaces. Short surveys may be found in [Schm87] and in [Ama95], III, 4.1, 4.2, pp. 128–135. As before, $S(\mathbb{R}^n)$ is the Schwartz space of all $\mathbb{C}$-valued rapidly decreasing, infinitely differentiable functions on $\mathbb{R}^n$. Then the space of E-valued tempered distributions $S'(\mathbb{R}^n, E)$ consists of all linear mappings f from $S(\mathbb{R}^n)$ into E for which there exists a constant $c > 0$ and a number $k \in \mathbb{N}_0$ with

$$\|f(\varphi) \mid E\| \leq c \sum_{|\alpha| \leq k} \sup_{x \in \mathbb{R}^n} (1 + |x|^2)^{\frac{k}{2}} |D^\alpha \varphi(x)| \qquad (15.4)$$

for all $\varphi \in S(\mathbb{R}^n)$. The well-known properties for $S'(\mathbb{R}^n)$ have their expected E-valued counterparts, see [Schm87]. We mention in particular the Fourier transform. Let $\hat{\varphi}$ be the scalar Fourier transform of $\varphi \in S(\mathbb{R}^n)$ given by (10.2). Let $f \in S'(\mathbb{R}^n, E)$, then

$$(Ff)(\varphi) = \hat{f}(\varphi) = f(\hat{\varphi}) \quad \text{for all} \quad \varphi \in S(\mathbb{R}^n). \qquad (15.5)$$

In particular $\hat{f} \in S'(\mathbb{R}^n, E)$. In a similar way the inverse Fourier transform $F^{-1}f$ or $\check{f}$ is given. We use the letters F and F^{-1} both for the scalar case and the E-valued case. As in the scalar case we have

$$\hat{f}(x) = (2\pi)^{-\frac{n}{2}} \int_{\mathbb{R}^n} e^{-ix\xi} f(\xi) \, d\xi \quad \text{if} \quad f \in L_1(\mathbb{R}^n, E). \qquad (15.6)$$

For further discussions we refer again to [Schm87] and [Ama95], III, 4.2, pp. 130–135. If $f \in S'(\mathbb{R}^n, E)$ and $\varphi \in S(\mathbb{R}^n)$ then

$$(f * \varphi)(x) = f(\varphi(x - \cdot)), \quad x \in \mathbb{R}^n, \tag{15.7}$$

denotes the convolution of f and φ. Extending a well-known property from the scalar case to the vector-valued case we have for $f \in S'(\mathbb{R}^n, E)$ and $\varphi \in S(\mathbb{R}^n)$

$$\left(\varphi \hat{f}\right)^\vee(x) = (2\pi)^{-\frac{n}{2}} f(\check{\varphi}(x - \cdot)), \quad x \in \mathbb{R}^n, \tag{15.8}$$

which is the convolution of f and $\check{\varphi}$. If φ has compact support, then $\check{\varphi}$ is an entire analytic function on $\mathbb{R}^n$, and, after extension, on $\mathbb{C}^n$. But then the regular E-valued distribution $(\varphi \hat{f})^\vee(x)$ is also an E-valued entire analytic function on $\mathbb{R}^n$. In particular if $f \in S'(\mathbb{R}^n, E)$ and *supp* $\hat{f}$ is compact in $\mathbb{R}^n$, then

$$f = (\varphi \hat{f})^\vee$$

for $\varphi \in S(\mathbb{R}^n)$ with compact support and $\varphi(x) = 1$ in a neighbourhood of *supp* $\hat{f}$. Hence f is an E-valued entire analytic function. This is the vector-valued version of the Paley-Wiener-Schwartz theorem, see [Tri83], p. 13, for the scalar case and related references. This is the basis for extending some assertions of the L_p-theory of analytic functions developed in [Tri83], Ch. 1, from the scalar case to the E-valued case.

15.3 L_p-spaces of vector-valued analytic functions
As always E stands for a complex Banach space. Let K be a compact set in $\mathbb{R}^n$ and let $0 < p \le \infty$; then

$$L_p^K(\mathbb{R}^n, E) = \left\{ f \in S'(\mathbb{R}^n, E) : \ \text{supp} \, \hat{f} \subset K, \ \|f \mid L_p(\mathbb{R}^n, E)\| < \infty \right\}. \tag{15.9}$$

This is the vector-valued counterpart of [Tri83], 1.4.1, p. 22. By 15.2 and the E-valued counterparts of [Tri83], Ch. 1, $L_p^K(\mathbb{R}^n, E)$ are quasi-Banach spaces (Banach spaces if $p \ge 1$) of E-valued entire analytic functions. We describe some properties extending the technique developed in [Tri83] from the scalar case to the vector-valued case. But it is not our aim, and also beyond the intentions of this book, to give a self-contained systematic treatment. Just on the contrary, we use these comparatively simple spaces to demonstrate the surprisingly close interplay between integral representations, maximal inequalities, and the Hahn-Banach theorem, in order to extend scalar assertions to E-valued ones. First we comment on E-valued maximal inequalities and Plancherel-Polya-Nikol'skij inequalities. By the approximation arguments in [Tri83], 1.4.1, p. 22, we may assume that $f \in L_p^K(\mathbb{R}^n, E)$ decays sufficiently strongly if $|x| \to \infty$. Let $\psi \in S(\mathbb{R}^n)$ with $\hat{\psi}(x) = 1$ if $x \in K$. Then

$$f(x) = (\hat{\psi} \hat{f})^\vee(x) = c \int_{\mathbb{R}^n} \psi(x - y) f(y) \, dy, \quad x \in \mathbb{R}^n. \tag{15.10}$$

Let $e' \in E'$ with $\|e'\| \leq 1$, then

$$\left| \left(\frac{\partial f}{\partial x_1}(x), e' \right) \right| = c \left| \int_{\mathbb{R}^n} \frac{\partial \psi}{\partial x_1}(x-y)\,(f(y), e')\,dy \right|$$
$$\leq c' \int_{\mathbb{R}^n} \|f(y)\,|\,E\|\,(1 + |x-y|)^{-\lambda}\,dy, \quad x \in \mathbb{R}^n, \tag{15.11}$$

where $\lambda > 0$ may be chosen arbitrarily large. By the Hahn-Banach theorem there is for any $x \in \mathbb{R}^n$ a functional $e' = e'(x)$ of norm 1 such that the left-hand side of (15.11) equals $\left\| \frac{\partial f}{\partial x_1}(x)\,|\,E \right\|$. Slightly generalized, we have

$$\|\nabla f(x)\,|\,E\| \leq c \int_{\mathbb{R}^n} \|f(y)\,|\,E\|\,(1 + |x-y|)^{-\lambda}\,dy, \quad x \in \mathbb{R}^n. \tag{15.12}$$

But this is just the (scalar) starting point in [Tri83], p. 16–17. Now the method, also in the more complicated cases $B^s_{pq}(\mathbb{R}^n, E)$ and $F^s_{pq}(\mathbb{R}^n, E)$, is clear:

First one looks for identities of type (15.10) in E for fixed $x \in \mathbb{R}^n$, then one applies $e' \in E'$ with $\|e'\| \leq 1$ and reduce the problem to the scalar case, and chooses afterwards $e' = e'(x)$ in an optimal way.

In other words: There is a good chance to extend those properties of $L^K_p(\mathbb{R}^n)$, $B^s_{pq}(\mathbb{R}^n)$ and $F^s_{pq}(\mathbb{R}^n)$ from the scalar case to the vector-valued case which are derived by the above outlined scheme. This applies especially to all assertions obtained via maximal inequalities.

Returning to $L^K_p(\mathbb{R}^n, E)$ we obtain in that way the following typical assertions which generalize what has been derived in [Tri83], 1.3.1, 1.4.1, pp. 14–17, 22–23:

(i) *Let $0 < r < \infty$. There is a constant $c > 0$ such that for all $f \in L^K_p(\mathbb{R}^n, E)$ and all $x \in \mathbb{R}^n$*

$$\sup_{z \in \mathbb{R}^n} \frac{\|f(x-z)\,|\,E\|}{1 + |z|^{\frac{n}{r}}} \leq c\,[(M\|f\,|\,E\|^r)(x)]^{\frac{1}{r}}. \tag{15.13}$$

Here M is the Hardy-Littlewood maximal function

$$Mg(x) = \sup |B|^{-1} \int_B |g(y)|\,dy, \quad x \in \mathbb{R}^n, \tag{15.14}$$

where the supremum is taken over all balls B in $\mathbb{R}^n$, centred at $x \in \mathbb{R}^n$.

(ii) *As a consequence of (15.13) and the Hardy-Littlewood maximal inequality one obtains for $0 < r < p$*

$$\left\| \sup_{z \in \mathbb{R}^n} \frac{\|f(\cdot - z)\,|\,E\|}{1 + |z|^{\frac{n}{r}}} \,|\,L_p(\mathbb{R}^n) \right\| \leq c\,\|f\,|\,L_p(\mathbb{R}^n, E)\| \tag{15.15}$$

for a suitable $c > 0$ and all $f \in L^K_p(\mathbb{R}^n, E)$.

(iii) *(Plancherel-Polya-Nikol'skij inequalities). Let $0 < p \leq q \leq \infty$ and let α be a multi-index. There is a constant $c > 0$ such that for all $f \in L_p^K(\mathbb{R}^n, E)$*

$$\|D^\alpha f \,|\, L_q(\mathbb{R}^n, E)\| \leq c \,\|f \,|\, L_p(\mathbb{R}^n, E)\|. \tag{15.16}$$

Furthermore, the Fourier multiplier assertions derived in [Tri83], 1.5.1, 1.5.2, pp. 25–28, can also be extended to the E-valued case. But instead of doing this we formulate the E-version of the more general $L_p(l_q)$ multiplier theorem proved in [Tri83], 1.6.3, pp. 31–32. The related proofs in [Tri83], 1.6.2 and 1.6.3, use exclusively maximal inequalities of the above type and the Fefferman-Stein $L_u(l_v)$ maximal inequality, see [Tri83], p. 15, or [Tri92], p. 89, for a formulation.

(iv) *Let K_j be compact subsets of $\mathbb{R}^n$ with $d_j = \operatorname{diam} K_j > 0$ where $j \in \mathbb{N}$. Let*

$$0 < p < \infty, \; 0 < q \leq \infty, \; \varkappa > \frac{n}{2} + \frac{n}{\min(p,q)} \quad \text{and} \quad M_j \in H_2^\varkappa(\mathbb{R}^n), \tag{15.17}$$

where $H_2^\varkappa(\mathbb{R}^n)$ are the spaces introduced in (10.13). Then

$$\left\| \left(\sum_{j=1}^\infty \|(M_j \hat{f}_j)^\vee(\cdot) \,|\, E\|^q \right)^{\frac{1}{q}} \,\Big|\, L_p(\mathbb{R}^n) \right\|$$

$$\leq c \sup_l \|M_l(d_l \cdot) \,|\, H_2^\varkappa(\mathbb{R}^n)\| \left\| \left(\sum_{j=1}^\infty \|f_j(\cdot) \,|\, E\|^q \right)^{\frac{1}{q}} \,\Big|\, L_p(\mathbb{R}^n) \right\| \tag{15.18}$$

for some $c > 0$, all $\{M_l\} \subset H_2^\varkappa(\mathbb{R}^n)$ and all $f_j \in L_p^{K_j}(\mathbb{R}^n, E)$ (with the usual modification if $q = \infty$).

In [Tri83], 1.6.4, 2.4.9, and in [Tri92], 2.2.4, we added a comment about how to improve the above number $\varkappa$, which was based on the Michlin-Hörmander Fourier multiplier theorem for $L_p(\mathbb{R}^n)$ with $1 < p < \infty$. But the corresponding assertion for $L_p(\mathbb{R}^n, E)$ is not true for general E. In other words it may well happen that an improvement of $\varkappa$ in (15.17), (15.18) depends on the geometry of the underlying Banach space E.

15.4 Definition *Let $\{\varphi_j\}_{j=0}^\infty$ be the dyadic resolution of unity introduced in (10.3)–(10.5) and let again E be a complex Banach space.*

(i) *Let $s \in \mathbb{R}$, $0 < p \leq \infty$, and $0 < q \leq \infty$. Then*

$$B_{pq}^s(\mathbb{R}^n, E) = \{ f \in S'(\mathbb{R}^n, E) :$$

$$\left(\sum_{j=0}^\infty 2^{jsq} \left\| (\varphi_j \hat{f})^\vee |\, L_p(\mathbb{R}^n, E) \right\|^q \right)^{\frac{1}{q}} < \infty \} \tag{15.19}$$

(with the usual modification if $q = \infty$).

(ii) *Let* $s \in \mathbb{R}$, $0 < p < \infty$, *and* $0 < q \leq \infty$. *Then*

$$F^s_{pq}(\mathbb{R}^n, E) = \{f \in S'(\mathbb{R}^n, E) :$$

$$\left\| \left(\sum_{j=0}^{\infty} 2^{jsq} \|(\varphi_j \hat{f})^{\vee}(\cdot) \mid E\|^q \right)^{\frac{1}{q}} \mid L_p(\mathbb{R}^n) \right\| < \infty\} \tag{15.20}$$

(with the usual modifications if $q = \infty$).

15.5 Theorem *Under the indicated restrictions for* s, p, *and* q, *both* $B^s_{pq}(\mathbb{R}^n, E)$ *and* $F^s_{pq}(\mathbb{R}^n, E)$ *are quasi-Banach spaces (Banach spaces if* $p \geq 1$ *and* $q \geq 1$). *They are independent of the chosen dyadic resolution of unity* $\{\varphi_j\}_{j=0}^{\infty}$.

Proof. Outline The crucial point is the independence of $B^s_{pq}(\mathbb{R}^n, E)$ and $F^s_{pq}(\mathbb{R}^n, E)$ on the dyadic resolution of unity. But this follows from (15.18) with $d_l \sim 2^l$ and

$$K_0 = \{x : |x| \leq c\}, \quad K_j = \{x : c_0 2^j \leq |x| \leq c_1 2^j\}, \quad j \in \mathbb{N}, \tag{15.21}$$

for some $c > 0$ and $0 < c_0 < c_1$, see [Tri83], 2.3.2 or [Tri92], 2.3.2, for details. As for the completeness of these spaces we refer to [Tri92], 2.3.2. There we used the scalar version of (15.18) with the same maximal function of $(M_j \hat{f}_j)^{\vee}$ as in (15.15) in place of $(M_j \hat{f}_j)^{\vee}$ itself. This can also be done in the E-valued case based on the same references as above. Then one can follow [Tri92], pp. 94–95.

15.6 Fourier multipliers
For given s, p, q, there are numbers $c > 0$ and $N \in \mathbb{N}$ such that

$$\left\| \left(m \hat{f} \right)^{\vee} \mid F^s_{pq}(\mathbb{R}^n, E) \right\|$$

$$c \sup_{x \in \mathbb{R}^n, |\alpha| \leq N} \langle x \rangle^{|\alpha|} |D^\alpha m(x)| \|f \mid F^s_{pq}(\mathbb{R}^n, E)\| \tag{15.22}$$

for all m and $f \in F^s_{pq}(\mathbb{R}^n, E)$. Of course, there is a corresponding assertion with B^s_{pq} in place of F^s_{pq}. We put $\langle x \rangle = (1 + |x|^2)^{\frac{1}{2}}$. This is a multiplier assertion of Michlin-Hörmander type. To prove this claim we replace f in (15.20) by $(m \hat{f})^{\vee}$ and obtain $(\varphi_j m \hat{f})^{\vee}$. By the support properties of φ_j it is sufficient to handle $(\varphi_j m \hat{f}_k)^{\vee}$ with $f_k = (\varphi_k \hat{f})^{\vee}$ where k is near j. But then one can apply (15.18), where the first factor on the right-hand side for $l \in \mathbb{N}$ is of type

$$\|\varphi_l(2^l \cdot) m(2^l \cdot) \mid H^N_2(\mathbb{R}^n)\| \leq c \sup_{c_1 \leq |x| \leq c_2} \sum_{|\alpha| \leq N} D^\alpha [m(2^l x)] \tag{15.23}$$

for some $c > 0$, $0 < c_1 < c_2 < \infty$ and $N > \varkappa$. Together with a modified estimate in case of $l = 0$ we obtain (15.22). In other words, all the spaces $B^s_{pq}(\mathbb{R}^n, E)$ and $F^s_{pq}(\mathbb{R}^n, E)$ enjoy Fourier multiplier assertions of Michlin-Hörmander type without any restriction for the Banach spaces E. This is in sharp contrast to the situation for $L_p(\mathbb{R}^n, E)$ with $1 < p < \infty$, where one needs additional properties of E to have a corresponding assertion, see 15.14 for some comments.

15.7 Equivalent quasi-norms

We outlined the general philosophy of how to handle the E-valued spaces

$$L_p^K(\mathbb{R}^n, E) \quad \text{and} \quad B_{pq}^s(\mathbb{R}^n, E), \quad F_{pq}^s(\mathbb{R}^n, E)$$

in comparison with their scalar ancestors:

Pointwise estimates, based on integral representations and maximal inequalities, combined with the Hahn-Banach principle, pave the way to carry over those assertions from the scalar case to the vector case which can be obtained in that way.

Fortunately, substantial parts of the theory as developed in [Tri92] are based on such principles. This applies in particular to the main theorems 2.4.1 and 2.5.1 in [Tri92], pp. 100–101, 131–132. In rough terms, the main aim of these two theorems, which (together with their subsequent complements) are the corner stones in the theory as presented in [Tri92], can be described as follows: Let φ_0 and φ be two C^∞ functions in $\mathbb{R}^n$ or $\mathbb{R}^n \setminus \{0\}$. Under which conditions are

$$\|\varphi_0(D)f \mid L_p(\mathbb{R}^n)\| + \left\| \left(\int_0^1 t^{-sq} |\varphi(tD) f(\cdot)|^q \frac{dt}{t} \right)^{\frac{1}{q}} \mid L_p(\mathbb{R}^n) \right\| \quad (15.24)$$

and

$$\left\| \left(\sum_{j=0}^{\infty} 2^{jsq} |\varphi_j(D) f(\cdot)|^q \right)^{\frac{1}{q}} \mid L_p(\mathbb{R}^n) \right\| \quad (15.25)$$

equivalent quasi-norms in $F_{pq}^s(\mathbb{R}^n)$, where

$$\varphi(tD)f = (\varphi(t\cdot)\hat{f})^\vee \quad \text{and} \quad \varphi_j(x) = \varphi(2^{-j}x). \quad (15.26)$$

Of course there is a B_{pq}^s-counterpart. It turns out that this can be done under rather weak assumptions for φ_0, φ in dependence on s, p in the B-case and s, p, q in the F-case, covering many desirable characterizations by classical means. This applies in particular to descriptions in terms of differences Δ_h^m, harmonic and thermic extensions. Some of the relevant scalar assertions may be found in [Tri92], Theorem 2.6.1, p. 140, Theorem 2.6.2, p. 144, Theorem 2.6.4, p. 152, which have E-valued counterparts. We do not go into detail but restrict ourselves to two examples.

(i) As usual, let

$$\Delta_h^M f(x) = \sum_{j=0}^{M} (-1)^{M-j} \binom{M}{j} f(x + hj), \quad M \in \mathbb{N}, \ h \in \mathbb{R}^n, \ x \in \mathbb{R}^n, \quad (15.27)$$

be the differences of a $\mathbb{C}$-valued or E-valued function in $\mathbb{R}^n$. Let $1 \le p \le \infty$, $1 \le q \le \infty$ and $0 < s < M \in \mathbb{N}$. Then $B_{pq}^s(\mathbb{R}^n, E)$ is the collection of all $f \in L_p(\mathbb{R}^n, E)$, such that

$$
\|f \mid B_{pq}^s(\mathbb{R}^n, E)\|_M = \|f \mid L_p(\mathbb{R}^n, E)\|
$$
$$
+ \left(\int_{|h| \le 1} |h|^{-sq} \left\| \Delta_h^M f \mid L_p(\mathbb{R}^n, E) \right\|^q \frac{dh}{|h|^n} \right)^{\frac{1}{q}} \qquad (15.28)
$$

is finite (modification if $q = \infty$). Furthermore, $\|f \mid B_{pq}^s(\mathbb{R}^n, E)\|_M$ is for any $M > s$ an equivalent norm. This assertion is covered by the above references and the indicated technique. Of course, we arrive at the E-valued counterpart of the classical Besov spaces (complemented by $p = 1$ and $p = \infty$) as described in 10.5(v). By [Tri92], p. 140, this assertion can be extended to $0 < p < 1$ and/or $0 < q \le 1$. But there are some minor technical complications which will not be discussed here. There is a counterpart for some spaces $F_{pq}^s(\mathbb{R}^n, E)$, see [Tri92], p. 144, for the scalar case. Of peculiar interest, also for possible applications of vector-valued and operator-valued spaces in the theory of parabolic equations, are the Hölder-Zygmund spaces which can be defined by

$$
\mathscr{C}^s(\mathbb{R}^n, E) = B_{\infty, \infty}^s(\mathbb{R}^n, E), \quad s \in \mathbb{R}, \qquad (15.29)
$$

generalizing (10.16). If $0 < s < M \in \mathbb{N}$, then $\mathscr{C}^s(\mathbb{R}^n, E)$ consists of all $f \in L_\infty(\mathbb{R}^n, E)$ where

$$
\|f \mid \mathscr{C}^s(\mathbb{R}^n, E)\|_M
$$
$$
= \sup_{x \in \mathbb{R}^n} \|f(x) \mid E\| + \sup |h|^{-s} \|\Delta_h^M f(x) \mid E\| \qquad (15.30)
$$

is finite (equivalent norms). The second supremum in (15.30) is taken over all $x \in \mathbb{R}^n$, $h \in \mathbb{R}^n$ with $0 < |h| \le 1$. This is the E-version of (10.18). As in the scalar case, some differences in (15.30) can be replaced by derivatives. This is also covered in the way indicated above. A collection of different approaches in connection with E-valued Besov spaces may be found in [Schm87].

(ii) Let

$$
\varphi_0(\xi) = e^{-|\xi|} \quad \text{and} \quad \varphi(\xi) = |\xi|^k e^{-|\xi|} \quad \text{with} \quad k \in \mathbb{N}.
$$

Then we have, say, in the scalar case

$$
\varphi(tD) f(x) = t^k \left(|\xi|^k e^{-t|\xi|} \hat{f} \right)^{\vee}(x)
$$
$$
= t^k \frac{\partial^k}{\partial t^k} \left(e^{-t|\xi|} \hat{f} \right)^{\vee}(x), \quad x \in \mathbb{R}^n. \qquad (15.31)
$$

But this is just the harmonic extension as treated in 12.2 and Theorem 12.5. By [Tri92], pp. 152, 153, this fits in the above described scheme. By [Tri92], Corollaries 2 on pp. 108 and 134 this applies also to related maximal functions, also that ones needed in the second terms in (12.19) and (12.22). Hence, at least for the second terms in (12.19) and (12.22) one has E-valued counterparts. As for the E-valued versions of the first terms in (12.19) and (12.22) we refer to 15.3 where we discussed the necessary assertions. In other words, there is an E-valued counterpart of Theorem 12.5 at least as far as the representation in part (i) and the equivalent quasi-norms in (ii) and (iii) are concerned. We put as in (12.1)

$$u(x,t) = \left(e^{-t|\xi|}\, \hat{f}\right)^{\vee}(x) \tag{15.32}$$

where f is now E-valued. Let

$$s \in \mathbb{R},\ 0 < q \leq \infty \quad \text{and} \quad \frac{n}{n+1} < p \leq \infty$$

($p < \infty$ in the F-case), and let $k \in \mathbb{N}$ be sufficiently large; then for any $c > 0$

$$\sum_{l=0}^{k-1} \left\| \sup_{|\cdot - y| \leq c} \left\| \frac{\partial^l u}{\partial t^l}(y,1)\, |E\right\| \ |L_p(\mathbb{R}^n)\right\|$$

$$+ \left(\int_0^1 t^{(k-s)q} \left\| \sup \left\| \frac{\partial^k u}{\partial t^k}(y,\tau)\, |E\right\| \ |L_p(\mathbb{R}^n)\right\|^q \frac{dt}{t}\right)^{\frac{1}{q}} \tag{15.33}$$

(modification if $q = \infty$) is an equivalent quasi-norm in $B_{pq}^s(\mathbb{R}^n, E)$ and

$$\sum_{l=0}^{k-1} \left\| \sup_{|\cdot - y| \leq c} \left\| \frac{\partial^l u}{\partial t^l}(y,1)\, |E\right\| \ |L_p(\mathbb{R}^n)\right\|$$

$$+ \left\| \left(\int_0^1 t^{(k-s)q} \sup \left\| \frac{\partial^k u}{\partial t^k}(y,\tau)\, |E\right\|^q \frac{dt}{t}\right)^{\frac{1}{q}} \ |L_p(\mathbb{R}^n)\right\| \tag{15.34}$$

(modification if $q = \infty$) is an equivalent quasi-norm in $F_{pq}^s(\mathbb{R}^n, E)$. Both in (15.33) and (15.34) sup is the supremum taken over $|x - y| \leq ct$ and $t \leq \tau \leq 2t$. But these assertions give the possibility to extend quarkonial and atomic decompositions from the scalar to the vector case.

First we formulate the counterpart of Theorem 14.4. Recall

$$\sigma_p = n\left(\frac{1}{p} - 1\right)_+ \quad \text{and} \quad \sigma_{pq} = n\left(\frac{1}{\min(p,q)} - 1\right)_+. \tag{15.35}$$

Furthermore, the $(s,p)_L - \gamma$-quarks $(\gamma qu)_{\nu m}^L$ have the same (scalar) meaning as in Definition 14.2. As there we write $(\gamma qu)_{\nu m}$ if $L = -1$, see also 14.3 where we gave some explanations. In particular, the sequence spaces b_{pq} and f_{pq} introduced in Definition 13.5 will be applied to the (scalar) sequences

$$\lambda^\gamma = \{\lambda^\gamma_{\nu m} \in \mathbb{C} : \nu \in \mathbb{N}_0, m \in \mathbb{Z}^n\},$$

where again $\gamma \in \mathbb{N}_0^n$ is a multi-index. We give a precise formulation repeating 14.4 as far as possible.

15.8 Theorem (i) *Let* $\frac{n}{n+1} < p \le \infty$, $0 < q \le \infty$ *and* $s \in \mathbb{R}$. *Let* $M \in \mathbb{N}$ *with* $M > \sigma_p$ *and* $M > s$; *and* L *with* $\frac{L+1}{2} \in \mathbb{N}_0$ *and* $L \ge \max(-1, [\sigma_p - s])$ *be fixed. Let* $(\gamma qu)_{\nu m}$ *be* $(M, p)_{-1} - \gamma$-*quarks and let* $(\gamma qu)_{\nu m}^L$ *be* $(s, p)_L - \gamma$-*quarks. There is a number* $\varkappa > 0$ *with the following property. Let* $\mu > \varkappa$. *Let* E *be a complex Banach space. Then* $f \in S'(\mathbb{R}^n, E)$ *belongs to* $B^s_{pq}(\mathbb{R}^n, E)$ *if, and only if, it can be represented as*

$$f = \sum_{\gamma \in \mathbb{N}_0^n} \sum_{\nu=0}^{\infty} \sum_{m \in \mathbb{Z}^n} \left(\varrho^\gamma_{\nu m} (\gamma qu)_{\nu m}(x) e^\gamma_{\nu m} \right.$$

$$\left. + \lambda^\gamma_{\nu m} (\gamma qu)_{\nu m}^L(x) e^{\gamma, L}_{\nu m} \right), \tag{15.36}$$

convergence being in $S'(\mathbb{R}^n, E)$ *(first* m, *then* ν, *then* γ), *with*

$$\left\{ e^\gamma_{\nu m}, e^{\gamma, L}_{\nu m} \right\} \subset U_E \quad (\textit{unit ball in } E) \tag{15.37}$$

and

$$\sup_\gamma 2^{\mu|\gamma|} \left(\|\varrho^\gamma \mid b_{pq}\| + \|\lambda^\gamma \mid b_{pq}\| \right) < \infty. \tag{15.38}$$

Furthermore, the infimum in (15.38) *over all admissible representations in* (15.36) *with* (15.37) *is an equivalent quasi-norm in* $B^s_{pq}(\mathbb{R}^n, E)$.

(ii) *Let* $\frac{n}{n+1} < p < \infty$, $0 < q \le \infty$ *and* $s \in \mathbb{R}$. *Let* $M \in \mathbb{N}$ *with* $M > \sigma_{pq}$ *and* $M > s$; *and* L *with* $\frac{L+1}{2} \in \mathbb{N}_0$ *and* $L \ge \max(-1, [\sigma_{pq} - s])$ *be fixed. The quarks and* $\varkappa$ *have the same meaning as in part (i) (now, perhaps, with a different value of* L). *Let* $\mu > \varkappa$. *Then* $f \in S'(\mathbb{R}^n, E)$ *belongs to* $F^s_{pq}(\mathbb{R}^n, E)$ *if, and only if, it can be represented by* (15.36) *with* (15.37) *and*

$$\sup_\gamma 2^{\mu|\gamma|} \left(\|\varrho^\gamma \mid f_{pq}\| + \|\lambda^\gamma \mid f_{pq}\| \right) < \infty. \tag{15.39}$$

Furthermore, the infimum in (15.39) *over all admissible representations* (15.36) *with* (15.37) *is an equivalent quasi-norm in* $F^s_{pq}(\mathbb{R}^n, E)$.

Proof. (Outline)

Step 1. We prove the «if-parts» of the above theorem. Under the assumptions (15.37) and (15.38) the right-hand side of (15.36) converges in $S'(\mathbb{R}^n, E)$. This follows from the arguments given in Step 1 of the proof of Theorem 14.4 with the respective references to Corollary 13.9 and Theorem 13.8. In the proof of Theorem 13.8 we used local means. But equivalent quasi-norms based on local means also fit in the scheme described in 15.7. Hence, there is an E-valued counterpart. Then the reasoning in Step 1 of the proof of Theorem 14.4 gives the desired result. In particular, *we have a corresponding assertion for arbitrary related atoms in place of the above quarks and for all $0 < p \le \infty$ ($p < \infty$ in the F-case), $0 < q \le \infty$ and $s \in \mathbb{R}$.*

Step 2. To prove the «only-if-parts» one can follow Steps 2 and 3 in the proof of Theorem 14.4 based on the example (ii) in 15.7, especially (15.33) and (15.34). As remarked there we have the E-version of (12.7) and (13.76) and hence optimal decompositions in E-*valued harmonic atoms* according to (13.82)–(13.85), where the counterpart of (13.79) is now given by

$$\lambda_{\nu m} = 2^{\nu(s-\frac{n}{p})} 2^{-\nu k} \sup \left\| \frac{\partial^k u}{\partial t^k}(y,t) \,|\, E \right\|, \quad \nu > \mu, \; m \in \mathbb{Z}^n, \tag{15.40}$$

where the supremum is taken over

$$|2^{-\nu} m - y| \le c\, 2^{-\nu+\mu-1} \quad \text{and} \quad c^{-1} 2^{-\nu+\mu} \le t \le c\, 2^{-\nu+\mu+1}$$

for sufficiently large $c \ge 1$. There is a corresponding counterpart of (13.80). But now one is in the same position as in 13.10 with the counterpart of (13.81), which follows from (15.40) and (15.34), and the related E-valued harmonic atoms $a_{\nu m}(x)$. There is an E-valued representation formula (13.88) for harmonic functions. This follows from the scalar case and the Hahn-Banach theorem. Then we have the E-valued counterpart of (13.90). Armed with these observations we follow (14.22)–(14.25), where now $c_\gamma \in E$ and $|c_\gamma|$ in (14.25) becomes $\|c_\gamma \,|\, E\|$. In other words, the elements $e_{\nu m}^\gamma$ originate from the normalized E-coefficients c_γ (depending on ν and m). The rest remains unchanged, now E-valued. This applies also to Step 3 of the proof of Theorem 14.4, including the E-version of (14.33).

15.9 Corollary (i) *Let $\frac{n}{n+1} < p \le \infty$, $0 < q \le \infty$ and $s > \sigma_p$. Let $\mu > \varkappa$ as in Theorem 15.8. Then $f \in S'(\mathbb{R}^n, E)$ belongs to $B_{pq}^s(\mathbb{R}^n, E)$ if, and only if, it can be represented as*

$$f = \sum_{\gamma \in \mathbb{N}_0^n} \sum_{\nu=0}^{\infty} \sum_{m \in \mathbb{Z}^n} \lambda_{\nu m}^\gamma \, (\gamma q u)_{\nu m}(x) \, e_{\nu m}^\gamma, \tag{15.41}$$

convergence being in $S'(\mathbb{R}^n, E)$ (first m, then ν, then γ), where $(\gamma q u)_{\nu m}$ are $(s, p)_{-1} - \gamma$-quarks,

$$\sup_\gamma 2^{\mu|\gamma|} \|\lambda^\gamma \,|\, b_{pq}\| < \infty, \tag{15.42}$$

and

$$\{e_{\nu m}^{\gamma}\} \subset U_E.$$ (15.43)

Furthermore, the infimum in (15.42) *over all admissible representations* (15.41) *with* (15.43) *is an equivalent quasi-norm in* $B_{pq}^s(\mathbb{R}^n, E)$.

(ii) *Let* $\frac{n}{n+1} < p < \infty$, $0 < q \le \infty$, *and* $s > \sigma_{pq}$. *Let* $\mu > \varkappa$. *Then* $f \in S'(\mathbb{R}^n, E)$ *belongs to* $F_{pq}^s(\mathbb{R}^n, E)$ *if, and only if, it can be represented by* (15.41), *where the quarks and* $e_{\nu m}^{\gamma}$ *have the same meaning as in part (i) and*

$$\sup_{\gamma} 2^{\mu |\gamma|} \|\lambda^{\gamma} \mid f_{pq}\| < \infty.$$ (15.44)

Furthermore, the infimum in (15.44) *over all admissible representations* (15.41) *with* (15.43) *is an equivalent quasi-norm in* $F_{pq}^s(\mathbb{R}^n, E)$.

Proof. This is the counterpart of Corollary 14.7 and as there covered by the outlined proof of Theorem 15.8.

15.10 Remarks and comments In applications it might well be the case that Hölder-Zygmund spaces $\mathcal{C}^s(\mathbb{R}^n, E)$ introduced in (15.29), (15.30) are of peculiar interest. This comes also from the fact that for general Banach spaces E there are no preferences among the spaces $B_{pq}^s(\mathbb{R}^n, E)$ and $F_{pq}^s(\mathbb{R}^n, E)$ including $p = q = 2$. In case of $s > 0$ one has both (15.41), (15.42) (where $b_{\infty\infty} = l_{\infty}$) and (15.36) with the related conditions. Both representations have their advantages. Whereas (15.41) is simpler, (15.36) splits $f \in \mathcal{C}^s(\mathbb{R}^n, E)$ into a first term with an arbitrarily large smoothness, M, and a second term where arbitrarily many moment conditions are available, L, which is very useful in connection with (singular) integrals and pseudodifferential operators. So far we restricted p by $p > \frac{n}{n+1}$. But there are no problems to extend 14.10 and 14.11 from the scalar case to the vector case and to include in that way $0 < p \le \frac{n}{n+1}$, see also the end of Step 1 of the proof of Theorem 15.8 marked in italics. Maybe the most surprising assertion of Theorem 15.8 and Corollary 15.9 is the total decoupling of E and $\mathbb{R}^n$. This comes from the E-version of the representation formula (13.88) for harmonic functions. This makes clear why we did not try to prove first more general atomic or even molecular decompositions in the vector case. Just on the contrary, having the total decoupling in E and in (scalar) quarks, it is now essentially a scalar matter to replace the quarks by more general atoms according to Theorem 13.8 and the subsequent considerations in Section 13. Coming that way one now has, at least at first glance, the restriction $p > \frac{n}{n+1}$. Furthermore the subtle interplay between E and $\mathbb{R}^n$ comes in according to Theorem 15.8. Otherwise one has the same (scalar) reasoning as in Step 2 of the proof of Theorem 13.8. In that way one arrives at the E-version of Theorem 13.8 which we formulate now.

15.11 Theorem (i) *Let* $\frac{n}{n+1} < p \le \infty$, $0 < q \le \infty$, *and* $s \in \mathbb{R}$. *Let* $M \in \mathbb{N}$ *with* $M > \sigma_p$ *and* $M > s$, $K \in \mathbb{N}_0$, *and* $L + 1 \in \mathbb{N}_0$ *with*

$$K \ge (1 + [s])_+ \quad and \quad L \ge \max(-1, [\sigma_p - s])$$ (15.45)

be fixed. There is a number $\varkappa > 0$ with the following property. Let $\mu > \varkappa$. Then $f \in S'(\mathbb{R}^n, E)$ belongs to $B_{pq}^s(\mathbb{R}^n, E)$ if, and only if, it can be represented as

$$f = \sum_{\gamma \in \mathbb{N}_0^n} \sum_{\nu=0}^{\infty} \sum_{m \in \mathbb{Z}^n} \left(\varrho_{\nu m}^\gamma a_{\nu m}^\gamma(x) e_{\nu m}^\gamma + \lambda_{\nu m}^\gamma a_{\nu m}^{\gamma,L}(x) e_{\nu m}^{\gamma,L} \right), \qquad (15.46)$$

convergence being in $S'(\mathbb{R}^n, E)$, (first m, then ν, then γ), where $a_{\nu m}^\gamma(x)$ are $(M, p)_{K, -1}$-atoms and $a_{\nu m}^{\gamma,L}(x)$ are $(s, p)_{K,L}$-atoms according to Definition 13.3 with (13.21),

$$\left\{ e_{\nu m}^\gamma, \; e_{\nu m}^{\gamma,L} \right\} \subset U_E \qquad (15.47)$$

and

$$\sup_\gamma 2^{\mu|\gamma|} \left(\|\varrho^\gamma \mid b_{pq}\| + \|\lambda^\gamma \mid b_{pq}\| \right) < \infty. \qquad (15.48)$$

The infimum in (15.48) over all admissible representations is an equivalent quasi-norm in $B_{pq}^s(\mathbb{R}^n, E)$.

  (ii) *Let $\frac{n}{n+1} < p < \infty$, $0 < q \le \infty$, and $s \in \mathbb{R}$. Let $M \in \mathbb{N}$ with $M > \sigma_{pq}$ and $M > s$, $K \in \mathbb{N}_0$, and $L + 1 \in \mathbb{N}_0$ with*

$$K \ge (1 + [s])_+ \quad and \quad L \ge \max(-1, [\sigma_{pq} - s]) \qquad (15.49)$$

be fixed. Again let $\mu > \varkappa$. Then $f \in S'(\mathbb{R}^n, E)$ belongs to $F_{pq}^s(\mathbb{R}^n, E)$ if, and only if, it can be represented by (15.46), where the atoms have the same meaning as in part (i) (now, perhaps, with a different value of L), with (15.47) and the counterpart of (15.48) with f_{pq} in place of b_{pq} (including that the related infimum is an equivalent quasi-norm in $F_{pq}^s(\mathbb{R}^n, E)$).

15.12 Remark As remarked in 15.10, the theorem is covered by Theorems 15.8 and 13.8. The infimum of (15.48) of all admitted representations is an equivalent quasi-norm in $B_{pq}^s(\mathbb{R}^n, E)$ (similarly for $F_{pq}^s(\mathbb{R}^n, E)$). This follows from the corresponding assertions in 15.8 since the quarks, up to constants, are special atoms. The alternative proof of (scalar) subatomic decompositions outlined in 14.15 and commented in 14.16 can also be extended to the vector-valued case.

15.13 Weak spaces
Quarkonial and atomic decompositions of vector-valued function spaces as presented in the Theorems 15.8 and 15.11, respectively, might be useful not only in applications but also to prove properties of these spaces: embeddings, traces, (Fourier) multipliers, duality etc. We will not do this here. But we mention an almost immediate consequence of these theorems compared with their scalar counterparts. Simply to avoid some technical difficulties we assume $f \in \mathscr{C}^{-\infty}(\mathbb{R}^n, E)$, which is the E-valued version of (12.16), and we have also the E-valued counterpart of (12.17). In particular there is a representation of type (15.36), and $(f, e') \in \mathscr{C}^{-\infty}(\mathbb{R}^n)$ makes sense for any $e' \in E'$. Let $s \in \mathbb{R}$, $0 < p < \infty$,

and $0 < q \leq \infty$. Then the *weak space* $F_{pq}^s(\mathbb{R}^n, E)^*$ is the collection of all $f \in \mathscr{C}^{-\infty}(\mathbb{R}^n, E)$ such that $(f, e') \in F_{pq}^s(\mathbb{R}^n)$ for all $e' \in E'$ and

$$\|f \mid F_{pq}^s(\mathbb{R}^n, E)^*\| = \sup_{e' \in U_{E'}} \|(f, e') \mid F_{pq}^s(\mathbb{R}^n)\| < \infty. \tag{15.50}$$

Recall that $U_{E'}$ is the unit ball in E', the dual of E. Similarly one can define $B_{pq}^s(\mathbb{R}^n, E)^*$. As an immediate consequence of Theorems 15.8 and 14.4 it follows

$$\|f \mid F_{pq}^s(\mathbb{R}^n, E)^*\| \leq c \|f \mid F_{pq}^s(\mathbb{R}^n, E)\| \quad \text{for all} \quad f \in F_{pq}^s(\mathbb{R}^n, E) \tag{15.51}$$

and some $c > 0$. We mention an application of this observation. The E-valued counterparts of (10.9), (10.14), (10.15) cannot be expected in general, see also what follows in 15.14. Let $1 < p < \infty$ and $s \in \mathbb{R}$. Let $f \in F_{p,2}^s(\mathbb{R}^n, E)$, then (10.14) and (15.51) yield

$$(f, e') \in H_p^s(\mathbb{R}^n) \quad \text{for all} \quad e' \in E'. \tag{15.52}$$

Recall that $H_p^s(\mathbb{R}^n)$ with $s \in \mathbb{N}_0$ are the classical Sobolev spaces, see (10.15), (10.10).

15.14   The UMD-property and Sobolev spaces

The spaces $B_{pq}^s(\mathbb{R}^n, E)$ and $F_{pq}^s(\mathbb{R}^n, E)$ introduced in Definition 15.4 have many good properties, for example, Fourier multipliers of Michlin-Hörmander type and equivalent quasi-norms by classical means as described in 15.6 and 15.7, respectively. Furthermore, despite the three sums we obtained in Theorem 15.8 a rather simple description of these spaces by very elementary building blocks. Hence they might well be the *right* E-valued spaces, both from the point of view of their theory and of possible applications. On the other hand it is natural to introduce E-valued Sobolev spaces and to ask for E-valued counterparts of the properties (i)–(iii) in 10.5. We give a very brief description. Let $1 < p < \infty$ and let I_σ be the lift defined in (10.11). In analogy to (10.13) we introduce the (fractional) Sobolev spaces

$$H_p^s(\mathbb{R}^n, E) = I_{-s} L_p(\mathbb{R}^n, E), \quad s \in \mathbb{R}, 1 < p < \infty. \tag{15.53}$$

The classical Sobolev spaces are defined by

$$W_p^m(\mathbb{R}^n, E) = \{f \in L_p(\mathbb{R}^n, E) : D^\alpha f \in L_p(\mathbb{R}^n, E), |\alpha| \leq m\}, \tag{15.54}$$

where $m \in \mathbb{N}_0$ and $1 < p < \infty$. Both $H_p^s(\mathbb{R}^n, E)$ and $W_p^m(\mathbb{R}^n, E)$ are normed in the obvious way. The lifts in (15.53) and the derivatives in (15.54) must be understood in $S'(\mathbb{R}^n, E)$. It turns out that E-valued versions of the Littlewood-Paley assertions (10.8) and (10.14), Fourier multiplier theorems of Michlin-Hörmander type, and assertions of type (10.15) cannot be expected for general complex Banach spaces E. Problems of that type have attracted a lot of attention. A rather careful collection of relevant publications may be found in [Schm87] and, as far as more recent papers are concerned, in [Ama95], 4.4, 4.5, pp. 141–147. A decisive role in this connection has been played by the so-called UMD-property (**U**nconditionality of **M**artingale **D**ifferences) of a Banach space E. At least for an analyst the simplest way to define what is meant is the following:

A Banach space E is called a UMD-space if the Hilbert transform is bounded in $L_2(\mathbb{R}, E)$.

Then it is also bounded in $L_p(\mathbb{R}, E)$ with $1 < p < \infty$, see [Ama95], pp. 139–147, also for more background information and properties of UMD-spaces. In our context the following two assertions shed some light on the indicated problems:

(i) *E is a UMD-space if, and only if, $L_p(\mathbb{R}^n, E)$ with $1 < p < \infty$ has the Michlin-Hörmander Fourier multiplier property.*

(ii) *E is a UMD-space if, and only if,*

$$W_p^m(\mathbb{R}^n, E) = H_p^m(\mathbb{R}^n, E), \quad m \in \mathbb{N}, \quad 1 < p < \infty. \tag{15.55}$$

For details, references, and in case of (i) also for explanations, we refer to [Schm87] and [Ama95].

15.15 Taylor expansions for E-valued tempered distributions
We outlined in 14.13 how to get Taylor expansions of type (14.51) for tempered distributions. The arguments there are supported by the (rather non-trivial) mapping properties $f \mapsto w(\cdot)f$ of weighted spaces onto corresponding unweighted spaces. Taking for granted that the same mapping properties hold also in the E-valued case (what has not been done so far) then one can argue as in 14.13. One obtains Taylor expansions of type (14.51) but now modified according to (15.41), or, more general, (15.36). Again one has a total decoupling of elements $e_{\nu m}^\gamma \in E$ and the γ-quarks. In 14.13 we stressed the analogy with Taylor expansions for analytic functions. It is quite interesting that even in case of E-valued analytic functions there is a corresponding analogy in connection with complex interpolation. We do not go into detail and refer to [Tri78], Theorem 1.9.1, pp. 65–58.

16 Graphs of functions

16.1 Introduction
This section is also somewhat outside the main stream of the book. But it illustrates the intimate connection between atomic decompositions of functions on the one hand, and fractal geometry on the other hand. Let, say,

$$f \in \mathscr{C}^s(\mathbb{R}^{n-1}) = B_{\infty,\infty}^s(\mathbb{R}^{n-1}), \quad s > 0, \quad n \geq 2,$$

be a real function on $\mathbb{R}^{n-1}$. Then the Hausdorff dimension of the graph of f, restricted, say, to the unit cube Q in $\mathbb{R}^{n-1}$,

$$graph\, f = \{(x, f(x)),\ x \in Q\}, \tag{16.1}$$

in $\mathbb{R}^n$ is of interest. If $s > 1$, then *graph f* is a smooth (Lipschitzian) $(n-1)$-dimensional manifold in $\mathbb{R}^n$, and, hence

$$\dim_H(graph\, f) = n - 1, \qquad (16.2)$$

where we use the notation introduced in Definition 2.6. If $0 < s < 1$ then the graph becomes non-smooth and the relation between s and $\dim_H(graph\, f)$ attracted some attention. The first candidate of a non-smooth function to be studied in this context is the famous Weierstrass function

$$f(x) = \sum_{\nu=1}^{\infty} \varrho^{-s\nu} \sin(\varrho^{\nu} x), \quad x \in [0,1], \qquad (16.3)$$

$\varrho > 1$ and $0 < s < 1$. We refer to [Fal85], 8.2, pp. 111–117, and [Fal90], Ch. 11, in particular pp. 147–148. If ϱ is large then

$$\dim_H(graph\, f) = 2 - s. \qquad (16.4)$$

In the two books mentioned one finds further results and references about this subject. In the context of function spaces the Hausdorff dimension and the nearby fractal dimension of graphs have been studied in [DeJ92] (one dimension) and [Sol96] (also several dimensions). It is not our aim to give a systematic treatment of the subject, which presumably could be based on atomic decompositions according to Theorem 13.8, or, maybe even better, on *qu*-decompositions according to Corollary 14.7. We restrict ourselves to the simplest case where f belongs to the Hölder spaces $\mathscr{C}^s$ with $0 < s < 1$. In what follows Q stands for the open unit cube in $\mathbb{R}^{n-1}$, $n \geq 2$, centred at the origin, with side-length 1. Furthermore, if f is the restriction to $\overline{Q}$ of a real function belonging to $\mathscr{C}^s(\mathbb{R}^{n-1})$ with $0 < s < 1$, then f is continuous on $\overline{Q}$ and

$$|f(x) - f(y)| \leq c\,|x - y|^s, \quad x \in \overline{Q},\, y \in \overline{Q}, \qquad (16.5)$$

for some $c > 0$, see (10.16) and (10.18).

16.2 Theorem *Let* $0 < s < 1$, $n \geq 2$, *and let* f *be the restriction to* $\overline{Q}$ *of a real function belonging to* $\mathscr{C}^s(\mathbb{R}^{n-1})$.

(i) *Let graph f be given by (16.1); then*

$$\dim_H(graph\, f) \leq n - s. \qquad (16.6)$$

(ii) *There are functions f of this type such that*

$$\dim_H(graph\, f) = n - s. \qquad (16.7)$$

Proof.
Step 1. We prove (i). As before $Q_{\nu m}$ with $\nu \in \mathbb{N}_0$ and $m \in \mathbb{Z}^{n-1}$ stands for a cube (with sides parallel to the axes of coordinates) centred at $2^{-\nu}m$ and with side-length $2^{-\nu}$. We may assume

$$Q_{\nu m} \subset cQ \quad \text{for some} \quad c > 1. \tag{16.8}$$

If $x \in Q_{\nu m}$ and $y \in Q_{\nu m}$ then by (16.5) $|f(x) - f(y)| \leq c\, 2^{-\nu s}$. Hence to cover the graph of f over $Q_{\nu m}$ one needs at most $c\, 2^{\nu(1-s)}$ cubes in $\mathbb{R}^n$ of side-length $2^{-\nu}$. Hence the graph of f over $\overline{Q}$ can be covered by

$$c\, 2^{\nu(n-1)}\, 2^{\nu(1-s)} = c\, 2^{\nu(n-s)}$$

cubes of side-length $2^{-\nu}$ in $\mathbb{R}^n$. With $d = n - s$ and $\delta = c\, 2^{-\nu}$ we obtain by (2.3)

$$\mathcal{H}_\delta^d(\text{graph}\, f) \leq c_1 \sum (\text{diam}\, Q_{\nu m})^{n-s} \leq c_2\, 2^{\nu(n-s)}\, 2^{-\nu(n-s)} \leq c_3 < \infty \tag{16.9}$$

where the sum is taken over those m with (16.8). Now (16.6) follows from (2.5).

Step 2. We outline the proof of (ii). Let

$$\omega(x) = e^{-\frac{1}{1-|2x|^2}} \text{ if } |x| < \frac{1}{2} \quad \text{and} \quad \omega(x) = 0 \text{ if } |x| \geq \frac{1}{2}, \quad x \in \mathbb{R}^{n-1}, \tag{16.10}$$

be the classical standard function belonging to $C_0^\infty(\mathbb{R}^{n-1})$. Let $\nu_j = 2^{\varkappa j}$ for some $\varkappa \in \mathbb{N}$ and all $j \in \mathbb{N}$. We choose $\varkappa$ later on. Let

$$f = \sum_{j=1}^\infty f_j \quad \text{with} \quad f_j(x) = 2^{-\nu_j s} \sum_m {}^j \omega(2^{\nu_j} x - m), \tag{16.11}$$

where we sum over those $m \in \mathbb{Z}^{n-1}$ such that $Q_{\nu_j m} \subset Q$. By $\mathscr{C}^s = B_{\infty\infty}^s$, Theorem 13.8 and (13.18) it follows that this lacunary function f belongs to $\mathscr{C}^s(\mathbb{R}^{n-1})$. We claim that *graph* f is a d-set with $d = n - s$ according to Definition 3.1, at least if $\varkappa$ is large. We fix $j \in \mathbb{N}$ and look at the surface

$$x_n = 2^{-\nu_j s} \omega(2^{\nu_j} x) + \sum_{l<j} f_l(x), \quad |x| \leq 2^{-\nu_j-1}, \; x \in \mathbb{R}^{n-1}. \tag{16.12}$$

We have

$$|\nabla f_l(x)| \leq c_1 2^{\nu_l(1-s)}, \quad x \in \mathbb{R}^{n-1}, \tag{16.13}$$

and for some $c_2 > 0$ and $0 < \eta_0 < \eta_1 < \frac{1}{2}$,

$$2^{-\nu_j s} |\nabla \omega(2^{\nu_j} x)| \geq c_2\, 2^{\nu_j(1-s)} \quad \text{if} \quad 0 < \eta_0 < 2^{\nu_j}|x| < \eta_1. \tag{16.14}$$

Hence we may assume that

$$\left| \nabla \sum_{l=1}^{j} f_l(x) \right| \geq c \, 2^{\nu_j(1-s)}, \quad c > 0, \tag{16.15}$$

if x belongs to the annulus in (16.14) and its admitted translates in Q where c is independent of j. On the other hand we have

$$\left| f(x) - \sum_{l=1}^{j} f_l(x) \right| \leq c \, 2^{-\nu_{j+1}s} \leq 2^{-\nu_j} \tag{16.16}$$

since $s\nu_{j+1} = \nu_j \, 2^{\varkappa}s > \nu_j$ if $2^{\varkappa}s > 1$. We divide the neighbourhood

$$T_j = \left\{ Y \in \mathbb{R}^n : \, \exists X \in graph \, f, \, |X - Y| \leq 2^{-\nu_j} \right\} \tag{16.17}$$

such that

$$T_j = \bigcup_{k=1}^{N_j} E_k^j, \quad E_k^j \cap E_l^j = \emptyset \text{ if } l \neq k, \, diam \, E_l^j \sim 2^{-\nu_j}. \tag{16.18}$$

By (16.16) and (16.15) it follows

$$N_j \geq c \, 2^{\nu_j(n-1)} \, 2^{\nu_j(1-s)} = 2^{\nu_j(n-s)} \tag{16.19}$$

for some $c > 0$ which is independent of j. We may assume $N_j \sim 2^{\nu_j(n-s)}$ and that any set E_k^{j+1} is a subset of some E_l^j. Now we use the mass distribution procedure described in the proof of Theorem 4.15 with a reference to [Fal90], pp. 13/14. We distribute the unit mass uniformly and have

$$\mu(E_l^j) \sim 2^{-\nu_j d} \sim (diam \, E_l^j)^d \quad \text{with} \quad d = n - s. \tag{16.20}$$

But then we arrive at a Radon measure μ on $graph \, f$ with

$$supp \, \mu = graph \, f, \quad \mu\left(B(X,r) \cap graph \, f\right) \sim r^d \tag{16.21}$$

where $B(X,r)$ is a ball in $\mathbb{R}^n$ with radius $0 < r < 1$ centred at $X \in graph \, f$. By Definition 3.1 it follows that $graph \, f$ is a d-set. By Theorem 3.4 and Corollary 3.6 we obtain in particular, $\dim_H(graph \, f) = d$.

16.3 Remark We proved a bit more than stated. Under the rather restrictive assumptions about f, given by (16.11), it comes out that $graph \, f$ is even a d-set with $d = n - s$. One obtains as a by-product the well-known assertion that for any d with $0 < d < n$ there are d-sets in $\mathbb{R}^n$: If $d \in \mathbb{N}$ then smooth manifolds will do it. If $k - 1 < d < k$ one has d-sets in $\mathbb{R}^k$ by the above construction. These are also d-sets in $\mathbb{R}^n$ with $\mathbb{R}^k \subset \mathbb{R}^n$.

16.4 d-sets as boundaries

The construction in Step 2 of the proof of Theorem 16.2 results in (16.18) and (16.21). The subdivision of E_l^j in (16.18) and the selection of those sets belonging to $\{E_k^{j+1}\}$ is, at leat in qualitative terms, the same for every admitted l. This imitates in qualitative terms the similarity constructions in 4.5–4.7 combined with the above described mass distribution procedure which results in the distinguished measure μ in (16.21). However this qualitative similarity construction can be applied to modified situations. For example, the $(n-1)$-dimensional unit cube Q in 16.1–16.3 can be replaced by the $(n-1)$-dimensional unit sphere S_{n-1} in $\mathbb{R}^n$. Instead of the x_n-direction in (16.1) one may use the direction of the outer normal. There is an obvious counterpart of (16.11) now with S_{n-1} instead of Q where one starts from. The resulting *graph f* is a d-set in $\mathbb{R}^n$ with $n-1 < d = n-s < n$. It is the boundary $\partial\Omega$ of a star-like bounded domain Ω (with respect to the origin) in $\mathbb{R}^n$. At least in the plane, which means $n = 2$, there are many beautiful constructions resulting in bounded (not necessarily star-like) domains Ω with a d-set as boundary $\partial\Omega$, where $1 < d < 2$. One of the most prominent examples is the snowflake curve, see e.g. [Fal90], p. XV.

Chapter IV
Function spaces on and of fractals

17 The distributional dimension

17.1 Introduction

It is the main aim of this Chapter to discuss the seminal interrelation between fractals and function spaces. This paves the way to a substantial spectral theory of fractal (pseudo)differential operators which will be developed in Chapter V. We begin in Section 17 with a rather final Fourier analytical characterization of the Hausdorff dimension of arbitrary Borel (or even Suslin) sets in $\mathbb{R}^n$ with vanishing Lebesgue measure. L_p-spaces on fractals and their connection with B_{pq}^s spaces on $\mathbb{R}^n$ will be studied in Section 18. On that basis we introduce B_{pq}^s spaces on fractals in Section 20 and calculate the entropy numbers of compact embeddings between them. The remaining sections complement these results.

The spaces $B_{pq}^s(\mathbb{R}^n)$ and $F_{pq}^s(\mathbb{R}^n)$ have been introduced in Definition 10.3 with some special cases listed in 10.5. Of special interest for us will be the Hölder-Zygmund scale $\mathcal{C}^s(\mathbb{R}^n) = B_{\infty,\infty}^s(\mathbb{R}^n)$ with $s \in \mathbb{R}$, see (10.16). We denote the Lebesgue measure of a measurable set Γ in $\mathbb{R}^n$ by $|\Gamma|$. We collected in Section 1–3 the measure-theoretical facts needed in the sequel. Some results which will be stated here for Borel sets in $\mathbb{R}^n$ actually hold for the more general Suslin sets (see [Fal85], p. 6, for a definition of Suslin sets, and also for related assertions in connection with Hausdorff measures in terms of Suslin sets in $\mathbb{R}^n$).

17.2 Definition
Let Γ be a non-empty closed subset of $\mathbb{R}^n$ with $|\Gamma| = 0$. Suppose that $s \in \mathbb{R}$, $0 < p \le \infty$, and $0 < q \le \infty$. Then

$$B_{pq}^{s,\Gamma}(\mathbb{R}^n) = \{f \in B_{pq}^s(\mathbb{R}^n) : f(\varphi) = 0 \quad if \quad \varphi \in S(\mathbb{R}^n), \varphi|\Gamma = 0\}. \quad (17.1)$$

17.3 Discussion
In Sections 17 and 18 we follow partly [TrW96b], which in turn is based on [TrW95] and [Win95]. Of special interest for us are

$$\mathcal{C}^{s,\Gamma}(\mathbb{R}^n) = B_{\infty,\infty}^{s,\Gamma}(\mathbb{R}^n). \quad (17.2)$$

Of course $\varphi|\Gamma$ stands for the restriction of φ to Γ. We use also $tr_\Gamma \varphi = \varphi|\Gamma$, but we reserve tr_Γ preferably for the *trace operator*, mapping spaces of type B_{pq}^s or F_{pq}^s on $\mathbb{R}^n$ into or onto suitable function spaces on Γ. But whether we use $\varphi|\Gamma$ or $tr_\Gamma \varphi$ is sometimes simply a matter of convenience or clarity. Let

$$0 < p \le \infty, \quad 0 < q \le \infty, \quad s > \sigma_p = n\left(\frac{1}{p} - 1\right)_+. \quad (17.3)$$

H. Triebel, *Fractals and Spectra*, Modern Birkhäuser Classics,
DOI 10.1007/978-3-0348-0034-1_4, © Birkhäuser Verlag 1997

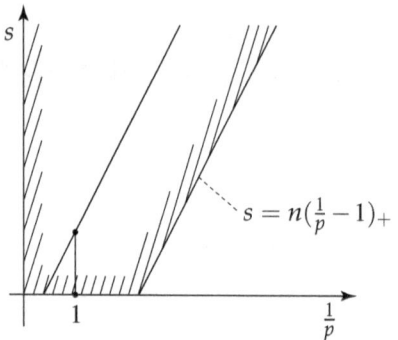

Fig. 17.1

Then we have

$$B_{pq}^s(\mathbb{R}^n) \subset L_1^{loc}(\mathbb{R}^n), \tag{17.4}$$

what corresponds to the shaded area in Fig. 17.1 and is covered by (11.16), (11.17), which means embeddings along lines of slope n in the $(\frac{1}{p}, s)$-diagram, in connection, say, with (10.14), (10.13). A final clarification of (17.4), including limiting situations may be found in [SicT95] and [RuS96], 2.2.4. In particular $B_{pq}^s(\mathbb{R}^n)$ with (17.3) consists of regular distributions. Hence we have $B_{pq}^{s,\Gamma}(\mathbb{R}^n) = \{0\}$ in that case. This follows from $|\Gamma| = 0$ and

$$\operatorname{supp} f \subset \Gamma \quad \text{if} \quad f \in B_{pq}^{s,\Gamma}(\mathbb{R}^n). \tag{17.5}$$

Of course, (17.5) holds for all admitted s, p, q. In other words, in any case, $B_{pq}^{s,\Gamma}(\mathbb{R}^n)$ consists of singular distributions (with exception of the zero-distribution) and only points $(\frac{1}{p}, s)$ in Fig. 17.1 outside the shaded area are of interest. Furthermore the support assertion in (17.5) is necessary for $f \in B_{pq}^{s,\Gamma}(\mathbb{R}^n)$, but not sufficient. For example, let $\Gamma = [-1, 1]$, considered as a subset of $\mathbb{R}^2$. Then one has for the δ-distribution in $\mathbb{R}^2$,

$$\delta(\varphi) = \frac{\partial \delta}{\partial x_1}(\varphi) = 0, \quad \text{but, in general,} \quad \frac{\partial \delta}{\partial x_2}(\varphi) \neq 0 \tag{17.6}$$

if $\varphi \in S(\mathbb{R}^2)$ and $\varphi|\Gamma = 0$. In other words, whereas δ and $\frac{\partial \delta}{\partial x_1}$ belong to suitable spaces $B_{pq}^{s,\Gamma}(\mathbb{R}^2)$, this is not the case for $\frac{\partial \delta}{\partial x_2}$. Finally we remark that one can introduce corresponding spaces $F_{pq}^{s,\Gamma}(\mathbb{R}^n)$, but they will not be needed later on (what a pity).

17.4 Definition (distributional dimension). *Let Γ be a non-empty Borel set in $\mathbb{R}^n$ with $|\Gamma| = 0$. Then*

$$\dim_D \Gamma$$

$$= \sup \left\{ d \ : \ \mathscr{C}^{-n+d,\Lambda}(\mathbb{R}^n) \text{ is non-trivial for some compact } \Lambda \subset \Gamma \right\}. \tag{17.7}$$

17.5 Discussion

Of course, *non-trivial* means $\mathscr{C}^{-n+d,\Lambda}(\mathbb{R}^n) \neq \{0\}$. Furthermore, $\dim_D \Gamma$ *is monotone in* Γ. In particular, if Γ is compact then

$$\dim_D \Gamma = \sup \left\{ d : \mathscr{C}^{-n+d,\Gamma}(\mathbb{R}^n) \text{ is non-trivial} \right\}. \tag{17.8}$$

As we said we restrict our attention to Borel sets in $\mathbb{R}^n$ although it is quite clear that the above definition can be extended to more general sets, at least if $|\Gamma| = 0$. Furthermore as we shall see in Theorem 17.8 one can replace $\mathscr{C}^{-n+d}$ in (17.7) by some suitable spaces B_{pq}^s. Finally we remark that

$$0 \leq \dim_D \Gamma \leq n, \tag{17.9}$$

as it should be. The right-hand side follows from the observation in 17.3 that $\mathscr{C}^{-n+d,\Lambda}(\mathbb{R}^n)$ is trivial if $-n + d > 0$. To prove the left-hand side of (17.9) we remark that the δ-distribution belongs to $\mathscr{C}^{-n}(\mathbb{R}^n)$. This follows from (10.6) with $s = -n$, $p = q = \infty$. Hence, if $\{0\} = \Lambda \subset \Gamma$ then $\mathscr{C}^{-n,\Lambda}(\mathbb{R}^n)$ is non-trivial. Of course, the off-point 0 can be replaced by any other point in $\mathbb{R}^n$ belonging to Λ. As in Definition 2.6 we denote the Hausdorff dimension of Γ by $\dim_H \Gamma$.

17.6 Theorem *Let* Γ *be a non-empty Borel set in* $\mathbb{R}^n$ *with* $|\Gamma| = 0$; *then*

$$\dim_H \Gamma = \dim_D \Gamma. \tag{17.10}$$

Proof.

Step 1. Assume that (17.10) holds for every compact $\Lambda \subset \Gamma$. Then (17.10) for Γ follows from (17.7) and its counterpart

$$\dim_H \Gamma = \sup \{ \dim_H \Lambda : \Lambda \text{ compact with } \Lambda \subset \Gamma \}. \tag{17.11}$$

The latter assertion is a consequence of Theorem 8.13 in [Mat95], p. 117, or Theorem 5.6 in [Fal85], p. 69. Hence it is sufficient to prove (17.10) for non-empty compact sets Γ in $\mathbb{R}^n$ with $|\Gamma| = 0$.

Step 2. Let Γ be non-empty and compact with $|\Gamma| = 0$. We prove

$$\dim_D \Gamma \geq \dim_H \Gamma. \tag{17.12}$$

By (17.9) and (2.6) we may assume $\dim_H \Gamma > 0$. By (2.5) we have

$$\mathscr{H}^\delta(\Gamma) = \infty \tag{17.13}$$

for every

$$0 < \delta < \dim_H \Gamma. \tag{17.14}$$

By Theorem 2.7 there are a compact set $\Lambda \subset \Gamma$ and a constant $c > 0$ such that $\mathcal{H}^\delta(\Lambda) > 0$ and

$$\mathcal{H}^\delta(\Lambda \cap B(x,r)) \leq c r^\delta, \quad x \in \mathbb{R}^n, \quad 0 < r \leq 1, \tag{17.15}$$

where again $B(x,r)$ stands for the closed ball in $\mathbb{R}^n$ centred at x and of radius r. We define $f \in S'(\mathbb{R}^n)$ by

$$f(\varphi) = \int_\Lambda \varphi(\lambda) \, \mathcal{H}^\delta(d\lambda), \quad \varphi \in S(\mathbb{R}^n). \tag{17.16}$$

Obviously, $\mathcal{H}^\delta(\Lambda) < \infty$ and $f(\varphi) = 0$ if $\varphi|\Gamma = 0$. We claim $f \in \mathcal{C}^{-n+\delta}(\mathbb{R}^n)$. We use the characterization (11.10) via local means. Let $k(y)$ be a suitable kernel with a support in the unit ball, then (11.3) is given by

$$k(t,f)(x) = t^{-n} \int_\Lambda k\left(\frac{\lambda - x}{t}\right) \mathcal{H}^\delta(d\lambda), \quad x \in j\mathbb{R}^n, \, t > 0. \tag{17.17}$$

By (17.15) we have

$$|k(t,f)(x)| \leq c t^{-n} \mathcal{H}^\delta(\Lambda \cap B(x,t)) \leq c t^{-n+\delta}. \tag{17.18}$$

By (11.10) with (11.8) or (11.9) where $s = -n + \delta$, $p = q = \infty$, it follows $f \in \mathcal{C}^{-n+\delta}(\mathbb{R}^n)$, and consequently $f \in \mathcal{C}^{-n+\delta,\Gamma}(\mathbb{R}^n)$. By construction f is non-trivial. Hence by (17.8) we have $\dim_D \Gamma \geq \delta$ and (17.12).

Step 3. The proof of the converse of (17.12) relies on the following observation. *Let Γ be as in Step 2, let $s \leq 0$, then $\mathcal{C}^{s,\Gamma}(\mathbb{R}^n)$ is non-trivial if, and only if,*

$$\{\varphi \in S(\mathbb{R}^n): \ \varphi|\Gamma = 0\} \ \text{is not dense in} \ B_{1,1}^{-s}(\mathbb{R}^n). \tag{17.19}$$

Recall the duality assertion

$$\left(B_{1,1}^{-s}(\mathbb{R}^n)\right)' = B_{\infty,\infty}^s(\mathbb{R}^n) = \mathcal{C}^s(\mathbb{R}^n), \tag{17.20}$$

see [Tri83], 2.11.2, p. 178. First assume that (17.19) holds. Then by (17.20) and the Hahn-Banach theorem there exists a non-trivial $f \in \mathcal{C}^{s,\Gamma}(\mathbb{R}^n)$. See [Rud91], 3.5, p. 60, for the used consequence of the Hahn-Banach theorem. Conversely if there is a non-trivial $f \in \mathcal{C}^{s,\Gamma}(\mathbb{R}^n)$ then (17.19) holds: if not, then we get a contradiction.

Step 4. Let again Γ as in Step 2. We prove

$$\dim_D \Gamma \leq \dim_H \Gamma. \tag{17.21}$$

By (17.9) and (2.6) we may assume $\dim_H \Gamma < n$. By (2.5) we have $\mathcal{H}^\delta(\Gamma) = 0$ if

$$\dim_H \Gamma < \delta < n. \tag{17.22}$$

We wish to prove that

$$\{\varphi \in S(\mathbb{R}^n) : \varphi|\Gamma = 0\} \quad \text{is dense in} \quad B_{1,1}^{n-\delta}(\mathbb{R}^n). \tag{17.23}$$

By (2.3) and the compactness of Γ, for every $\varrho > 0$ and $\varepsilon > 0$ there is a finite covering of Γ by open balls B_j centred at Γ and with diameters less than ε such that

$$\sum_{j=1}^{N} (\text{diam } B_j)^\delta < \varrho. \tag{17.24}$$

The union $\bigcup_{j=1}^{N} B_j$ also covers the closure $\overline{\Gamma_\lambda}$ of some neighbourhood

$$\Gamma_\lambda = \{x \in \mathbb{R}^n : \text{dist}(x, \Gamma) < \lambda\}$$

where λ depends on ϱ. Let $\{\varphi_j\}_{j=1}^{N}$ be a smooth resolution of unity of $\overline{\Gamma_\lambda}$ subordinated to $\{B_j\}_{j=1}^{N}$, in particular

$$\varphi(x) = \sum_{j=1}^{N} (\text{diam } B_j)^\delta (\text{diam } B_j)^{-\delta} \varphi_j(x) = 1 \quad \text{if} \quad x \in \overline{\Gamma_\lambda}, \tag{17.25}$$

and $\text{supp } \varphi_j \subset B_j$. Up to immaterial constants we may assume that the functions $(\text{diam } B_j)^{-\delta} \varphi_j$ are $(n - \delta, 1)$ atoms according to Definition 13.3 without moment conditions. By Theorem 13.8 we have

$$\|\varphi \mid B_{1,1}^{n-\delta}(\mathbb{R}^n)\| \le c \sum_{j=1}^{N} (\text{diam } B_j)^\delta \le c \varrho, \tag{17.26}$$

where c is independent of ϱ. Let $\psi \in S(\mathbb{R}^n)$ have compact support and choose some $\eta \in S(\mathbb{R}^n)$ with $\eta(x) = 1$ if either $x \in \text{supp } \varphi$ or $x \in \text{supp } \psi$, where φ has the above meaning. Then

$$\psi(x) (\eta(x) - \varphi(x)) = 0 \quad \text{if} \quad x \in \Gamma \tag{17.27}$$

and

$$\|\psi - \psi(\eta - \varphi) \mid B_{1,1}^{n-\delta}(\mathbb{R}^n)\| = \|\psi\varphi \mid B_{1,1}^{n-\delta}(\mathbb{R}^n)\| \le c \|\varphi \mid B_{1,1}^{n-\delta}(\mathbb{R}^n)\|, \tag{17.28}$$

where we used that ψ is a pointwise multiplier in $B_{1,1}^{n-\delta}(\mathbb{R}^n)$, see [Tri92], 4.2.2, p. 205. In particular, c in (17.28) may depend on ψ but not on φ. Recall that $S(\mathbb{R}^n)$ is dense in $B_{1,1}^{n-\delta}(\mathbb{R}^n)$, see [Tri83], p. 48. By easy arguments it follows that also the compactly supported functions belonging to $S(\mathbb{R}^n)$ are dense in $B_{1,1}^{n-\delta}(\mathbb{R}^n)$. Together with (17.26)–(17.28) we obtain (17.23). Now we can easily finish the proof. By (17.23) and Step 3 it follows that $\mathscr{C}^{-n+\delta,\Gamma}(\mathbb{R}^n)$ is trivial. Hence we obtain by (17.7) that $\dim_D \Gamma \le \delta$ and finally (17.21).

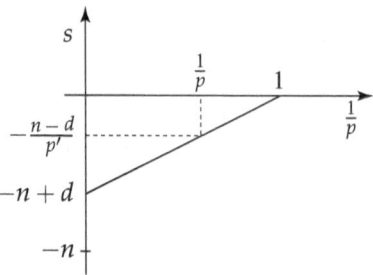

Fig. 17.2

17.7 Remark By the same arguments as in the preceding proof one can extend (17.10) to Suslin sets. Discussing (17.10) it is quite clear that $\dim_H \Gamma$ stands for the *geometric approach to measure the massiveness of a Borel set* Γ *with* $|\Gamma| = 0$, whereas $\dim_D \Gamma$ *reflects the other side of the coin, the analytic approach to measure the ability of* Γ *to carry non-trivial singular distributions as smooth as possible.* It is useful for our later purposes to generalize (17.7) and (17.10) replacing $\mathscr{C}^s = B_{\infty,\infty}^s$ by B_{pq}^s. The correct way to do this is indicated by the line in Fig. 17.2, which will also play a crucial role later on in connection with the spectral theory of fractal pseudodifferential operators. Recall that $B_{pq}^{s,\Gamma}(\mathbb{R}^n)$ has been introduced in (17.1). As usual we put $\frac{1}{p} + \frac{1}{p'} = 1$ if $1 \le p \le \infty$.

17.8 Theorem *Let* Γ *be a non-empty Borel set in* $\mathbb{R}^n$ *with* $|\Gamma| = 0$. *Let* $1 < p \le \infty$ *and* $0 < q \le \infty$. *Then*

$$\dim_H \Gamma = \sup\left\{ d : B_{pq}^{-\frac{n-d}{p'},\Lambda}(\mathbb{R}^n) \right.$$

$$\left. \text{is non-trivial for some compact } \Lambda \subset \Gamma \right\}. \tag{17.29}$$

Proof.
Step 1. As in Step 1 of the proof of Theorem 17.6 we may assume by (17.11) and (17.29) that Γ is compact. Hence we have to prove

$$\dim_H \Gamma = \sup\left\{ d : B_{pq}^{-\frac{n-d}{p'},\Gamma}(\mathbb{R}^n) \text{ is non-trivial} \right\}. \tag{17.30}$$

As in Step 2 of the proof of Theorem 17.6, in order to prove

$$\dim_H \Gamma \le \sup\left\{ d : B_{pq}^{-\frac{n-d}{p'},\Gamma}(\mathbb{R}^n) \text{ is non-trivial} \right\} \tag{17.31}$$

we may assume $\dim_H \Gamma > 0$. We choose δ as in (17.14) and have (17.15)–(17.17). We apply Hölder's inequality with $\frac{1}{p} + \frac{1}{p'} = 1$ to (17.17) and obtain

$$|k(t,f)(x)|$$

$$\leq t^{-n} \left(\int_\Lambda \left| k\left(\frac{\lambda - x}{t}\right) \right| \mathcal{H}^\delta(d\lambda) \right)^{\frac{1}{p}} \left(\int_\Lambda \left| k\left(\frac{\lambda - x}{t}\right) \right| \mathcal{H}^\delta(d\lambda) \right)^{\frac{1}{p'}}$$

$$\leq c\, t^{-n} \mathcal{H}^\delta \left(\Lambda \cap B(x,t) \right)^{\frac{1}{p'}} \left(\int_\Lambda \left| k\left(\frac{\lambda - x}{t}\right) \right| \mathcal{H}^\delta(d\lambda) \right)^{\frac{1}{p}}$$

$$\leq c'\, t^{-n+\frac{\delta}{p'}} \left(\int_\Lambda \left| k\left(\frac{\lambda - x}{t}\right) \right| \mathcal{H}^\delta(d\lambda) \right)^{\frac{1}{p}}.$$

$$(17.32)$$

Taking the p-th power and integrating over $x \in \mathbb{R}^n$ we obtain

$$\| k(t,f) \,|\, L_p(\mathbb{R}^n) \| \leq c\, t^{-n+\frac{\delta}{p'}+\frac{n}{p}} = c\, t^{-\frac{n-\delta}{p'}}. \qquad (17.33)$$

Now by (11.10) and (11.8) it follows $f \in B_{p,\infty}^{-\frac{n-\delta}{p'}}(\mathbb{R}^n)$. By (17.1) and (17.16) we have $f \in B_{p,\infty}^{-\frac{n-\delta}{p'},\Gamma}(\mathbb{R}^n)$. Furthermore, f is non-trivial. By (17.14) we arrive at (17.31) with $q = \infty$. By (11.14) this includes also $0 < q \leq \infty$.

Step 2. Again we assume that Γ is a compact, non-empty set with $|\Gamma| = 0$. We wish to prove the converse of (17.31), which means

$$\sup \left\{ d : B_{pq}^{-\frac{n-d}{p'},\Gamma}(\mathbb{R}^n) \text{ is non-trivial} \right\} \leq \dim_H \Gamma. \qquad (17.34)$$

First we observe that in analogy to (17.9) the left-hand side of (17.34) is smaller than or equal to n. This follows from (17.3), (17.4) and $|\Gamma| = 0$. Hence to prove (17.34) we may assume (17.22). The counterpart of Step 3 of the proof of Theorem 17.6 reads as follows: *Let $s \leq 0$, then $B_{pp}^{s,\Gamma}(\mathbb{R}^n)$ is non-trivial if, and only if,*

$$\{ \varphi \in S(\mathbb{R}^n) : \varphi|\Gamma = 0 \} \quad \text{is not dense in} \quad B_{p'p'}^{-s}(\mathbb{R}^n). \qquad (17.35)$$

The proof is the same as in Step 3 in 17.6 based on the duality

$$\left(B_{p'p'}^{-s}(\mathbb{R}^n) \right)' = B_{pp}^{s}(\mathbb{R}^n), \quad 1 < p \leq \infty, \qquad (17.36)$$

see [Tri83], 2.11.2, p. 178. Now we are in the same position as in Step 4 in 17.6 and we wish to prove the counterpart of (17.23),

$$\{ \varphi \in S(\mathbb{R}^n) : \varphi|\Gamma = 0 \} \quad \text{is dense in} \quad B_{p'p'}^{\frac{n-\delta}{p'}}(\mathbb{R}^n) \qquad (17.37)$$

under the assumption (17.22). We have again (17.24) and we use once more the resolution of unity $\{\varphi_j\}_{j=1}^N$ of $\overline{\Gamma_\lambda}$. We replace (17.25) by

$$\varphi(x) = \sum_{j=1}^N (diam\, B_j)^{\frac{\delta}{p'}} \, (diam\, B_j)^{-\frac{\delta}{p'}} \, \varphi_j(x) = 1 \text{ if } x \in \overline{\Gamma_\lambda} \qquad (17.38)$$

and $supp\, \varphi_j \subset B_j$. Again we may assume that $(diam\, B_j)^{-\frac{\delta}{p'}} \varphi_j(x)$ are $(\frac{n-\delta}{p'}, p')$-atoms according to Definition 13.3. Then we have by Theorem 13.8

$$\left\| \varphi \mid B_{p'p'}^{\frac{n-\delta}{p'}} (\mathbb{R}^n) \right\|^{p'} \le c \sum_{j=1}^N (diam\, B_j)^\delta < c\, \varrho. \qquad (17.39)$$

The rest is the same as in Step 4 in 17.6 and we have (17.34) with $p = q$. The corresponding assertion for arbitrary q follows once more from (11.14).

17.9 Comment By (17.2), Theorem 17.6 with (17.7), is a special case of Theorem 17.8. The main aim of Theorem 17.8 is not so much to generalize Theorem 17.6 but to emphasize the importance of the line in Fig. 17.2 which will play a decisive role in our later considerations. In addition, Theorem 17.8 gives us the possibility to compare the above results with related considerations in literature.

17.10 The Fourier dimension
The dominating mathematician in the theory of *sets of points* in $\mathbb{R}^n$ from the twenties up to the sixties was A. S. Besicovitch. This is well documented in [Fal85]. On the other hand R. S. Strichartz expressed in his paper [Str90a] some doubts as to whether Besicovitch imagined the close connection between the theory developed by him and Fourier analysis. It is quite clear that the results presented in this chapter and in what follows may also be considered to illustrate the intimate interrelation between fractal geometry and the theory of function spaces which, in turn, is based on Fourier analytical principles, see Definition 10.3. Fourier analytical aspects with an emphasis on self-similarity, related fractal measures, and L_2-theory have been studied on a large scale by R. S. Strichartz, we refer to [Str90a], [Str90b], [Str91], [Str93a], [Str93b] and, in particular, to the survey [Str94]. Further references concerning the connection between Fourier analysis and fractal geometry may be found in [Mat95], p. 168–169. We modify slightly the approach given in [Mat95], pp. 110, 168 and show how closely the related notations are connected with what we have done so far. Let

$$A_0 = \{\xi \in \mathbb{R}^n : |\xi| \le 2\} \qquad (17.40)$$

and

$$A_j = \left\{\xi \in \mathbb{R}^n : 2^{j-1} \le |\xi| \le 2^{j+1}\right\}, \quad j \in \mathbb{N}, \qquad (17.41)$$

be the annuli used in the resolution of unity $\{\varphi_k\}_{k \in \mathbb{N}_0}$ in (10.5). Let μ be a Radon measure in $\mathbb{R}^n$ with a compact support and $\mu(\mathbb{R}^n) < \infty$. Let $0 \le d \le n$. *Then*

$$\mu \in B_{2,\infty}^{-\frac{n-d}{2}}(\mathbb{R}^n) \tag{17.42}$$

if, and only if,

$$\|\hat{\mu} \mid L_2(A_0)\| + \sup_{j \in \mathbb{N}} \left(\int_{A_j} |\xi|^d |\hat{\mu}(\xi)|^2 \frac{d\xi}{|\xi|^n} \right)^{\frac{1}{2}} < \infty \tag{17.43}$$

(equivalent norms). The proof follows easily from (10.6) with $p = 2$ and $q = \infty$: We have (17.42) if, and only if,

$$\|\varphi_0 \hat{\mu} \mid L_2(\mathbb{R}^n)\| + \sup_j 2^{-\frac{n-d}{2} j} \|\varphi_j \hat{\mu} \mid L_2(\mathbb{R}^n)\| < \infty. \tag{17.44}$$

But this coincides with (17.43). This observation suggests, in agreement with [Mat95], p. 168, and in analogy with (17.7), to measure the massiveness of sets in $\mathbb{R}^n$ as follows. Let Γ be a non-empty Borel set in $\mathbb{R}^n$ with $|\Gamma| = 0$. Then the *Fourier dimension of Γ is defined as*

$$\dim_F \Gamma = \sup \Big\{ d : \text{ there exists a non-zero Radon measure } \mu$$

$$\text{with a compact support, } supp\,\mu \subset \Gamma, \ \mu(\mathbb{R}^n) < \infty, \tag{17.45}$$

$$\text{and } |\hat{\mu}(\xi)| \le |\xi|^{-\frac{d}{2}} \text{ for all } 0 \ne \xi \in \mathbb{R}^n \Big\}.$$

We claim

$$\dim_F \Gamma \le \dim_D \Gamma = \dim_H \Gamma, \tag{17.46}$$

where the latter assertion is covered by Theorem 17.6. To prove the first inequality in (17.46) we assume that μ has the above property with $d < \dim_F \Gamma$ and $\Lambda = supp\,\mu$ compact, and $\Lambda \subset \Gamma$. By (17.43) we have (17.42) where $\mu \in S'(\mathbb{R}^n)$ is interpreted in analogy to (17.16) and hence

$$\mu \in B_{2,\infty}^{-\frac{n-d}{2}, \Lambda}(\mathbb{R}^n). \tag{17.47}$$

Now (17.46) follows from the Theorems 17.8 and 17.6. The question arises whether $\dim_F \Gamma$ coincides with $\dim_H \Gamma$. The answer is negative. Sets having this property are called *Salem sets*, see [Mat95], p. 168, for details and references.

17.11 Capacities and capacitary dimension

Let μ be a Radon measure in $\mathbb{R}^n$; then for $d \geq 0$

$$I_d(\mu) = \int \int |x - y|^{-d} \, \mu(dx) \, \mu(dy) \qquad (17.48)$$

is called the *d-energy*, a notion which had been studied by Frostman some sixty years ago. It is another way to measure the massiveness of μ. In analogy to (17.45) the *capacitary dimension* of a Borel set Γ with $|\Gamma| = 0$ is defined by

$$\dim_C \Gamma = \sup \Big\{ d : \text{ there exists a non-zero Radon measure } \mu$$

$$\text{with a compact support, } supp\, \mu \subset \Gamma, \ \mu(\mathbb{R}^n) < \infty, \qquad (17.49)$$

$$\text{and } I_d(\mu) < \infty \Big\}.$$

Let Γ be a Borel set in $\mathbb{R}^n$ with $|\Gamma| = 0$ then

$$\dim_C \Gamma = \dim_H \Gamma. \qquad (17.50)$$

This is one of the standard facts in fractal geometry. We refer to [Mat95], pp. 110–114, or [Fal85], 6.2. The close connection of $I_d(\mu)$ with the Fourier analytical considerations in 17.10 may be found in [Mat95], p. 162. The assertion (17.50) will not be needed in the sequel. We mentioned it for sake of completeness on the one hand. But on the other hand it is closely related with the capacity of a Borel set in terms of function spaces, which, in turn, is directly linked with some of the above arguments: In Step 2 of the proof of Theorem 17.8 we used the duality (17.36) and asked for the minimal norm in $B_{p'p'}^{\frac{n-\delta}{p'}}(\mathbb{R}^n)$ of smooth functions which are identically 1 on Γ. But this is the well-known and extensively studied *capacity of a set* Γ with respect to a given function space. Let $s > 0$, say, $1 \leq p \leq \infty$ and $0 < q \leq \infty$, then the *capacity* $C(\Gamma, B_{pq}^s)$ of a compact set Γ in $\mathbb{R}^n$ is defined by

$$C(\Gamma, B_{pq}^s) = \inf \left(\|\varphi \,|\, B_{pq}^s(\mathbb{R}^n)\| : \varphi \in S(\mathbb{R}^n) \text{ real}, \varphi \geq 1 \text{ on } \Gamma \right), \qquad (17.51)$$

extending [AdH96], p. 105, see also [Maz85], 7.2.1, p. 351. But now it is quite clear that we asked in Step 2 of the proof of Theorem 17.8 whether Γ has capacity zero with respect to $B_{p'p'}^{\frac{n-\delta}{p'}}(\mathbb{R}^n)$. In other words, assertions of zero-capacities of compact sets with respect to suitable function spaces are dual to the way described above. The study of capacities of compact and more general sets in $\mathbb{R}^n$ in connection with function spaces attracted much attention. As for Sobolev-Besov spaces we refer to [Maz85], see also [AdH96] as far as the extensive work of V. G. Mazja in this connection is concerned. In [HKM93], p. 43, results about H_p^1-capacities, $1 < p \leq n$, are obtained, which, after suitable reformulations, are near to the above assertions. As we mentioned the nearest to us is [AdH96], in particular Section 4.3 and Ch. 5. We do not go into detail, references may be found in the Note-sections 4.9 and 5.7 in [AdH96], pp. 125–127 and 152–153. Far-reaching refinements both of capacities in B_{pq}^s and F_{pq}^s spaces and also for Hausdorff measures and Hausdorff dimensions have been obtained recently by Yu. V. Netrusov in [Net88], [Net90], [Net92a] and [Net92b].

18 The spaces $L_p(\Gamma)$

18.1 Introduction
Let $x = (x', x_n) \in \mathbb{R}^n$ with $x' \in \mathbb{R}^{n-1}$. As usual

$$tr_{\mathbb{R}^{n-1}} : f(x) \mapsto f(x', 0) \tag{18.1}$$

is called the trace of f on $\mathbb{R}^{n-1}$. The following assertions are well-known:

(i) *Let $0 < p \le \infty$, $0 < q \le \infty$, and $s - \frac{1}{p} > (n-1)(\frac{1}{p} - 1)_+$; then*

$$tr_{\mathbb{R}^{n-1}} B_{pq}^s(\mathbb{R}^n) = B_{pq}^{s-\frac{1}{p}}(\mathbb{R}^{n-1}). \tag{18.2}$$

(ii) *Let $0 < p < \infty$, $0 < q \le \infty$, and $s - \frac{1}{p} > (n-1)(\frac{1}{p} - 1)_+$; then*

$$tr_{\mathbb{R}^{n-1}} F_{pq}^s(\mathbb{R}^n) = B_{pp}^{s-\frac{1}{p}}(\mathbb{R}^{n-1}). \tag{18.3}$$

We refer to [Tri92], 4.4.1 and 4.4.2, pp. 212–220, where we listed also the necessary references. Of interest is the limiting case $s = \frac{1}{p}$. We discussed this in [Tri92], 4.4.3, pp. 220–221, as a curiosity. In particular, if $p < 1$, then $L_p(\mathbb{R}^{n-1})$ cannot be interpreted as a subspace of $S'(\mathbb{R}^{n-1})$. But nevertheless one has the following complements of (i) and (ii):

(iii) *Let $n \ge 2$, $0 < p < \infty$, and $0 < q \le \min(1, p)$. Then*

$$tr_{\mathbb{R}^{n-1}} B_{pq}^{\frac{1}{p}}(\mathbb{R}^n) = L_p(\mathbb{R}^{n-1}). \tag{18.4}$$

(iv) *Let $n \ge 2$, $0 < p \le 1$, and $0 < q \le \infty$. Then*

$$tr_{\mathbb{R}^{n-1}} F_{pq}^{\frac{1}{p}}(\mathbb{R}^n) = L_p(\mathbb{R}^{n-1}). \tag{18.5}$$

Comments and references to the related papers by J. Peetre, M. L. Gol'dman, V. I. Burenkov, M. Frazier and B. Jawerth may be found in [Tri92], 4.4.3, pp. 220–221. It is one aim of Section 18 to extend (18.4) and (18.5) from $\mathbb{R}^{n-1}$ to fractals Γ in $\mathbb{R}^n$. In case of d-sets according to Definition 3.1 we obtain in 18.6 and 18.12 rather final answers. This brings us in the position to study later on embeddings of type (18.2), (18.3) with a compact d-set Γ in place of $\mathbb{R}^{n-1}$. However instead of trying to introduce intrinsically spaces of B_{pq}^s-type on Γ we define corresponding spaces via traces and study their properties. This task will be shifted to Section 20. The present section deals with a second problem related to spaces $L_p(\Gamma)$ where Γ is a d-set according to Definition 3.1 or a compact anisotropic or nonisotropic d-set according to 5.2 and 5.3. Any $f^\Gamma \in L_p(\Gamma)$ with $1 \le p \le \infty$ can be interpreted as a tempered distribution $f \in S'(\mathbb{R}^n)$ given by

$$f(\varphi) = \int_\Gamma f^\Gamma(\gamma)\,(\varphi|\Gamma)(\gamma)\,\mu(d\gamma), \quad \varphi \in S(\mathbb{R}^n), \tag{18.6}$$

where $\varphi|\Gamma$ is the pointwise trace of φ on Γ as we used it in 17.2, 17.3, and μ is the respective measure on Γ. It is our first aim in this section to identify $L_p(\Gamma)$ via the interpretation (18.6) with some spaces $B_{pq}^{s,\Gamma}(\mathbb{R}^n)$ introduced in Definition 17.2. Recall that, as usual, $\frac{1}{p} + \frac{1}{p'} = 1$, where $1 \le p \le \infty$. First we consider the (isotropic) d-sets. The anisotropic and the nonisotropic cases will be treated later on in 18.15 and 18.17.

18.2 Theorem *Let Γ be a d-set in $\mathbb{R}^n$ according to Definition 3.1 with $0 < d < n$. Let $1 < p \le \infty$; then*

$$L_p(\Gamma) = B_{p,\infty}^{-\frac{n-d}{p'},\Gamma}(\mathbb{R}^n). \tag{18.7}$$

Proof.

Step 1. Let $f^\Gamma \in L_p(\Gamma)$ and let f be given by (18.6). We prove $f \in B_{p,\infty}^{-\frac{n-d}{p'}}(\mathbb{R}^n)$ and

$$\|f \mid B_{p,\infty}^{-\frac{n-d}{p'}}(\mathbb{R}^n)\| \le c\,\|f^\Gamma \mid L_p(\Gamma)\| \tag{18.8}$$

for some $c > 0$ which is independent of f^Γ. We wish to use (11.10) and (11.9). Let k be a suitable kernel then we modify (17.32) and obtain

$$|k(t,f)(x)| \le c\,t^{-n+\frac{d}{p'}} \left(\int_\Gamma |f^\Gamma(\gamma)|^p \left| k\left(\frac{\gamma - x}{t}\right) \right| \mu(d\gamma) \right)^{\frac{1}{p}} \tag{18.9}$$

and

$$\|k(t,f) \mid L_p(\mathbb{R}^n)\| \le c\,t^{-n+\frac{d}{p'}+\frac{n}{p}} \|f^\Gamma \mid L_p(\Gamma)\|. \tag{18.10}$$

By (11.10) and (11.9) we have (18.8), and by (18.6) also

$$f \in B_{p,\infty}^{-\frac{n-d}{p'},\Gamma}(\mathbb{R}^n).$$

Step 2. To prove the converse we need a preparation. By Theorem 3.8 we know that $D|\Gamma$ is dense in $L_{p'}(\Gamma)$. Let $\varphi \in D(\mathbb{R}^n)$ and let $\{\varphi_j\}_{j=1}^N$ be the smooth resolution

of unity in a neighbourhood Λ of $\Gamma \cap supp\, \varphi$ used in (17.25). We assume now that
the involved balls B_j have the same radius r. Let $\lambda_j = \max_{x \in B_j} |\varphi(x)|$. Then we have

$$
\begin{aligned}
\varphi(x) &= \sum_{j=1}^{N} \varphi(x)\varphi_j(x) \\
&= \sum_{j=1}^{N} \lambda_j r^{\frac{d}{p'}} \left[r^{-\frac{d}{p'}} \lambda_j^{-1} \varphi(x)\, \varphi_j(x) \right], \quad x \in \Lambda.
\end{aligned}
\tag{18.11}
$$

Of course, terms with $\lambda_j = 0$ are simply omitted. Suppose that $\varphi|\Gamma$ does not vanish
identically on Γ. In dependence on such a given φ we choose $r > 0$ small. Hence
also $\{\varphi_j\}$ is chosen in dependence on φ. Then we may assume by Definition 13.3
that the functions in brackets are $(\frac{n-d}{p'}, p')$ atoms, and by Theorem 13.8 we have

$$
\left\| \varphi \sum \varphi_j \mid B_{p',1}^{\frac{n-d}{p'}}(\mathbb{R}^n) \right\| \le c \left(\sum_{j=1}^{N} |\lambda_j|^{p'} r^d \right)^{\frac{1}{p'}} \le c' \, \|\varphi \mid L_{p'}(\Gamma)\|, \tag{18.12}
$$

where the last inequality comes from the assumption that Γ is a d-set according
to 3.1 and that $r > 0$ can be chosen arbitrarily small. Now we assume $f \in B_{p,\infty}^{-\frac{n-d}{p'},\Gamma}(\mathbb{R}^n)$. Then we have for all $\varphi \in D(\mathbb{R}^n)$

$$
\begin{aligned}
|f(\varphi)| &= |f(\varphi \sum \varphi_j)| \\
&\le \left\| f \mid B_{p,\infty}^{-\frac{n-d}{p'}}(\mathbb{R}^n) \right\| \left\| \varphi \sum \varphi_j \mid B_{p',1}^{\frac{n-d}{p'}}(\mathbb{R}^n) \right\| \\
&\le c \left\| f \mid B_{p,\infty}^{-\frac{n-d}{p'}}(\mathbb{R}^n) \right\| \, \|\varphi \mid L_{p'}(\Gamma)\|,
\end{aligned}
\tag{18.13}
$$

where we used the duality assertion

$$
\left(B_{p',1}^{\frac{n-d}{p'}}(\mathbb{R}^n) \right)' = B_{p,\infty}^{-\frac{n-d}{p'}}(\mathbb{R}^n), \tag{18.14}
$$

see [Tri83], 2.11.2, p. 178. The last estimate in (18.13) comes from (18.12). Since

$$
f(\varphi) = f(\psi) \quad \text{if} \quad \varphi \in D(\mathbb{R}^n), \psi \in D(\mathbb{R}^n) \quad \text{and} \quad \varphi|\Gamma = \psi|\Gamma, \tag{18.15}
$$

it follows that $f(\varphi)$ is a linear functional on $D|\Gamma$. On the other hand, by Theorem
3.8 this set is dense in $L_{p'}(\Gamma)$. Hence by (18.13) there is a uniquely determined
function $f^\Gamma \in L_p(\Gamma)$ such that

$$
f(\varphi) = \int_\Gamma f^\Gamma(\gamma)\, (\varphi|\Gamma)\, \mu(d\gamma) \tag{18.16}
$$

with

$$\|f^\Gamma \mid L_p(\Gamma)\| \le c \left\| f \mid B_{p',\infty}^{-\frac{n-d}{p'}} (\mathbb{R}^n) \right\|.$$ (18.17)

The proof is complete.

18.3 Remark This theorem coincides with Theorem 2 in [TrW96b]. In this paper we gave a slightly different but more detailed proof, especially as far as the somewhat tricky arguments in Step 2 of the above proof are concerned. The question arises whether (18.7) can be extended to $p = 1$. But this is not so. Let $0 \in \Gamma$ and let δ be the δ-distribution. Then we have the desirable assertions $\delta(\varphi) = \varphi(0)$ and $\delta \in B_{1,\infty}^{0,\Gamma}(\mathbb{R}^n)$, but $\delta \notin L_1(\Gamma)$ in the interpretation (18.16), since $d > 0$. On the other hand, Step 1 of the above proof can be extended to $p = 1$. Hence we have

$$L_1(\Gamma) \subset B_{1,\infty}^{0,\Gamma}(\mathbb{R}^n).$$ (18.18)

This assertion can be generalized. Let $M(\mathbb{R}^n)$ be the space of all complex-valued finite Radon measures in $\mathbb{R}^n$ normed in an obvious way. Then we have

$$M(\mathbb{R}^n) \subset B_{1,\infty}^{0}(\mathbb{R}^n).$$ (18.19)

This is an easy consequence of (11.10) and (11.8).

18.4 Remark It is by no means surprising that spaces of type

$$B_{pq}^{-\frac{n-d}{p'},\Gamma}(\mathbb{R}^n)$$

play a similar role both in Theorem 18.2 and in Theorem 17.8. It underlines the close connection between spaces of this type and fractal geometry, at least as far as d-sets are concerned. In case of $p = 2$ this is related to fractal Plancherel theorems as discussed in [Str90a] and [Str94]. In these papers are also some assertions for $1 < p \le \infty$, connected with convolutions by approximate identities, which result in estimates of type (18.10).

18.5 Traces
As mentioned in the Introduction 18.1 traces of spaces $B_{pq}^s(\mathbb{R}^n)$ or $F_{pq}^s(\mathbb{R}^n)$ on a fractal Γ on the one hand and interpretations of L_p-spaces on Γ as spaces on $\mathbb{R}^n$ as we did it in Theorem 18.2 on the other hand are two sides of the same coin. We are looking now for counterparts of (18.4) and (18.5). Let Γ be a closed set in $\mathbb{R}^n$, say, with $|\Gamma| = 0$. Assume that there is a Radon measure μ in $\mathbb{R}^n$ with $supp\, \mu = \Gamma$, and let $L_p(\Gamma)$ be the related L_p-spaces on Γ. Then $tr_\Gamma \varphi = \varphi|\Gamma$ makes sense pointwise for any $\varphi \in S(\mathbb{R}^n)$. Suppose that for some $s > 0, 0 < p < \infty, 0 < q < \infty$ there is a constant $c > 0$ such that

$$\| tr_\Gamma \varphi \mid L_p(\Gamma)\| \le c \,\|\varphi \mid B_{pq}^s(\mathbb{R}^n)\|, \quad \varphi \in S(\mathbb{R}^n).$$ (18.20)

Recall that $S(\mathbb{R}^n)$ is dense in $B^s_{pq}(\mathbb{R}^n)$, see [Tri83], p. 48. Then (18.20) can be extended by completion to any $f \in B^s_{pq}(\mathbb{R}^n)$ and the resulting function on Γ is denoted by $tr_\Gamma f$. By standard arguments it is independent of the approximation of f in $B^s_{pq}(\mathbb{R}^n)$ by $S(\mathbb{R}^n)$-functions. In addition, the equality

$$tr_\Gamma B^s_{pq}(\mathbb{R}^n) = L_p(\Gamma) \tag{18.21}$$

must be understood that any $f^\Gamma \in L_p(\Gamma)$ is the trace in the above sense of a suitable $g \in B^s_{pq}(\mathbb{R}^n)$ on Γ and $\|f^\Gamma \,|\, L_p(\Gamma)\|$ is equivalent to

$$\inf \left\{ \|g \,|\, B^s_{pq}(\mathbb{R}^n)\| : \; tr_\Gamma g = f^\Gamma \right\}. \tag{18.22}$$

In case of $\Gamma = \mathbb{R}^{n-1}$ one might consider (18.4) and (18.5) as examples of these interpretations. We mention explicitly that (18.21) does not mean that there is a linear extension operator from $L_p(\Gamma)$ into $B^s_{pq}(\mathbb{R}^n)$. This is not so, even not for $\Gamma = \mathbb{R}^{n-1}$, which means in the case of (18.4), see [Tri83], Remark 5 on p. 139, or [Tri92], p. 220, and the references given there. However this is also not necessary for us later on since compact embeddings between function spaces will be considered in terms of entropy numbers which are non-linear by nature.

18.6 Theorem *Let Γ be a d-set according to Definition 3.1 with $0 < d < n$. Let $\frac{d}{n} < p < \infty$ and $0 < q \leq \min(1,p)$. Then*

$$tr_\Gamma B^{\frac{n-d}{p}}_{pq} (\mathbb{R}^n) = L_p(\Gamma) \tag{18.23}$$

with the interpretation given in 18.5.

Proof.
Step 1. Let $0 < p < \infty$ and let $0 < q \leq \min(1,p)$. We prove

$$tr_\Gamma B^{\frac{n-d}{p}}_{pq} (\mathbb{R}^n) \subset L_p(\Gamma). \tag{18.24}$$

We use an optimal atomic decomposition of $\varphi \in S(\mathbb{R}^n)$ in $B^{\frac{n-d}{p}}_{pq} (\mathbb{R}^n)$ according to Theorem 13.8 and obtain

$$\varphi = \sum_{\nu=0}^{\infty} \sum_{m \in \mathbb{Z}^n} \lambda_{\nu m} \, a_{\nu m}(x) \tag{18.25}$$

where the sequence

$$\lambda = \{\lambda_{\nu m} \in \mathbb{C} : \nu \in \mathbb{N}_0, \; m \in \mathbb{Z}^n\}$$

and the $(\frac{n-d}{p}, p)$-atoms $a_{\nu m}(x)$ have the same meaning as in 13.8. In particular
we have

$$\|\lambda \,|\, b_{pq}\| \leq c \,\left\|\varphi \,|\, B_{pq}^{\frac{n-d}{p}} (\mathbb{R}^n)\right\|, \qquad (18.26)$$

where $c > 0$ is independent of φ and

$$|a_{\nu m}(x)| \leq 2^{-\nu(\frac{n-d}{p} - \frac{n}{p})} = 2^{\frac{d\nu}{p}}, \quad \nu \in \mathbb{N}_0, \ m \in \mathbb{Z}^n. \qquad (18.27)$$

Let $0 < p \leq 1$. By the support property (13.21) of $a_{\nu m}(x)$ and (3.1) it follows that

$$\|tr_\Gamma \,\varphi \,|\, L_p(\Gamma)\|^p \leq \sum_{\nu=0}^{\infty} \int_\Gamma \left| \sum_{m \in \mathbb{Z}^n} \lambda_{\nu m} \, a_{\nu m}(\gamma) \right|^p \mu(d\gamma)$$

$$\leq c \sum_{\nu=0}^{\infty} \sum_{m \in \mathbb{Z}^n} |\lambda_{\nu m}|^p \, 2^{d\nu} \, 2^{-d\nu} \qquad (18.28)$$

$$\leq c \, \|\lambda \,|\, b_{pq}\|^p,$$

where we used $q \leq p$. Together with (18.26) we obtain (18.24) by completion. If
$p > 1$, then the first inequality in (18.28) must be replaced by the usual triangle
inequality, and one has to use $q \leq 1$.

Step 2. Let $\frac{d}{n} < p < \infty$ and $0 < q \leq \min(1, p)$. We prove

$$L_p(\Gamma) \subset tr_\Gamma \, B_{pq}^{\frac{n-d}{p}} (\mathbb{R}^n). \qquad (18.29)$$

We use the same arguments as in Step 2 of the proof of Theorem 18.2. As there
we assume that $\varphi \in D(\mathbb{R}^n)$ does not vanish identically on Γ. With a suitable
choice of $\{\varphi_j\}$ (in dependence of φ) we obtain in the same way as in (18.12)

$$\left\|\varphi \sum \varphi_j \,|\, B_{pq}^{\frac{n-d}{p}} (\mathbb{R}^n)\right\| \leq c \, \|tr_\Gamma \,\varphi \,|\, L_p(\Gamma)\|, \qquad (18.30)$$

where $c > 0$ is independent of φ. To justify the application of Theorem 13.8 we
remark that

$$\frac{n - d}{p} > n(\frac{1}{p} - 1)_+, \qquad (18.31)$$

since $p > \frac{d}{n}$, and hence by (13.29) no moment conditions for the related atoms
are needed. By Theorem 3.8 we know that $D|\Gamma$ is dense in $L_p(\Gamma)$. Hence any
$f \in L_p(\Gamma)$ can be represented as

$$f(\gamma) = \sum_{j=1}^{\infty} f_j(\gamma), \quad \gamma \in \Gamma, \quad f_j \in D(\mathbb{R}^n), \qquad (18.32)$$

with

$$0 < \|tr_\Gamma \, f_j \,|\, L_p(\Gamma)\| \leq c \, 2^{-j} \, \|f \,|\, L_p(\Gamma)\|, \quad j \in \mathbb{N}, \tag{18.33}$$

where $c > 0$ is independent of j and f. If necessary we modify $f_j \in D(\mathbb{R}^n)$ in (18.32) such that we can apply (18.30) to each f_j separately. We have by (18.32), (18.33), and by 18.5

$$ext \, f = \sum_{j=1}^{\infty} f_j \in B_{pq}^{\frac{n-d}{p}} (\mathbb{R}^n), \quad tr_\Gamma \, ext \, f = f \tag{18.34}$$

and

$$\left\| ext \, f \,|\, B_{pq}^{\frac{n-d}{p}} (\mathbb{R}^n) \right\| \leq c \, \|f \,|\, L_p(\Gamma)\|. \tag{18.35}$$

But (18.34) and (18.35) prove (18.29).

18.7 Comment With $d = n - 1$ we obtain (18.4) as a special case of (18.23) at least if $p > \frac{n-1}{n}$. By Step 1 we have (18.24) for all $0 < p < \infty$. The restriction $p > \frac{d}{n}$ comes from Step 2. There we needed (18.31) to avoid moment conditions for the respective atoms according to (13.29). We can circumvent this problem in case of $\Gamma = \mathbb{R}^{n-1}$. We will do this in a more general context in 18.12.

18.8 Duality
The two Theorems 18.2 and 18.6 are dual to each other. To make clear what is meant we give a new proof of (18.24) restricted to $1 < p < \infty$ and $q \leq 1$ by dualizing (18.7). Let $\varphi \in S(\mathbb{R}^n)$ then we obtain by the $L_p(\Gamma) - L_{p'}(\Gamma)$ duality

$$\|tr_\Gamma \, \varphi \,|\, L_p(\Gamma)\| = \sup \left| \int_\Gamma (tr_\Gamma \, \varphi)(\gamma) \, f^\Gamma(\gamma) \, \mu(d\gamma) \right|, \tag{18.36}$$

where the supremum is taken over all

$$f^\Gamma \in L_{p'}(\Gamma) \quad \text{with} \quad \|f^\Gamma \,|\, L_{p'}(\Gamma)\| \leq 1.$$

By (18.6) and (18.7) it follows

$$\| tr_\Gamma \, \varphi \,|\, L_p(\Gamma)\| \leq c \, \sup |f(\varphi)| \tag{18.37}$$

where now the supremum is taken over all

$$f \in B_{p',\infty}^{-\frac{n-d}{p},\,\Gamma} (\mathbb{R}^n) \quad \text{with} \quad \left\| f \,|\, B_{p',\infty}^{-\frac{n-d}{p}} (\mathbb{R}^n) \right\| \leq c'.$$

By the duality assertion in [Tri83], 2.11.2, p. 178, and elementary embedding as far as $q \leq 1$ is concerned, we obtain

$$\| tr_\Gamma \, \varphi \,|\, L_p(\Gamma)\| \leq c \, \left\| \varphi \,|\, B_{p,q}^{\frac{n-d}{p}} (\mathbb{R}^n) \right\|, \quad \varphi \in S(\mathbb{R}^n), \tag{18.38}$$

what is just what we want. We again followed [TrW96b].

18.9 Independence from dimensions

Again let Γ be a d-set in $\mathbb{R}^n$ with $0 < d < n$. Let $m \in \mathbb{N}$ with $m > n$ and let $\mathbb{R}^n \subset \mathbb{R}^m$ be interpreted as a hyperplane in the usual way. Since $\Gamma \subset \mathbb{R}^n$ we have also $\Gamma \subset \mathbb{R}^m$ and (18.23) can be complemented by

$$tr_\Gamma \; B_{pq}^{\frac{m-d}{p}} (\mathbb{R}^m) = L_p(\Gamma), \tag{18.39}$$

now even for $\frac{d}{m} < p < \infty$ (and $0 < q \leq \min(1, p)$). By (18.31) and its counterparts with $n + 1, \dots, m$ and (18.2) it follows

$$tr_{\mathbb{R}^n} \; B_{pq}^{\frac{m-d}{p}} (\mathbb{R}^m) = B_{pq}^{\frac{n-d}{p}} (\mathbb{R}^n) \tag{18.40}$$

which again sheds light on (18.23) and (18.39). As we shall see in 18.10 the d-set $\Gamma \subset \mathbb{R}^n$ considered in $\mathbb{R}^{n+1}$ satisfies the so-called *ball condition*. This allows not only an extension of $p > \frac{d}{n}$ to $p > \frac{d}{n+1}$ but to all $0 < p < \infty$.

18.10 Definition (Ball condition) *Let Γ be a non-empty Borel set in $\mathbb{R}^n$ with $|\Gamma| = 0$. Then Γ satisfies the ball condition if there is a number $0 < \eta < 1$ with the following property: For any ball $B(x, r)$ centred at $x \in \Gamma$ and of radius $0 < r < 1$ there is a ball $B(y, \eta r)$ centred at some $y \in \mathbb{R}^n$, depending on x, and of radius ηr, such that*

$$B(y, \eta r) \subset B(x, r) \quad and \quad B(y, \eta r) \cap \overline{\Gamma} = \emptyset. \tag{18.41}$$

18.11 Remark This is a slightly modified version of Definition 4 in [TrW96b]. It is quite clear that conditions of this type are related to the *open set condition* introduced in Definition 4.5. One may assume in addition

$$dist(B(y, \eta r), \overline{\Gamma}) \geq \eta r. \tag{18.42}$$

This follows from (18.41) where one replaces, if necessary, η by $\frac{\eta}{2}$. Conditions of the above type have been used in the literature in connection with non-smooth boundaries of bounded domains, see [TrW96a], 3.2, and [ET96], 2.5, pp. 58–59, and the references given there.

18.12 Corollary *Let Γ be a d-set in $\mathbb{R}^n$ according to Definition 3.1 with $0 < d < n$ satisfying the ball condition introduced in 18.10.*

(i) *Let $0 < p < \infty$ and $0 < q \leq \min(1, p)$. Then*

$$tr_\Gamma \; B_{pq}^{\frac{n-d}{p}} (\mathbb{R}^n) = L_p(\Gamma) \tag{18.43}$$

with the interpretation given in 18.5.

(ii) *Let $0 < p \leq 1$ and $0 < q \leq \infty$. Then*

$$tr_\Gamma \; F_{pq}^{\frac{n-d}{p}} (\mathbb{R}^n) = L_p(\Gamma) \tag{18.44}$$

with the interpretation given in 18.5 (complemented in the proof below in case of $q = \infty$).

Proof. (Outline)

Step 1. We prove (i). By Theorem 18.6 we may assume $0 < p \le \frac{d}{n}$. By Step 1 of the proof of Theorem 18.6 we have (18.24). It remains to prove (18.29) now under the assumption $0 < q \le p \le \frac{d}{n}$. Unfortunately (18.31) is no longer valid and hence by Theorem 13.8, in particular by (13.29), moment conditions for the respective $(\frac{n-d}{p}, p)$-atoms are needed which are not necessarily guaranteed for $\varphi\,\varphi_j$ in (18.30). This is the point where the ball condition comes in. Assume as in the left-hand sides of (18.30) and of the appropriately modified formula (18.12) that $\varphi \in D(\mathbb{R}^n)$ does not vanish identically on Γ, and

$$supp\, \varphi\,\varphi_j \subset B(x^j, r) \quad \text{for some} \quad x^j \in \Gamma, \quad \text{and some} \quad 0 < r < 1; \quad (18.45)$$

and let $B(y^j, \eta r)$ be a ball with the properties (18.41) and (18.42). Then we replace $\varphi\,\varphi_j$ by

$$\psi_j(x) = (\varphi\,\varphi_j)(x) + \chi_j(x) \quad (18.46)$$

such that

$$supp\, \chi_j \subset B(y^j, \eta r), \quad (18.47)$$

and ψ_j is an $(\frac{n-d}{p}, p)$-atom satisfying the necessary moment conditions

$$\int x^\beta \psi_j(x)\, dx = 0, \quad |\beta| \le L, \quad (18.48)$$

where L is given by (13.29) with $s = \frac{n-d}{p}$. The somewhat tricky explicit construction of χ_j and ψ_j in a similar situation may be found in [TrW96a], proof of Theorem 3.6. The atoms ψ_j and $\varphi\,\varphi_j$ coincide in a neighbourhood of Γ. Now we have the counterpart of (18.30) and we arrive at (18.29). The proof of (i) is complete.

Step 2. We prove (ii). Assume that $tr_\Gamma\, F_{pq}^{\frac{n-d}{p}}(\mathbb{R}^n)$ is independent of q. Then (18.44), including the equivalence of the two involved quasi-norms, follows from (18.43) with $p = q \le 1$ since $B_{pp}^s(\mathbb{R}^n) = F_{pp}^s(\mathbb{R}^n)$. To prove the independence of the trace on Γ of q we first remark that

$$tr_\Gamma\, F_{p,q_1}^{\frac{n-d}{p}}(\mathbb{R}^n) \subset tr_\Gamma\, F_{p,q_2}^{\frac{n-d}{p}}(\mathbb{R}^n) \quad \text{if} \quad 0 < q_1 \le q_2 \le \infty \quad (18.49)$$

follows simply by the monotonicity of the two spaces. To show the reverse inclusion we choose an optimal atomic decomposition for $f \in F_{p,q_2}^{\frac{n-d}{p}}(\mathbb{R}^n)$ according to Theorem 13.8,

$$f = \sum_{\nu=0}^{\infty} \sum_{m \in \mathbb{Z}^n} \lambda_{\nu m}\, a_{\nu m}(x) \quad (18.50)$$

with

$$\|\lambda \mid f_{pq_2}\| \sim \left\| f \mid F_{p,q_2}^{\frac{n-d}{p}}(\mathbb{R}^n) \right\|. \quad (18.51)$$

Let

$$f_\Gamma = \sum_{\nu=0}^{\infty} \sum_{m \in \mathbb{Z}^n}{}^\Gamma \lambda_{\nu m} \, a_{\nu m}(x), \qquad (18.52)$$

where the summation in (18.52) is restricted to those m, in dependence on ν, such that, say,

$$2 \, supp \, a_{\nu m} \cap \Gamma \neq \emptyset. \qquad (18.53)$$

Of course, f and f_Γ have the same trace on Γ. Furthermore we may assume that the atoms $a_{\nu m}(x)$ have sufficiently many moment conditions according to (13.32) not only with q_2 but also with q_1 in place of q. Using again the ball condition it follows that even $f_\Gamma \in F_{p,q_1}^{\frac{n-d}{p}}(\mathbb{R}^n)$. This follows from the construction detailed in [TrW96a], Step 2 of the proof of Theorem 3.5, and outlined in [TrW96b], Step 3 in 4.5. Denoting by λ^Γ that sub-sequence of λ collecting those $\lambda_{\nu m}$ which appear in (18.52), then it turns out that

$$\|\lambda^\Gamma \, | \, f_{pq_1}\| \sim \|\lambda^\Gamma \, | \, f_{pq_2}\|. \qquad (18.54)$$

The (outlined) proof is complete.

18.13 Comment We must admit that the proof of the last corollary, in particular its end, is rather sketchy. But what we mostly need later on is not so much Corollary 18.12 and even not the equality (18.23) but the inclusion (18.24). It ensures the existence of the trace. The assumption $p > \frac{d}{n}$ in Theorem 18.6 is not very restrictive. If Γ is a d-set in $\mathbb{R}^n$ and if $0 < p < \infty$ is given then we choose a natural number $m \geq n$ such that $\frac{d}{m} < p$. Afterwards by 18.9 we may replace $\Gamma \subset \mathbb{R}^n$ by $\Gamma \subset \mathbb{R}^m$. Then Theorem 18.6 can be applied and hence $L_p(\Gamma)$ is a trace spaces under any circumstances.

18.14 Anisotropic and nonisotropic fractals
In the remaining subsections in this Section 18 we ask for the counterparts of the Theorems 18.2 and 18.6 for anisotropic and nonisotropic d-sets. We introduced these fractals in *qualitative terms* in Section 5, in particular in Definitions 5.2 and 5.3, respectively, which by Proposition 5.6 generalize the *constructive* PXT-sets and OF-sets we dealt with in 4.17 and 4.19. The philosophy behind the distinction between *anisotropic* and *nonisotropic* had been discussed in 5.9. As far as fractals are concerned the notation used has the same meaning as in Section 5. The spaces $L_p(\Gamma)$ on the anisotropic or nonisotropic d-set Γ are taken with respect to the Radon measure μ according to Theorem 5.5. As in Theorem 18.2, if $1 \leq p \leq \infty$ then $f^\Gamma \in L_p(\Gamma)$ generates by (18.6) a tempered distribution f. The inclusion in the following theorem must be understood in that way. Finally how to understand traces, in particular the trace operator tr_Γ, had been discussed in 18.5.

18.15 Theorem *Let Γ be a compact anisotropic d-set according to Definition 5.2 with the (anisotropic) deviation $0 \le a \le 1$. Let $1 \le p \le \infty$; then*

$$L_p(\Gamma) \subset B_{p,\infty}^{-\frac{1}{p'}(2-\frac{d}{1+a})}(\mathbb{R}^2) \tag{18.55}$$

and

$$tr_\Gamma \, B_{p,1}^{\frac{1}{p}(2-\frac{d}{1+a})}(\mathbb{R}^2) \subset L_p(\Gamma). \tag{18.56}$$

Proof.
Step 1. We prove (18.55). We use the same arguments as in (17.32) and in Step 1 of the proof of Theorem 18.2 where again $\frac{1}{p} + \frac{1}{p'} = 1$. Let $Q(x,t)$ be a square centred at $x \in \mathbb{R}^2$, with sides parallel to the axes of coordinates and with side-length $t = 2^{-(1+a)j}$ for some $j \in \mathbb{N}$. By (5.7) and (5.10) it follows that $Q(x,t)$ has a non-empty intersection with at most 4 rectangles R_1^j. By (5.17) and (5.4) we obtain

$$\mu(Q(x,t)) \le 4\,\mu(R_1^j) \le c\,2^{-jd} = c\,t^{\frac{d}{1+a}}. \tag{18.57}$$

Now as in (17.32) and (18.9) we have for $f^\Gamma \in L_p(\Gamma)$

$$|k(t,f)(x)| \le c\,t^{-2+\frac{d}{p'(1+a)}} \left(\int_\Gamma |f^\Gamma(\gamma)|^p \left| k\left(\frac{\gamma - x}{t}\right) \right| \mu(d\gamma) \right)^{\frac{1}{p}} \tag{18.58}$$

and

$$\|k(t,f)\,|\,L_p(\mathbb{R}^2)\| \le c\,t^{-\frac{2}{p'}+\frac{d}{p'(1+a)}} \|f^\Gamma\,|\,L_p(\Gamma)\| \tag{18.59}$$

with $0 < t < 1$. The rest is now the same as at the end of Step 1 of the proof of Theorem 18.2 and we obtain (18.55) with an obvious *counterpart* of the *norm inequality* as in *(18.8)*.

Step 2. We prove (18.56). If $p = \infty$ we have

$$B_{\infty,1}^0(\mathbb{R}^2) \subset C(\mathbb{R}^2)$$

and (18.56) is obvious. Here $C(\mathbb{R}^2)$ is the space of all uniformly continuous bounded functions on $\mathbb{R}^2$. Let $p < \infty$. We use the duality procedure described in 18.8. Let $\varphi \in S(\mathbb{R}^2)$, then we have (18.36) with

$$f^\Gamma \in L_{p'}(\Gamma) \quad \text{and} \quad \|f^\Gamma\,|\,L_{p'}(\Gamma)\| \le 1.$$

By (18.55) we obtain

$$\left\| f\,|\,B_{p',\infty}^{-\frac{1}{p}(2-\frac{d}{1+a})}(\mathbb{R}^2) \right\| \le c\,\left\| f^\Gamma\,|\,L_{p'}(\Gamma) \right\| \le c'. \tag{18.60}$$

We have again (18.37) where we take the supremum over all f in, say, the unit ball in $B_{p',\infty}^{-\frac{1}{p}(2-\frac{d}{1+a})}(\mathbb{R}^2)$. By the duality assertion in [Tri83], 2.11.2, p. 178, we obtain

$$\|tr_\Gamma \, \varphi\,|\,L_p(\Gamma)\| \le c\,\left\| \varphi\,|\,B_{p,1}^{\frac{1}{p}(2-\frac{d}{1+a})}(\mathbb{R}^2) \right\|, \quad \varphi \in S(\mathbb{R}^2).$$

Completion yields the desired result.

18.16 Remark If one replaces the deviation a in (5.10) by the more natural deviation a' in (5.15), then one obtains (18.55) and (18.56) with $a' + \varepsilon$ in place of a for any $\varepsilon > 0$. The restriction $p \geq 1$ in connection with (18.56) and also with (18.62) below is not necessary and will be removed in the Corollaries 22.4 and 22.7, respectively.

18.17 Theorem *Let Γ be a compact nonisotropic d-set according to Definition 5.3 with the (nonisotropic) deviation $0 \leq a \leq 1$. Let $1 \leq p \leq \infty$; then*

$$L_p(\Gamma) \subset B_{p,\infty}^{-\frac{1}{p'}(\frac{5}{2}-\frac{d}{1+a})}(\mathbb{R}^2) \tag{18.61}$$

and

$$tr_\Gamma \, B_{p,1}^{\frac{1}{p}(\frac{5}{2}-\frac{d}{1+a})}(\mathbb{R}^2) \subset L_p(\Gamma). \tag{18.62}$$

Proof. We follow the proof of Theorem 18.15. Again let $Q(x,t)$ be a square centred at $x \in \mathbb{R}^2$ with side-length $t = 2^{-(1+a)j}$ for some $j \in \mathbb{N}$. We estimate the number of ellipses E_l^j having a non-empty intersection with $Q(x,t)$. By (5.10) and Fig. 5.1 the length of the minor axis of E_l^j is larger than or equal to t. If $Q(x,t) \cap E_l^j \neq \emptyset$ then it follows by elementary reasoning that

$$vol(2Q(x,t) \cap E_l^j) \geq c\, t^{\frac{5}{2}} \tag{18.63}$$

for some $c > 0$. Hence by (18.63) and (5.7) with E_l^j in place of R_l^j there are at most $c\, t^{-\frac{1}{2}}$ ellipses E_l^j having a non-empty intersection with $Q(x,t)$. In other words, the number 4 in (18.57) must be replaced by $c t^{-\frac{1}{2}}$ and hence, $\frac{d}{1+a}$ in the exponent in (18.57) must be modified by $\frac{d}{1+a} - \frac{1}{2}$. The rest is the same as in the proof of Theorem 18.15 and we obtain (18.55) and (18.56) with $\frac{5}{2}$ in place of the number 2. But this is just what we want.

18.18 Anisotropic versus nonisotropic
We return to our discussion in 5.9. For isotropic d-sets we obtained in the Theorems 18.2 and 18.6 final assertions. For anisotropic d-sets in the plane $\mathbb{R}^2$ we have Theorem 18.15 which seems to be rather satisfactory. At least at first glance it is hard to see how to improve (18.55) and (18.56) without additional assumptions for the fractal Γ. As for nonisotropic fractals in the plane Theorem 18.17 is less convincing. In particular, if $\frac{5}{2} - \frac{d}{1+a}$ is larger than 2, then both (18.61) and (18.62) are trivial since

$$B_{pq}^s(\mathbb{R}^2) \subset C(\mathbb{R}^2) \quad \text{when} \quad 1 \leq p \leq \infty \quad \text{and} \quad s > \frac{2}{p}, \tag{18.64}$$

where again $C(\mathbb{R}^2)$ is the spaces of all uniformly continuous bounded functions on $\mathbb{R}^2$. Here (18.64) follows from (11.16), (11.17) amd (10.16). But it is reasonable that the assertions for the anisotropic case are better than that for the nonisotropic case. For further improvements one needs presumably additional geometric assumptions.

19 A second digression: Distributional cascades and iterated function systems

19.1 Introduction

This section is somewhat outside of our main intentions and will not be used later on in this book. By what we have done so far it is quite clear that some fractals are closely connected with measures, distributions, and related function spaces. This applies both to the diverse isotropic, anisotropic and nonisotropic d-sets introduced in 3.1, 5.2 and 5.3, and to their more constructive counterparts generated via contractive affine maps (in recent literature also called *iterated function systems*) as studied in Section 4. In any case either a characteristic Radon measure was given or constructed, as in Theorem 4.15 in case of nonisotropic fractals Γ in $\mathbb{R}^2$. In Section 18 it turned out that the corresponding spaces $L_p(\Gamma)$ are closely related to some spaces $B_{pq}^s(\mathbb{R}^n)$ with $s < 0$. We continue this type of considerations in the remaining Sections 20–25 of this chapter. In the present section our intention is a little bit different. To get a feeling what type of conditions we needed so far we have a closer look at the self-affine fractals introduced in 4.10. Let A_l be given by (4.11), (4.12); then (4.13) and (4.15) seem to be reasonable, whereas (4.14) is a convenient but severe restriction. But it is just this assumption which was needed in Theorem 4.15 to construct the Radon measure μ by means of the mass distribution procedure. As it turned out in 18.14–18.18 we were even forced to discard some of the lovely fractals covered by 4.15 such as those ones depicted in the Fig.'s 4.2 and 4.3. The main aim of this section is to reveal the intimate and symbiotic relation between more general (self-affine) fractals on the one hand and Fourier analysis, tempered distributions, and related measures on the other hand. However the connection with function spaces is not so clear at this moment.

Let Γ be a, say, compact fractal in $\mathbb{R}^n$ with $|\Gamma| = 0$. Then we introduce

$$S'^{\Gamma}(\mathbb{R}^n) = \{f \in S'(\mathbb{R}^n) : f(\varphi) = 0 \quad \text{if} \quad \varphi \in S(\mathbb{R}^n), \varphi|\Gamma = 0\}. \qquad (19.1)$$

This is the counterpart of (17.1). A discussion was given in 17.3. In particular we have

$$\operatorname{supp} f \subset \Gamma \quad \text{if} \quad f \in S'^{\Gamma}(\mathbb{R}^n) \qquad (19.2)$$

as a necessary, but not sufficient, condition. We wish to construct characterizing distributions $f \in S'^{\Gamma}(\mathbb{R}^n)$ by means of atomic decompositions although there is no hope to do this in the isotropic way. Just on the contrary the atoms involved must be adapted to the fractal Γ and hence they must be nonisotropically distorted. On the other hand since a characterizing distribution $f \in S'^{\Gamma}(\mathbb{R}^n)$ is singular, Theorem 13.8 suggests that some moment conditions might be indispensable. This complicates the procedure. Fortunately Theorem 17.8 and (17.30) in connection with Theorem 13.8 hint at the possibility that first moment conditions might be sufficient. And so it is. Furthermore it turns out that the constructed characterizing distribution $f \in S'^{\Gamma}(\mathbb{R}^n)$ can be identified with a uniquely determined Radon measure μ on Γ such that

$$f(\varphi) = \int_{\Gamma} (\varphi|\Gamma)(\gamma) \mu(d\gamma), \quad \varphi \in S(\mathbb{R}^n). \qquad (19.3)$$

This is the usual interpretation of a finite measure as a distribution which we used several times, see e.g. (18.6).

19.2 The set-up, iterated function systems

We generalize 4.10 slightly. Let Q be the closed unit cube in $\mathbb{R}^n$, that means

$$Q = \{x \in \mathbb{R}^n : 0 \leq x_j \leq 1 \text{ where } j = 1, \ldots, n \text{ and } x = (x_1, \ldots, x_n)\}, \quad (19.4)$$

and let $A = (A_l)_{l=1}^N$ be $N \geq 2$ affine maps of $\mathbb{R}^n$ *onto* itself, hence

$$(A_l x)_j = \sum_{m=1}^n a_{jm}^l x_m + a_j^l \quad \text{where} \quad j = 1, \ldots, n, \quad (19.5)$$

and $x = (x_1, \ldots, x_n)$. We assume that the maps are *contractions* according to 4.1 and that

$$A_l Q \subset Q \quad (19.6)$$

and

$$\sum_{l=1}^N vol(A_l Q) < 1. \quad (19.7)$$

In contrast to 4.10 we do not suppose (4.14). In other words, overlappings of $A_l Q$ are now allowed. On the other hand, (19.6) and (19.7) are rather natural in our context. We add a discussion about these two properties.

Assume that $(A_l)_{l=1}^N$ are arbitrary contractive affine maps of $\mathbb{R}^n$ onto itself. Let $r_l < 1$ be the contraction of A_l according to (4.1). If $r = \max_{l=1,\ldots,N} r_l$ is small enough then there is always a suitable cube in $\mathbb{R}^n$ with the counterpart of (19.6). Hence the assumption that this cube is the unit cube Q in (19.4) is not a restriction. If r is, say, near 1, then we recall that the fractal Γ generated by $\{A_l : 1 \leq l \leq N\}$ can also be generated for any $m \in \mathbb{N}$ by $\{A_{l_1} \cdots A_{l_m} : 1 \leq l_j \leq N\}$. Choosing m sufficiently large then the contraction constants are as small as we want and we have the same situation as before. Hence in any case (19.6) is natural and convenient. The situation for (19.7) is slightly different. Now we assume that we have (19.6), (19.7) with (19.4) for the contractive affine maps (19.5) of $\mathbb{R}^n$ onto itself. As in 4.12 and 4.13 we introduce the uniquely determined number d with $0 < d < n$ and

$$\sum_{l=1}^N vol(A_l Q)^{\frac{d}{n}} = 1. \quad (19.8)$$

There is an obvious counterpart of (4.25) with $\frac{d}{n}$ in place of $\frac{d}{2}$. Let Γ be the resulting fractal and let $r = \max_{l=1,\ldots,N} r_l$ be the above number. We may assume

$\sqrt{n}r < 1$. Then we have by the counterparts of (4.17) and (4.25)

$$|\Gamma| \leq \sum_{l_1,\ldots,l_k}^{1,\ldots,N} vol\left(A_{l_1}\cdots A_{l_k}Q\right)$$

$$\leq \left(\sqrt{n}\,r\right)^{k\left(1-\frac{d}{n}\right)} \sum_{l_1,\ldots,l_k}^{1,\ldots,N} vol\left(A_{l_1}\cdots A_{l_k}Q\right)^{\frac{d}{n}} \tag{19.9}$$

$$= \left(\sqrt{n}r\right)^{k\left(1-\frac{d}{n}\right)}.$$

Hence the Lebesgue measure $|\Gamma| = 0$. We used this property throughout this book as far as fractals are concerned, sometimes it was indispensable, for instance in connection with the distributional dimension introduced in 17.4. Mainly for this reason we stick at this property and hence at (19.7) also in this section.

The creation of beautiful fractals via contractive affine maps (19.5) is nowadays very fashionable, mostly restricted to the plane $\mathbb{R}^2$. The respective key words are *encoding images* and (hyperbolic or contractive) *iterated function systems* (IFS). On that way one generates *ferns* and *grasses*, see [Fal90], p. 131. As for IFS we refer to [Bar88], [Edg90] and [Per93]. In [Per93], 2.3, pp. 16–27, one finds interesting fractals as attractors of explicitly given IFS: the *Sierpinski triangle*, the *twin dragon*, the *maple leaf*, the *black spleenwort fern*, a *gothic cathedral* and other examples. See also [Bar88] for further explicit examples. We compare the affine maps given there with our set-up. The simplest cases are the n-dimensional *Cantor set*, see Fig. 4.1, and the (two-dimensional) Sierpinski triangle, see Fig. 19.1. Both cases fit in our scheme (19.4)–(19.7), where, of course, one replaces in

Fig. 19.1 Sierpinski triangle

the latter case Q by an equilateral triangle. In both cases the related affine maps are (isotropic) dilations combined with suitable translations. By Theorem 4.7 it can be easily seen that we arrive at (isotropic) d-sets with $0 < d < n$ (with $n = 2$

in case of the Sierpinski triangle). The situation changes if one looks at the two generating affine maps of the *twin dragon* on p. 21 in [Per93]. In our notation we have just the case

$$vol(A_1 Q) + vol(A_2 Q) = 1, \qquad (19.10)$$

in contrast to our assumption (19.7). This might well be the case also for other IFS generating interesting fractals. But as we said we stick at (19.7) and (19.8) and $0 < d < n$ mostly for consistency with all the other sections in this book. In the next subsection we construct for any fractal Γ generated by the IFS given by (19.4)–(19.7) a *characterizing distribution* via a *cascade procedure*. Checking this procedure it works also in the limiting situation $d = n$ ($= 2$ in case of the twin dragon). Hence one can construct on that way a *twin dragon distribution* (belonging to $B^0_{1,\infty}(\mathbb{R}^2)$) and a related *twin dragon measure*. Of course, it would be of interest to follow these possibilities in greater detail. There is a second point which should be mentioned. We always assume that A_l are non-degenerate affine maps, that means in particular *vol* $A_l Q > 0$. But apparently it makes sense in the theory of IFS to admit also degenerate affine maps where the determinant of the related matrix in (19.5) is zero. In particular one of the four maps on p. 23 in [Per93] generating the *black spleenwort fern* is of that type. We summarize our intentions and somewhat cryptical discussion:

Given an IFS of non-degenerate affine contractive maps in $\mathbb{R}^n$ with (19.4)–(19.8) we are looking for a designer distribution $f \in S'^{\Gamma}(\mathbb{R}^n)$, see (19.1), characterizing the generated fractal Γ with $|\Gamma| = 0$, see (19.9), and a corresponding Radon measure.

Furthermore it seems to be desirable to extend this procedure to $d = n$ (mostly $n = 2$) and to degenerate affine maps: *twin dragons and other fractal beauties and monsters are waiting for their designer distributions (and optimally adapted designer function spaces ?)*. Fractals created by the above procedure with $d = n$ in (19.8) have in general Hausdorff dimension n. This follows from [Fal90], Theorem 9.12, p. 131.

19.3 Distributional cascades

Let the iterated function system be given by (19.4)–(19.8) and let Γ be the compact fractal generated. We have $|\Gamma| = 0$. Let $0 \leq b(x) \in C^\infty(\mathbb{R}^n)$ with

$$supp \, b \subset Q \quad \text{and} \quad \int_{\mathbb{R}^n} b(x)\, dx = 1. \qquad (19.11)$$

Let $k \in \mathbb{N}_0$ and let $j_1, \dots, j_k$ and l be natural numbers between 1 and N. We construct the *affine wavelets*

$$
\begin{aligned}
b_{j_1 \cdots j_k \, l}(x) = {} & \left(vol \, A_{j_1} \cdots A_{j_k} A_l Q\right)^{-\frac{n-d}{n}} b\left(A_l^{-1} A_{j_k}^{-1} \cdots A_{j_1}^{-1} x\right) \\
& - \left(vol \, A_l Q\right)^{\frac{d}{n}} \left(vol \, A_{j_1} \cdots A_{j_k} Q\right)^{-\frac{n-d}{n}} b\left(A_{j_k}^{-1} \cdots A_{j_1}^{-1} x\right)
\end{aligned}
$$

$$(19.12)$$

with an obvious modification in case of $k = 0$. Since Q is the unit cube it holds

$$vol\, A_l Q = |\det A_l|,$$

where $\det A_l$ is the determinant of the matrix in (19.5), and similarly for the iterated maps in (19.12). We have

$$supp\, b_{j_1 \cdots j_k l} \subset A_{j_1} \cdots A_{j_k} Q \quad \text{for any} \quad l = 1, \ldots, N, \tag{19.13}$$

and

$$\int_{\mathbb{R}^n} b_{j_1 \cdots j_k l}(x)\, dx \tag{19.14}$$
$$= \left(vol\, A_{j_1} \cdots A_{j_k} A_l Q\right)^{\frac{d}{n}} - \left(vol\, A_l Q\right)^{\frac{d}{n}} \left(vol\, A_{j_1} \cdots A_{j_k} Q\right)^{\frac{d}{n}} = 0.$$

Furthermore, by (19.8) it follows that

$$\sum_{l=1}^{N} b_{j_1 \cdots j_k l}(x) = \sum_{l=1}^{N} \left(vol\, A_{j_1} \cdots A_{j_k} A_l Q\right)^{-\frac{n-d}{n}} b\left(A_l^{-1} A_{j_k}^{-1} \cdots A_{j_1}^{-1} x\right)$$
$$- \left(vol\, A_{j_1} \cdots A_{j_k} Q\right)^{-\frac{n-d}{n}} b\left(A_{j_k}^{-1} \cdots A_{j_1}^{-1} x\right) \tag{19.15}$$

and consequently

$$f_m(x) = \sum_{k=1}^{m} \sum_{j_1, \ldots, j_k}^{1, \ldots, N} b_{j_1 \cdots j_k}(x) + b(x)$$
$$= \sum_{j_1, \ldots, j_m}^{1, \ldots, N} \left(vol\, A_{j_1} \cdots A_{j_m} Q\right)^{-\frac{n-d}{n}} b\left(A_{j_m}^{-1} \cdots A_{j_1}^{-1} x\right) \tag{19.16}$$

where $m \in \mathbb{N}$. These properties make clear why we call this procedure a *cascade*: On the one hand we have the same situation as in 4.10; the resulting fractal Γ is the limit

$$\Gamma = (AQ)^{\infty} = \bigcap_{r \in \mathbb{N}} (AQ)^r = \lim_{r \to \infty} (AQ)^r \tag{19.17}$$

of the sequence of monotonically decreasing sets

$$(AQ)^r = \bigcup_{1 \le j_l \le N} A_{j_1} \cdots A_{j_r} Q, \quad r \in \mathbb{N}. \tag{19.18}$$

On the other hand, stepping in (19.15) from level k to level $k + 1$, the involved functions cancel each other and result in the functions $f_m(x)$ in (19.16), following the cascade of the parallelograms $A_{j_1} \cdots A_{j_m} Q$ shrinking to Γ as indicated in Fig. 19.2. If $f_m(x)$ converges for $m \to \infty$ the outcome cannot be a function (regular distribution) since $|\Gamma| = 0$. We prove that f_m converges in $S'(\mathbb{R}^n)$ to a (singular) distribution f belonging to $S'^{\Gamma}(\mathbb{R}^n)$ given by (19.1) and that it can be identified with a Radon measure according to (19.3).

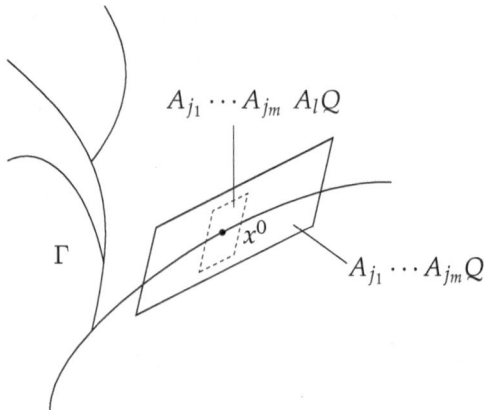

Fig. 19.2

19.4 Theorem *Let $A = (A_l)_{l=1}^N$ be the $N \geq 2$ non-degenerate contractive affine maps (19.5) in $\mathbb{R}^n$, satisfying (19.6) and (19.7). Let d with $0 < d < n$ be given by (19.8). Let Γ be the generated fractal according to (19.17). Let $0 \leq b(x) \in C^\infty(\mathbb{R}^n)$, the atoms $b_{j_1 \cdots j_r}(x)$ and $f_m(x)$ be given by (19.11), (19.12), and (19.16), respectively. Then f_m converges in $S'(\mathbb{R}^n)$ to $f \in S'^\Gamma(\mathbb{R}^n)$ given by (19.1). Furthermore f is independent of $b(x)$ and it can be identified with (a uniquely determined) Radon measure μ according to (19.3) with $\operatorname{supp}\mu = \Gamma$ and $\mu(\Gamma) = 1$.*

Proof.
Step 1. Let $\varphi \in S(\mathbb{R}^n)$ and let $x^{j_1 \cdots j_k}$ be the centres of $A_{j_1} \cdots A_{j_k} Q$. By (19.13) and (19.14) we have

$$\left| \int_{\mathbb{R}^n} b_{j_1 \cdots j_k j_{k+1}}(x)\, \varphi(x)\, dx \right|$$

$$= \left| \int_{\mathbb{R}^n} b_{j_1 \cdots j_{k+1}}(x) \left(\varphi(x) - \varphi(x^{j_1 \cdots j_k}) \right) dx \right| \tag{19.19}$$

$$\leq c \operatorname{vol}\left(A_{j_1} \cdots A_{j_{k+1}} Q \right)^{\frac{d}{n}} \operatorname{diam}(A_{j_1} \cdots A_{j_k} Q)$$

where c depends on the first derivatives of φ. The last factor can be estimated from above by a^k for some $0 < a < 1$, the largest contractivity constant of the maps A_l. Iteration of (19.8) yields

$$\sum_{j_1, \ldots, j_k}^{1, \ldots, N} \operatorname{vol}(A_{j_1} \cdots A_{j_k} Q)^{\frac{d}{n}} = 1. \tag{19.20}$$

Let $m > l$, then we have by (19.16), (19.19) and (19.20)

$$|f_m(\varphi) - f_l(\varphi)| = \left| \int_{\mathbb{R}^n} (f_m(x) - f_l(x))\, \varphi(x)\, dx \right| \leq c \sum_{r=l+1}^{m} a^r.$$

Hence,

$$\lim_{m\to\infty} f_m(\varphi) = f(\varphi) \tag{19.21}$$

for some $f \in S'(\mathbb{R}^n)$.

Step 2. Let $\varphi \in S(\mathbb{R}^n)$ and $\varphi(x) = 1$ near Γ. Then we have by (19.16) and (19.20)

$$
\begin{aligned}
f(\varphi) &= \lim_{m\to\infty} f_m(\varphi) \\
&= \lim_{m\to\infty} \sum_{j_1,\dots,j_m}^{1,\dots,N} (vol\, A_{j_1} \cdots A_{j_m} Q)^{-\frac{n-d}{n}} \int_{\mathbb{R}^n} b(A_{j_m}^{-1} \cdots A_{j_1}^{-1} x)\, dx \\
&= \lim_{m\to\infty} \sum_{j_1,\dots,j_m}^{1,\dots,N} vol(A_{j_1} \cdots A_{j_m} Q)^{\frac{d}{n}} = 1.
\end{aligned}
\tag{19.22}
$$

If one replaces Q in (19.8) by $A_{j_1} \cdots A_{j_m} Q$ then one has (19.8) with $vol(A_{j_1} \cdots A_{j_m} Q)^{\frac{d}{n}}$ in place of 1 on the right-hand side and a corresponding assertion in (19.20). Let $\varphi \in S(\mathbb{R}^n)$ with $\varphi(x) \geq 0$ and, say, $\varphi(x) = 1$ if $x \in A_{j_1} \cdots A_{j_m} Q$, then we obtain by the same arguments as in (19.22)

$$f(\varphi) \geq (vol\, A_{j_1} \cdots A_{j_m} Q)^{\frac{d}{n}} > 0. \tag{19.23}$$

Hence,

$$supp\, f \supset \Gamma. \tag{19.24}$$

Finally, let $\varphi \in S(\mathbb{R}^n)$ and $\varphi|\Gamma = 0$. Then $|\varphi(x)| \leq \varepsilon$ in a sufficiently small neighbourhood of Γ. In modification of (19.22) we obtain by (19.16) and (19.17)

$$|f(\varphi)| \leq \varepsilon \quad \text{for any} \quad \varepsilon > 0. \tag{19.25}$$

Hence, $f \in S'^{\Gamma}(\mathbb{R}^n)$ and

$$supp\, f = \Gamma. \tag{19.26}$$

Step 3. Let $\varphi_1 \in S(\mathbb{R}^n)$ and $\varphi_2 \in S(\mathbb{R}^n)$ with $\varphi_1|\Gamma = \varphi_2|\Gamma$. Then we have

$$f(\varphi_1) = f(\varphi_2). \tag{19.27}$$

Hence f is a positive linear functional on $S|\Gamma$. However this is sufficient to apply the Riesz representation theorem, see [Mat95], p. 15, and some sketchy remarks in the proof of the above Theorem 4.15. Hence by (19.22) there exists a Radon measure μ with

$$supp\, \mu = \Gamma, \quad \mu(\Gamma) = 1, \tag{19.28}$$

and

$$f(\varphi) = \int_{\Gamma} (\varphi|\Gamma)(\gamma)\, \mu(d\gamma), \quad \varphi \in S(\mathbb{R}^n). \tag{19.29}$$

It remains to prove that μ, and consequently also f, is independent of $b(x)$. For that purpose we fix an open cube $Q(x,t)$ centred at $x \in \Gamma$ and with side-length $t > 0$. Let $m \in \mathbb{N}$ and

$$Q(x,t)_m = \sum vol(A_{j_1} \cdots A_{j_m} Q)^{\frac{d}{n}} \tag{19.30}$$

where the sum is taken over the closed parallelograms

$$A_{j_1} \cdots A_{j_m} Q \subset Q(x,t). \tag{19.31}$$

By the monotonicity of the parallelograms $A_{j_1} \cdots A_{j_m} Q$ and the same arguments as in connection with (19.22) and (19.23) it follows that $Q(x,t)_m$ is monotonically increasing and

$$Q(x,t)_\infty = \lim_{m \to \infty} Q(x,t)_m \le \mu(Q(x,t)). \tag{19.32}$$

Let $0 < r < 1$ and let $rQ(x,t)$ be the open cube centred at $x \in \Gamma$ and with sides parallel to the sides of $Q(x,t)$ and of length rt. By the above construction we have

$$\mu(rQ(x,t)) \le Q(x,t)_\infty. \tag{19.33}$$

Since μ is Radon the left-hand side of (19.33) tends to $\mu(Q(x,t))$ if $r \to 1$. Hence we obtain

$$\mu(Q(x,t)) = Q(x,t)_\infty. \tag{19.34}$$

Consequently both μ and f are independent of $b(x)$.

19.5 Comment The above theorem is quite satisfactory. For any fractal Γ generated by the iterated function system $A = (A_l)_{l=1}^N$ in the above way we found a rather natural distribution f and a related measure μ. As for any finite Radon measure in $\mathbb{R}^n$, we have

$$f = \mu \in B_{1,\infty}^0(\mathbb{R}^n),$$

see also (18.19). Together with $f \in S'^\Gamma(\mathbb{R}^n)$ it follows

$$f = \mu \in B_{1,\infty}^{0,\Gamma}(\mathbb{R}^n), \tag{19.35}$$

see Definition 17.2. However in our context this does not say very much. For the later considerations one would need assertions of type (18.7) or (18.55). But the admitted overlappings of $A_{j_1} \cdots A_{j_m} Q$ and the highly nonisotropically distorted atoms $b_{j_1 \cdots j_k}$, see (19.12), cause a lot of trouble. We discuss a few examples and add a few more comments.

19.6 Examples (i) Let

$$A_l x = r_l x + a^l, \quad 0 < r_l < 1, \quad a^l \in \mathbb{R}^n, \tag{19.36}$$

be similarities, see 4.1, where $l = 1, \ldots, N$ with $N \geq 2$. Besides (19.6) and (19.7) we assume in addition (4.14). Then we can apply Theorem 4.7 with $0 < d < n$ where the above constructed measure μ is just the measure needed there. This follows from the arguments in Step 3 of the proof in 19.4, see in particular (19.34). By Theorem 18.2 we have in addition to (19.35)

$$f = \mu \in B_{p,\infty}^{-\frac{n-d}{p'}}(\mathbb{R}^n). \tag{19.37}$$

A direct proof, at least for values $\frac{n-d}{p'} < 1$, follows from the (now isotropic) atomic decomposition (19.16) with $m \to \infty$ and Theorem 13.8. One has to use a volume argument of the same type as in the proof of Proposition 5.6.

(ii) The anisotropic counterpart of Example (i) are the PXT fractals introduced in 4.17 with the properties (4.13)–(4.15). Of course there is no problem to extend these definitions from two to n dimensions. There is hardly any doubt that f_m in (19.16), based on (19.11)–(19.15), is an anisotropic atomic decomposition. By Theorem 19.4 the functions f_m converge in $S'(\mathbb{R}^2)$ to the uniquely determined PXT-distribution $f \in S'^\Gamma(\mathbb{R}^2)$ and the related measure μ. There should be natural counterparts of Theorems 18.2 and 18.6 in the context of anisotropic spaces and Theorem 18.15 should be a consequence of these assertions in the considered cases. But this theory has not yet been worked out so far. If one switches from isotropic cases or anisotropic PXT cases to genuine nonisotropic cases, one runs presumably into considerable difficulties, even under the non-overlapping assumption (4.14). One gets a faint impression if one compares the two Theorems 18.15 and 18.17 complemented by the remarks in 18.18.

(iii) In Theorem 19.4 we did not use non-overlapping conditions as in (4.14) and in Theorem 4.15. Then the number d in (19.8) and hence in 19.3 and 19.4 does not say very much about the nature of the resulting fractal Γ. Furthermore as we mentioned several times also the restriction $d < n$ is not necessary in the present context. We discuss both points simultaneously by looking at simple examples. Now let $d = n$ in (19.8), then we have

$$\sum_{l=1}^{N} vol(A_l Q) = 1 \tag{19.38}$$

under the assumptions that the contractive non-degenerate affine maps $(A_l)_{l=1}^{N}$ with $N \geq 2$ satisfy (19.4)–(19.6). Let $0 \leq b(x) \in C^\infty(\mathbb{R}^n)$ be as in (19.11), then (19.12) becomes

$$b_{j_1 \cdots j_k l}(x) = b\left(A_l^{-1} A_{j_k}^{-1} \cdots A_{j_1}^{-1} x\right) - (vol\, A_l Q)\, b\left(A_{j_k}^{-1} \cdots A_{j_1}^{-1} x\right), \tag{19.39}$$

and we have (19.13)–(19.18) with $d = n$, in particular

$$f_m(x) = \sum_{j_1,\ldots,j_m}^{1,\ldots,N} b\left(A_{j_m}^{-1}\cdots A_{j_1}^{-1}x\right), \quad m \in \mathbb{N}. \tag{19.40}$$

Let

$$A_l x = \frac{1}{3}x + x^l, \quad l = 1,\ldots,3^n. \tag{19.41}$$

First we choose the translates x^l such that the $A_l Q$ coincide with the 3^n cubes occurring in the process of trisecting each side of Q, see Fig. 4.1. Then we have

$$Q = \bigcup_{l=1}^{N} A_l Q, \quad N = 3^n. \tag{19.42}$$

Now it follows easily: $\Gamma = Q$, f is the characteristic function of Q, and μ is the Lebesgue measure on Q. On the other hand, if we choose $x^l = 0$ for all $l = 1,\ldots,N$ in (19.41) then we have $\Gamma = \{0\}$ and

$$f_m(x) = 3^{nm} b(3^m x) \to \delta \quad \text{in } S'(\mathbb{R}^n) \text{ if } m \to \infty, \tag{19.43}$$

where δ is the δ-distribution. Hence $f = \mu = \delta$.

19.7 Remark The constructions in Theorem 19.4 rely on the iterated function systems $(A_l)_{l=1}^N$. But the idea to construct *cascades* (19.15), (19.16) converging to Γ may have a more qualitative counterpart. A first step was done in [TrW95], 3.2 and 3.3. But there are some shortcomings.

19.8 Image measures
The iterated function system of contractive affine maps as described in (19.4)–(19.8) is a *cascade algorithm* generating the *attractor* Γ. Instead of the distributional cascades as discussed in 19.3 and 19.4 one can try to construct a corresponding cascade of *probability measures*. This procedure is well known in literature and has been extensively treated in recent times by R. S. Strichartz in his papers quoted in 17.10. We refer also to the survey [Lau95] where one finds a thorough discussion of these problems and their consequences, complemented by many references. By Theorem 19.4 the distributional cascade in (19.16) results in the probability measure μ. It comes out that the self-similar structure of μ in the understanding of the above-mentioned papers can be obtained rather easily in our context. For that purpose we recall what is meant by an *image measure*. We stick at our situation and assume that μ is the probability measure constructed in Theorem 19.4 and that A_l are the non-degenerate contractive affine maps introduced at the beginning of 19.2. Then the *image measure* μA_l^{-1} is given by

$$\left(\mu A_l^{-1}\right)(B) = \mu(A_l^{-1}B), \quad B \text{ Borel set in } \mathbb{R}^n. \tag{19.44}$$

It turns out that μA_l^{-1} is also a Radon measure and

$$supp \, \mu A_l^{-1} = A_l \Gamma, \quad l = 1, \ldots, N. \tag{19.45}$$

Since Γ is the related attractor we have in particular $A_l \Gamma \subset \Gamma$ and hence

$$supp \, \mu A_l^{-1} \subset \Gamma. \tag{19.46}$$

Furthermore, let, say, $\varphi \in S(\mathbb{R}^n)$, then

$$\int_{\mathbb{R}^n} \varphi(x) \left(\mu A_l^{-1} \right) (dx) = \int_{\mathbb{R}^n} \varphi(A_l x) \, \mu(dx) \tag{19.47}$$

or, by (19.46) and $supp \, \mu = \Gamma$,

$$\int_{\Gamma} \varphi(\gamma) \left(\mu A_l^{-1} \right) (d\gamma) = \int_{\Gamma} \varphi(A_l \gamma) \mu(d\gamma), \tag{19.48}$$

where, of course, $\varphi(\gamma)$ means $(\varphi|\Gamma)(\gamma)$, the restriction of φ to Γ, see (19.3), and $\mu(d\gamma)$ is the restriction of $\mu(dx)$ to Γ. Having (19.48) one obtains (19.44). Hence (19.44) and (19.48) are equivalent. The above construction of image measures and also (19.47) and (19.48) is standard in measure theory, we refer to [Mat95], pp. 15–17.

19.9 Corollary *The probability measure μ in Theorem* 19.4 *has the self-similarity property*

$$\mu = \sum_{l=1}^{N} (vol \, A_l Q)^{\frac{d}{n}} \, \mu A_l^{-1}. \tag{19.49}$$

Proof. By (19.21) and (19.3) we have for $\varphi \in S(\mathbb{R}^n)$

$$\int_{\Gamma} \varphi(\gamma) \, \mu(d\gamma) = \lim_{m \to \infty} \int_{\mathbb{R}^n} f_{m+1}(x) \, \varphi(x) \, dx \tag{19.50}$$

and by (19.16)

$$f_{m+1}(x) = \sum_{j=1}^{N} (vol \, A_j Q)^{-\frac{n-d}{n}} \, f_m(A_j^{-1} x). \tag{19.51}$$

We insert (19.51) in (19.50) and obtain

$$\int_{\Gamma} \varphi(\gamma) \, \mu(d\gamma) = \lim_{m \to \infty} \sum_{j=1}^{N} (vol \, A_j Q)^{\frac{d}{n}} \int_{\mathbb{R}^n} f_m(x) \, \varphi(A_j x) \, dx$$

$$= \sum_{j=1}^{N} (vol \, A_j Q)^{\frac{d}{n}} \int_{\Gamma} \varphi(A_j \gamma) \, \mu(d\gamma). \tag{19.52}$$

By (19.48) this coincides with (19.49).

19.10 Remark Let $a_j > 0$ where $j = 1, \ldots, N$, and

$$\sum_{j=1}^{N} a_j = 1. \tag{19.53}$$

Let A_j be the above contractive non-degenerate affine maps in $\mathbb{R}^n$. Let μ be the above probability measure. Then the rigorous set-up of self-similar measures having the property

$$\mu = \sum_{j=1}^{N} a_j \, \mu A_j^{-1} \tag{19.54}$$

is due to J. Hutchinson [Hut81] who in turn used some general ideas of B. B. Mandelbrot [Man77]. By (19.8) and (19.49) our approach fits in this scheme. References to papers containing the slightly more general construction (19.54) may be found in 17.10 and 19.8.

20 The spaces $\mathbb{B}_{pq}^{s}(\Gamma)$

20.1 Introduction

Let Ω be a bounded C^∞-domain in $\mathbb{R}^n$. Then for any $s \in \mathbb{R}$, $0 < p \le \infty$ ($p < \infty$ for the F-scale), and $0 < q \le \infty$ one can introduce the spaces $B_{pq}^s(\Gamma)$ and $F_{pq}^s(\Gamma)$ on the compact $(n-1)$-dimensional C^∞ manifold $\Gamma = \partial\Omega$. For that purpose one needs pointwise multiplier and diffeomorphism properties of the related spaces on $\mathbb{R}^n$, defined in 10.3. We refer to [Tri83], 3.2.2, pp. 192–3. The basic idea is to reduce spaces on Γ via an atlas of finitely many local C^∞ charts to corresponding spaces on $\mathbb{R}^{n-1}$ in that case. This method works for any d-dimensional compact C^∞ manifold with $d \in \mathbb{N}$ and even for some non-compact C^∞ manifolds such as Riemannian manifolds with bounded geometry and Lie groups, see [Tri92], Ch. 7. The situation changes drastically if Γ is non-smooth. Let Γ be a closed set in $\mathbb{R}^n$, say, with Lebesgue measure $|\Gamma| = 0$. It attracted some attention to define intrinsically spaces of Hölder (Lipschitz) type, or, more general of B_{pq}^s, F_{pq}^s, and Hardy type and to ask for (linear and) bounded extension operators from these spaces into suitable spaces on $\mathbb{R}^n$. This theory began in the thirties with Whitney's construction of a linear and bounded extension operator of intrinsically defined Hölder spaces on Γ into the corresponding spaces on $\mathbb{R}^n$, see [Ste70], Ch. 6, for details. Extending this procedure Jonsson and Wallin studied more general spaces, especially Besov spaces, but also Hardy spaces on Γ, see [JoW84] and [Wal88]. Recently it turned out that atomic decompositions as discussed in Section 13 and used afterwards in this book several times are useful to introduce in an intrinsic way and to study function spaces on closed sets in $\mathbb{R}^n$ and also in non-smooth bounded domains. We refer to [Jon93], [JoW95] and [TrW96a]. Slightly different but closely related is the study of spaces of B_{pq}^s and F_{pq}^s type on quasi-metric spaces equipped with a measure satisfying the so-called doubling condition in [HaS94]. In this paper, but also in [DaS93] the authors try to find out under which conditions for the underlying sets a substantial analysis around pseudodifferential operators and (Calderón-Zygmund) singular integral operators can be developed.

Our approach is somewhat different and guided by one of the main aims of this book, the spectral theory for fractal (pseudo)differential operators. According to Section 6, in particular Theorem 6.9 and Corollary 6.10, eigenvalues are closely related to entropy numbers of corresponding operators. These entropy numbers of *operators* can be reduced to entropy numbers of *compact embeddings* between function spaces on compact fractals or in neighbourhoods of these fractals. For that purpose we introduce in this Section 20 spaces of B_{pq}^s type on isotropic fractals Γ and estimate entropy numbers of compact embeddings between them. The following Section 21 deals with some limiting embeddings of that type. In case of anisotropic and nonisotropic fractals occur a few additional difficulties which will be discussed in Section 22. In Section 23 we compare the technique and the outcome here with corresponding results in [ET96]. To some extent the rest of this book might be considered as the *fractal twin of ET* (not the movie, but [ET96]).

Let Γ be a d-set in $\mathbb{R}^n$ according to Definition 3.1 with $0 < d < n$. By (18.24) we have for any $0 < p \le \infty$ and $\bar{p} = \min(1,p)$

$$tr_\Gamma \ B_{p\bar{p}}^{\frac{n-d}{p}} (\mathbb{R}^n) \subset L_p(\Gamma) \tag{20.1}$$

and

$$\| tr_\Gamma \ f \,|\, L_p(\Gamma)\| \le c \left\| f \,|\, B_{p\bar{p}}^{\frac{n-d}{p}} (\mathbb{R}^n) \right\| \tag{20.2}$$

for some $c > 0$ and all $f \in B_{p\bar{p}}^{\frac{n-d}{p}} (\mathbb{R}^n)$. Here we extended (18.24) to $p = \infty$: In that case $B_{\infty,1}^0(\mathbb{R}^n)$ consists of continuous functions and the trace is taken pointwise, see [Tri83], Remark 2.7.1/2, p. 130. Hence it makes sense to speak about traces on Γ for all spaces $B_{pq}^\sigma(\mathbb{R}^n)$ with $0 < p \le \infty$, $0 < q \le \infty$ and $\sigma > \frac{n-d}{p}$ as subspaces of $L_p(\Gamma)$. We remark that (20.1) is sharp, see Theorem 18.6, Corollary 18.12 and also the discussion in 18.9 and 18.13.

20.2 Definition *Let Γ be a d-set in $\mathbb{R}^n$ according to Definition 3.1 with $0 < d < n$. Let $s > 0$, $0 < p \le \infty$, and $0 < q \le \infty$; then*

$$\mathbb{B}_{pq}^s(\Gamma) = tr_\Gamma \ B_{pq}^{s+\frac{n-d}{p}} (\mathbb{R}^n), \tag{20.3}$$

equipped with the quasi-norm

$$\|f \,|\, \mathbb{B}_{pq}^s(\Gamma)\| = \inf \left\| g \,|\, B_{pq}^{s+\frac{n-d}{p}} (\mathbb{R}^n) \right\|, \tag{20.4}$$

where the infimum is taken over all $g \in B_{pq}^{s+\frac{n-d}{p}} (\mathbb{R}^n)$ with $tr_\Gamma \ g = f$.

20.3 Discussions

(i) By standard arguments, $\mathbb{B}^s_{pq}(\Gamma)$ is a quasi-Banach space (Banach space if $p \geq 1, q \geq 1$), always considered as a subspace of $L_p(\Gamma)$ according to (20.1) and (20.2) and what has been said in front of 20.2.

(ii) In Theorem 18.6 we assumed in addition $p > \frac{d}{n}$ and it turned out in (18.31) and in 18.9 that this restriction is not only useful but rather natural. In particular, let $m \in \mathbb{N}$ with $m > n$ and let $\mathbb{R}^n$ be interpreted as a hyperplane in $\mathbb{R}^m$. In analogy to (18.40) we have then for $s > 0$

$$tr_{\mathbb{R}^n} \ B_{pq}^{s+\frac{m-d}{p}} (\mathbb{R}^m) = B_{pq}^{s+\frac{n-d}{p}} (\mathbb{R}^n). \qquad (20.5)$$

Now it follows easily that $\mathbb{B}^s_{pq}(\Gamma)$ *in (20.3) in this case is independent of the dimension of the underlying space* $\mathbb{R}^n$. By 18.9 if Γ and p are given then for sufficiently large n we are always in the position $p > \frac{d}{n}$. In the applications in Chapter V we are only interested in values p with $1 \leq p \leq \infty$. Then (20.5) is always true, and hence we have the indicated independence without any further discussion.

(iii) In case $p < 1$ there might be another complication which explains why we wrote $\mathbb{B}^s_{pq}(\Gamma)$ instead of $B^s_{pq}(\Gamma)$. Let Γ be a, say, compact d-dimensional C^∞ manifold in $\mathbb{R}^n$, where $d \in \mathbb{N}$ and $d < n$. As we mentioned in 20.1 one may introduce spaces $B^s_{pq}(\Gamma)$ for any $s \in \mathbb{R}$, $0 < p \leq \infty$, and $0 < q \leq \infty$ via local charts, reducing these spaces to corresponding spaces on $\mathbb{R}^d$, maybe again interpreted as a hyperplane in $\mathbb{R}^n$. In other words we must compare the spaces $B^s_{pq}(\mathbb{R}^d)$ given by Definition 10.3 with d in place of n with the spaces $\mathbb{B}^s_{pq}(\mathbb{R}^d)$ according to Definition 20.2, where we assume $s > 0$. Let

$$s > d \left(\frac{1}{p} - 1 \right)_+ , \quad 0 < p \leq \infty, \quad 0 < q \leq \infty, \quad d \in \mathbb{N}. \qquad (20.6)$$

Then we have also

$$s + \frac{k}{p} > (d + k) \left(\frac{1}{p} - 1 \right)_+ \quad \text{where} \quad k \in \mathbb{N}. \qquad (20.7)$$

Iterative application of 18.1(i) yields

$$tr_{\mathbb{R}^d} \ B_{pq}^{s+\frac{n-d}{p}} (\mathbb{R}^n) = B^s_{pq}(\mathbb{R}^d) \qquad (20.8)$$

in the framework of tempered distributions on $\mathbb{R}^n$ and $\mathbb{R}^d$. Hence under the restriction (20.6) we have

$$B^s_{pq}(\mathbb{R}^d) = \mathbb{B}^s_{pq}(\mathbb{R}^d), \qquad (20.9)$$

where it should be noted that (20.6) is just the condition needed in 11.4, in particular in (11.17) combined with (10.8), which ensures that $B^s_{pq}(\mathbb{R}^d)$ is embedded in some spaces $L_r(\mathbb{R}^d)$ with $1 \leq r \leq \infty$ and, hence, consists of regular distributions. On the other hand, if

$$0 < p < 1, \quad 0 < q \leq \infty, \quad \text{and} \quad 0 < s < d\left(\frac{1}{p} - 1\right), \tag{20.10}$$

then $B^s_{pq}(\mathbb{R}^d)$ contains singular distributions, see [SicT95] or [RuS96], 2.2.4, whereas $\mathbb{B}^s_{pq}(\mathbb{R}^d)$ is by definition a subset of $L_p(\mathbb{R}^d)$. In other words, for (20.10) we have

$$B^s_{pq}(\mathbb{R}^d) \neq \mathbb{B}^s_{pq}(\mathbb{R}^d). \tag{20.11}$$

The two spaces are not comparable. However in all cases of interest in the later considerations in Chapter V we have always the situation (20.9).

(iv) In some cases the spaces $\mathbb{B}^s_{pq}(\Gamma)$ coincide with the spaces treated in the papers by Jonsson, Wallin and Han, Sawyer mentioned in 20.1. This gives the possibility to combine these different approaches. But this will not be done in this book.

20.4 Embeddings
We summarized in 11.4 embedding assertions for B^s_{pq} (and F^s_{pq}) spaces on $\mathbb{R}^n$. By Definition 20.2 some of these continuous embeddings can be carried over to the spaces $\mathbb{B}^s_{pq}(\Gamma)$, where Γ is a d-set. If the closed set Γ is, in addition, compact, then also the related embeddings between diverse spaces $\mathbb{B}^s_{pq}(\Gamma)$ are compact, at least in non-limiting situations. It is the main aim of Section 20 to estimate the entropy numbers of corresponding compact embeddings. We introduced entropy numbers in Definition 6.2 and collected some properties afterwards in Section 6. Two remarks seem to be in order.

(i) *First*, as we mentioned in 20.3(ii), in particular after (20.5), in all cases of interest, $\mathbb{B}^s_{pq}(\Gamma)$ given by (20.3) is independent of the underlying space $\mathbb{R}^n$. But the related restrictions for the parameters are immaterial for our reasoning below. Hence we always assume that Γ is a compact fractal in a given fixed space $\mathbb{R}^n$.

(ii) *Secondly*, the compactness of the embedding between suitable spaces $\mathbb{B}^s_{pq}(\Gamma)$ has nothing to do with the assumption that Γ is a fractal with $|\Gamma| = 0$ or a d-set with $d < n$. Let Ω be an arbitrary bounded domain in $\mathbb{R}^n$. Then $B^s_{pq}(\Omega)$ is, by definition, the restriction of $B^s_{pq}(\mathbb{R}^n)$ to Ω:

$$B^s_{pq}(\Omega)$$
$$= \{f \in D'(\Omega): \quad \text{there is a} \quad g \in B^s_{pq}(\mathbb{R}^n) \quad \text{with} \quad g|\Omega = f\}, \tag{20.12}$$
$$\|f \mid B^s_{pq}(\Omega)\| = \inf \|g \mid B^s_{pq}(\mathbb{R}^n)\|, \tag{20.13}$$

where the infimum is taken over all $g \in B^s_{pq}(\mathbb{R}^n)$ such that its restriction to Ω, denoted by $g|\Omega$, coincides in $D'(\Omega)$ with f. This definition makes sense whenever

$$s \in \mathbb{R}, \quad 0 < p \leq \infty, \quad 0 < q \leq \infty. \tag{20.14}$$

The embedding

$$id: \quad B^{s_1}_{p_1q_1}(\Omega) \rightarrow B^{s_2}_{p_2q_2}(\Omega) \tag{20.15}$$

is compact, provided that

$$-\infty < s_2 < s_1 < \infty,$$

$$0 < p_1 \le \infty, \; 0 < p_2 \le \infty, \; 0 < q_1 \le \infty, \; 0 < q_2 \le \infty, \tag{20.16}$$

and

$$\delta_+ = s_1 - s_2 - n\left(\frac{1}{p_1} - \frac{1}{p_2}\right)_+ > 0. \tag{20.17}$$

We use the same notation as in [ET96], 3.3.1, p. 105; furthermore $a_+ = \max(a, 0)$ if $a \in \mathbb{R}$. We proved in [ET96], 3.3.3, p. 118, and 3.5, p. 151, for the related entropy numbers

$$e_k\left(id: B^{s_1}_{p_1q_1}(\Omega) \rightarrow B^{s_2}_{p_2q_2}(\Omega)\right) \sim k^{-\frac{s_1-s_2}{n}}, \quad k \in \mathbb{N}. \tag{20.18}$$

We use «$\sim$» always in the following meaning: Let a_k and b_k be two sequences of positive numbers indexed by, say, $k \in \mathbb{N}$, then $a_k \sim b_k$ means that there are two positive numbers $0 < c_1 \le c_2 < \infty$ such that

$$c_1 a_k \le b_k \le c_2 a_k, \quad k \in \mathbb{N}. \tag{20.19}$$

Assertions of type (20.18) have a rather long history which will not be repeated here. We refer to [ET96], 3.3.5, p. 126–128, and the references given there. As far as spaces $B^s_{pq}(\Omega)$ in bounded, non-smooth domains Ω in connection with (20.18) are concerned we refer also to [TrW96a] and [ET96], 3.5, p. 151. The main aim of Section 20 is to ask for counterparts of (20.18) with $\mathbb{B}^s_{pq}(\Gamma)$ in place of $B^s_{pq}(\Omega)$. The methods developed here are completely different from those in [ET96]. In other words, applied to $B^s_{pq}(\Omega)$, where Ω stands for a bounded (non-smooth) domain in $\mathbb{R}^n$, instead of $\mathbb{B}^s_{pq}(\Gamma)$, one obtains a new proof of (20.18) with (20.16), (20.17), see also Section 23.

20.5 Proposition *Let Γ be a compact d-set in $\mathbb{R}^n$ with $0 < d < n$ according to Definition 3.1. Let $\mathbb{B}^s_{pq}(\Gamma)$ be the spaces introduced in Definition 20.2, notationally complemented by $\mathbb{B}^0_{pq}(\Gamma) = L_p(\Gamma)$ for any $0 < p \le \infty$ and $0 < q \le \infty$. Let*

$$0 \le s_2 < s_1 < \infty, \; 0 < p_1 \le \infty, \; 0 < p_2 \le \infty, \; 0 < q_1 \le \infty, \; 0 < q_2 \le \infty \tag{20.20}$$

and

$$\delta_+ = s_1 - s_2 - d\left(\frac{1}{p_1} - \frac{1}{p_2}\right)_+ > 0. \tag{20.21}$$

Then the embedding of $\mathbb{B}^{s_1}_{p_1q_1}(\Gamma)$ into $\mathbb{B}^{s_2}_{p_2q_2}(\Gamma)$ is compact and there is a constant $c > 0$ such that for the related entropy numbers

$$e_k\left(id: \mathbb{B}^{s_1}_{p_1q_1}(\Gamma) \rightarrow \mathbb{B}^{s_2}_{p_2q_2}(\Gamma)\right) \le c\, k^{-\frac{s_1-s_2}{d}} \text{ where } k \in \mathbb{N}. \tag{20.22}$$

Proof.
Step 1. Let $p_2 \geq p_1$. With

$$\sigma_1 = s_1 + \frac{n-d}{p_1}, \quad \sigma_2 = s_2 + \frac{n-d}{p_2} \quad \text{and} \quad \delta = \delta_+ \tag{20.23}$$

we have

$$\sigma_1 - \frac{n}{p_1} = s_1 - \frac{d}{p_1} = \delta + s_2 - \frac{d}{p_2} = \delta + \sigma_2 - \frac{n}{p_2}. \tag{20.24}$$

Let $f \in \mathbb{B}^{s_1}_{p_1 q_1}(\Gamma)$, then by (20.4) and (20.24) there is a (non-linear) bounded extension operator $g = ext\, f$ such that

$$tr_\Gamma\, g = f \quad \text{and} \quad \|g \mid B^{\sigma_1}_{p_1 q_1}(\mathbb{R}^n)\| \leq 2\|f \mid \mathbb{B}^{s_1}_{p_1 q_1}(\Gamma)\|. \tag{20.25}$$

To fix the imagination we may assume that g is zero outside of a fixed neighbourhood of Γ. We expand g according to the subatomic representation Theorem 14.4(i) and Corollary 14.11(i) in terms of $(M,p_1)_{-1} - \gamma$-quarks and $(\sigma_1, p_1)_L - \gamma$-quarks (with the modification in case of Corollary 14.11). Then we have (14.15) with g in place of f and the equivalent quasi-norm (14.16), or the modifications according to Corollary 14.11. The idea of the proof is to reduce everything to the building blocks as introduced in Definition 14.2 and to the knowledge that the involved sequences of complex numbers in (14.15) belong to the spaces $b_{p_1 q_1}$ according to (14.16). Hence it does not matter very much to look at one of the two terms in (14.15) and to assume that we can apply the somewhat simpler situation as described in Corollary 14.7 both to $B^{\sigma_1}_{p_1 q_1}(\mathbb{R}^n)$ and $B^{\sigma_2}_{p_2 q_2}(\mathbb{R}^n)$. The necessary technical modifications are clear in all other cases. Hence by Corollary 14.7 we have, without restriction of generality,

$$g = \sum_{\beta \in \mathbb{N}^n_0} \sum_{\nu=0}^{\infty} \sum_{m \in \mathbb{Z}^n} \lambda^\beta_{\nu m} 2^{-\nu(\sigma_1 - \frac{n}{p_1})} \psi^\beta(2^\nu x - m) \tag{20.26}$$

where $\psi(x)$ is given by (14.11), (14.12), $\psi^\beta(x) = x^\beta \psi(x)$, and

$$\sup_\beta 2^{\varrho_1 |\beta|} \|\lambda^\beta \mid b_{p_1 q_1}\| \sim \|g \mid B^{\sigma_1}_{p_1 q_1}(\mathbb{R}^n)\|, \tag{20.27}$$

with $\varrho_1 > 0$ large. Recall that

$$\lambda^\beta = \left(\lambda^\beta_{\nu m} : \nu \in \mathbb{N}_0 \text{ and } m \in \mathbb{Z}^n \right).$$

As before, $Q_{\nu m}$ denotes a cube in $\mathbb{R}^n$ with sides parallel to the axes of coordinates, centred at $2^{-\nu} m$ and with side-length $2^{-\nu}$ where $\nu \in \mathbb{N}_0$, $m \in \mathbb{Z}^n$. Let $cQ_{\nu m}$ be a cube concentric with $Q_{\nu m}$ and side length $c\, 2^{-\nu}$, see 13.2. Let

$$\lambda^{\beta,\Gamma} = \left(\lambda^\beta_{\nu m} : \nu \in \mathbb{N}_0, m \in \mathbb{Z}^n, C\, Q_{\nu m} \cap \Gamma \neq \emptyset \right), \tag{20.28}$$

where we may assume that $C > 1$ is fixed and sufficiently large such that all what follows is justified. Let M_ν for fixed $\nu \in \mathbb{N}_0$ be the number of cubes $Q_{\nu m}$ such that $C\, Q_{\nu m} \cap \Gamma \neq \emptyset$. Since Γ is compact it follows by Definition 3.1 and 3.5 that there are two positive numbers $0 < c_1 \leq c_2 < \infty$ with

$$c_1\, 2^{\nu d} \leq M_\nu \leq c_2\, 2^{\nu d}, \quad \nu \in \mathbb{N}_0. \tag{20.29}$$

This coincides with (8.1). Let

$$\ell_q \left(2^{\nu \delta}\, \ell_p^{M_\nu} \right) \quad \text{and} \quad \ell_\infty \left[2^{\varrho|\beta|}\, \ell_q \left(2^{\nu \delta}\, \ell_p^{M_\nu} \right) \right]$$

be the sequence spaces introduced in (8.2) and (9.1) (adapted to the present situation), where $\delta = \delta_+$, $\varrho = \varrho_1$, and M_ν have the same meaning as in (20.21), (20.27), and (20.29), respectively, and the ℓ_∞-norm is modified now by $\sup_{\beta \in \mathbb{N}_0^n}$. We introduce the (non-linear) operator S,

$$S: \quad B^{\sigma_1}_{p_1 q_1}(\mathbb{R}^n) \to \ell_\infty \left[2^{\varrho_1|\beta|}\, \ell_{q_1} \left(2^{\nu \delta}\, \ell_{p_1}^{M_\nu} \right) \right] \tag{20.30}$$

by

$$Sg = \eta = \left(\eta^{\beta, \Gamma} : \beta \in \mathbb{N}_0^n \right), \tag{20.31}$$

$$\eta^{\beta, \Gamma} = \left(2^{-\nu \delta}\, \lambda^\beta_{\nu m} : \nu \in \mathbb{N}_0,\, m \in \mathbb{Z}^n,\, C Q_{\nu m} \cap \Gamma \neq \emptyset \right), \tag{20.32}$$

where g is given by (20.26), (20.27). Recall that the expansion (20.26) is not unique, but this does not matter. By (20.27) it follows that S is a bounded map. Next we construct the linear map T,

$$T: \quad \ell_\infty \left[2^{\varrho_2|\beta|}\, \ell_{q_2}(\ell_{p_2}^{M_\nu}) \right] \to B^{\sigma_2}_{p_2 q_2}(\mathbb{R}^n), \tag{20.33}$$

given by

$$T \varkappa = \sum_{\beta \in \mathbb{N}_0^n} \sum_{\nu=0}^{\infty} \sum_{m} \varkappa^\beta_{\nu m}\, 2^{-\nu(\sigma_2 - \frac{n}{p_2})}\, \psi^\beta(2^\nu x - m), \tag{20.34}$$

where $\varkappa = (\varkappa^{\beta, \Gamma} : \beta \in \mathbb{N}_0^n)$ and the sum over m in (20.34) is taken according to (20.28) with $\varkappa^\beta_{\nu m}$ in place of $\lambda^\beta_{\nu m}$. As we said above we may assume without restriction of generality that Corollary 14.7 is applicable to $B^{\sigma_2}_{p_2 q_2}(\mathbb{R}^n)$ (instead of Theorem 14.4 or Corollary 14.11). Then it follows by Corollary 14.7(i) that T is a linear and bounded map. Now we complement the three bounded maps *ext*, S, and T by the identity

$$id: \quad \ell_\infty \left[2^{\varrho_1|\beta|}\, \ell_{q_1} \left(2^{\nu \delta}\, \ell_{p_1}^{M_\nu} \right) \right] \to \ell_\infty \left[2^{\varrho_2|\beta|}\, \ell_{q_2} \left(\ell_{p_2}^{M_\nu} \right) \right] \tag{20.35}$$

with $\varrho_1 > \varrho_2$ and the trace

$$tr_\Gamma: \quad B^{\sigma_2}_{p_2 q_2}(\mathbb{R}^n) \rightarrow \mathbb{B}^{s_2}_{p_2 q_2}(\Gamma). \tag{20.36}$$

By (9.4) and Theorem 9.2 the operator id is compact and tr_Γ is continuous by (20.23) and (20.3). For $s_2 = 0$ we have $\sigma_2 = \frac{n-d}{p_2}$ and may choose $q_2 = \overline{p_2}$ according to (20.1) and (20.2). We claim

$$id\left(\mathbb{B}^{s_1}_{p_1 q_1}(\Gamma) \rightarrow \mathbb{B}^{s_2}_{p_2 q_2}(\Gamma)\right) = tr_\Gamma \circ T \circ id \circ S \circ ext. \tag{20.37}$$

We follow the constructions. Let $f \in \mathbb{B}^{s_1}_{p_1 q_1}(\Gamma)$, then we have (20.25) and (20.26). Checking the coefficients in front of $\psi^\beta(2^\nu x - m)$ we arrive in (20.34) at

$$\varkappa^\beta_{\nu m} \, 2^{-\nu(\sigma_2 - \frac{n}{p_2})} = 2^{-\nu(\sigma_2 - \frac{n}{p_2}) - \nu\delta} \, \lambda^\beta_{\nu m} = 2^{-\nu(\sigma_1 - \frac{n}{p_1})} \, \lambda^\beta_{\nu m}, \tag{20.38}$$

where we used (20.24). But this is just what we need in (20.26). Hence taking finally tr_Γ we obtain f by (20.25), where we started from. This proves (20.37). In particular the final outcome is independent of ambiguities in the nonlinear constructions ext and S. The unit ball in $\mathbb{B}^{s_1}_{p_1 q_1}(\Gamma)$ is mapped by $S \circ ext$ into a bounded set in

$$\ell_\infty\left[2^{\varrho_1|\beta|} \, \ell_{q_1}\left(2^{\nu\delta} \, \ell^{M_\nu}_{p_1}\right)\right].$$

By (20.35) this bounded set is mapped into a pre-compact set in

$$\ell_\infty\left[2^{\varrho_2|\beta|} \, \ell_{q_2}(\ell^{M_\nu}_{p_2})\right]$$

which can be covered by 2^k balls of radius $c\,e_k(id)$ with

$$e_k(id) \leq c\,k^{-\frac{\delta}{d} + \frac{1}{p_2} - \frac{1}{p_1}}, \quad k \in \mathbb{N}. \tag{20.39}$$

This follows from Theorem 9.2 where we used $p_2 \geq p_1$. Afterwards the two linear and bounded maps T and tr_Γ do not change this covering assertion, not to speak about constants. Hence we arrive at a covering of the unit ball in $\mathbb{B}^{s_1}_{p_1 q_1}(\Gamma)$ by 2^k balls of radius $c e_k(id)$ in $\mathbb{B}^{s_2}_{p_2 q_2}(\Gamma)$. We insert $\delta = \delta_+$, given by (20.21) with $p_2 \geq p_1$ in the exponent in (20.39) and obtain (20.22).

Step 2. Let $p_2 < p_1$. We claim

$$\mathbb{B}^{s_2}_{p_1 q_2}(\Gamma) \subset \mathbb{B}^{s_2}_{p_2 q_2}(\Gamma). \tag{20.40}$$

By (20.3) and (20.23) this assertion is equivalent with

$$\{f \in B^{\sigma_2}_{p_1 q_2}(\mathbb{R}^n), \, f = 0 \text{ in } \mathbb{R}^n \backslash \Omega\} \subset B^{\sigma_2}_{p_2 q_2}(\mathbb{R}^n), \tag{20.41}$$

where Ω is a bounded open neighbourhood of Γ. However since $p_2 < p_1$, the embedding (20.41) follows from 11.2(ii) and the monotonicity of the L_p-spaces on bounded domains. Now (20.22) is a consequence of (20.40) and Step 1 applied to $p_1 = p_2$. The proof is complete.

20.6 Theorem *Let Γ be a compact d-set in $\mathbb{R}^n$ with $0 < d < n$ according to Definition 3.1. Let $\mathbb{B}^s_{pq}(\Gamma)$ be the spaces introduced in Definition 20.2, notationally complemented by $\mathbb{B}^0_{pq}(\Gamma) = L_p(\Gamma)$ for any $0 < p \le \infty$ and $0 < q \le \infty$. Let*

$$0 \le s_2 < s_1 < \infty,$$

$$0 < p_1 \le \infty, \ 02 < p_2 \le \infty, \ 0 < q_1 \le \infty, \ 0 < q_2 \le \infty, \tag{20.42}$$

and

$$\delta_+ 2 = s_1 - s_2 - d \left(\frac{1}{p_1} - \frac{1}{p_2} \right)_+ > 0. \tag{20.43}$$

Then the embedding of $\mathbb{B}^{s_1}_{p_1 q_1}(\Gamma)$ into $\mathbb{B}^{s_2}_{p_2 q_2}(\Gamma)$ is compact and for the related entropy numbers holds

$$e_k \left(id : \mathbb{B}^{s_1}_{p_1 q_1}(\Gamma) \to \mathbb{B}^{s_2}_{p_2 q_2}(\Gamma) \right) \sim k^{-\frac{s_1 - s_2}{d}}, \quad k \in \mathbb{N}. \tag{20.44}$$

Proof.
Step 1. By Proposition 20.5 it remains to prove that there is a constant $c > 0$ such that for all $k \in \mathbb{N}$,

$$e_k \left(id : \mathbb{B}^{s_1}_{p_1 q_1}(\Gamma) \to \mathbb{B}^{s_2}_{p_2 q_2}(\Gamma) \right) k^{\frac{s_1 - s_2}{d}} \ge c. \tag{20.45}$$

Assume that there is no such $c > 0$. Then we find a sequence $k_j \to \infty$ such that

$$e_{k_j} \left(id : \mathbb{B}^{s_1}_{p_1 q_1}(\Gamma) \to \mathbb{B}^{s_2}_{p_2 q_2}(\Gamma) \right) k_j^{\frac{s_1 - s_2}{d}} \to 0 \quad \text{when} \quad j \to \infty. \tag{20.46}$$

By Proposition 20.5 and the multiplication property (6.8) for entropy numbers we may assume in (20.46) that

$$\mathbb{B}^{s_2}_{p_2 q_2}(\Gamma) = L_{p_2}(\Gamma),$$

in particular, $s_2 = 0$, and $1 < p_1 \le \infty$. We have also

$$e_k \left(id : \mathbb{B}^{s_1}_{p_1 q_1}(\Gamma) \to L_{p_1}(\Gamma) \right) \le c \, k^{-\frac{s_1}{d}}, \quad k \in \mathbb{N}. \tag{20.47}$$

Assume $p_2 < 1$. For any $f \in L_{p_1}(\Gamma)$ we have

$$\|f \mid L_p(\Gamma)\| \le \|f \mid L_{p_1}(\Gamma)\|^{1-\Theta} \|f \mid L_{p_2}(\Gamma)\|^{\Theta}, \tag{20.48}$$

where

$$0 < \Theta < 1 \quad \text{and} \quad \frac{1}{p} = \frac{1-\Theta}{p_1} + \frac{\Theta}{p_2}. \tag{20.49}$$

By the interpolation property for entropy numbers, briefly mentioned in 6.7 and treated in detail in [ET96], 1.3.2, p. 13, we have

$$e_{2k_j}\left(id: \mathbb{B}_{p_1 q_1}^{s_1}(\Gamma) \to L_p(\Gamma)\right)$$
$$\leq c\, e_{k_j}\left(id: \mathbb{B}_{p_1 q_1}^{s_1}(\Gamma) \to L_{p_1}(\Gamma)\right)^{1-\Theta}\, e_{k_j}\left(id: \mathbb{B}_{p_1 q_1}^{s_1}(\Gamma) \to L_{p_2}(\Gamma)\right)^{\Theta}. \tag{20.50}$$

By (20.46) and (20.47) it follows

$$e_{2k_j}\left(id: \mathbb{B}_{p_1 q_1}^{s_1}(\Gamma) \to L_p(\Gamma)\right) k_j^{\frac{s_1}{d}} \to 0 \quad \text{when} \quad j \to \infty. \tag{20.51}$$

By (20.49) we may assume $p > 1$. In other words, to prove (20.45) it is sufficient to disprove the existence of a sequence $k_j \to \infty$ with

$$e_{k_j}\left(id: \mathbb{B}_{p_1 q_1}^{s_1}(\Gamma) \to L_{p_2}(\Gamma)\right) k_j^{\frac{s_1}{d}} \to 0, \tag{20.52}$$

where $1 < p_1 \leq \infty$ and $1 < p_2 \leq \infty$.

Step 2. By Step 1 it remains to prove the following assertion: Let $0 < q \leq \infty$,

$$1 < p_1 \leq \infty, \ 1 < p_2 \leq \infty, \ \text{and} \ s > d\left(\frac{1}{p_1} - \frac{1}{p_2}\right)_+, \tag{20.53}$$

then there is a constant $c > 0$ such that

$$e_k\left(id: \mathbb{B}_{p_1 q}^s(\Gamma) \to L_{p_2}(\Gamma)\right) \geq c\, k^{-\frac{s}{d}}, \quad k \in \mathbb{N}. \tag{20.54}$$

Since Γ is compact we may assume $\mu(\Gamma) = 1$. Furthermore by (3.1) and 3.5 we find $M_j \sim 2^{jd}$ disjoint balls $B_{j,r}$ in $\mathbb{R}^n$ of radius 2^{-j} and centred at $x^{j,r} \in \Gamma$, where $r = 1, \ldots, M_j$. Here $j \in \mathbb{N}$ and $M_j \sim 2^{jd}$ have the same meaning as in (20.29). Let φ and ψ be two non-negative C^∞ functions in $\mathbb{R}^n$ with supports in the unit ball and

$$c_{j,r}\, 2^{jd} \int_\Gamma \varphi\left(2^j(\gamma - x^{j,r})\right) \psi\left(2^j(\gamma - x^{j,r})\right) \mu(d\gamma) = 1, \tag{20.55}$$

where by (3.1) we may assume that there are two constants $0 < c_1 \leq c_2 < \infty$ such that

$$c_1 \leq c_{j,r} \leq c_2 \quad \text{for all} \quad j \in \mathbb{N} \quad \text{and} \quad r = 1, \ldots, M_j. \tag{20.56}$$

In the commutative diagram (20.57) let the operators A and B be

$$\begin{array}{ccc}
2^{(s-\frac{d}{p_1})}\ell_{p_1}^{M_j} & \xrightarrow{\quad A \quad} & \mathbb{B}_{p_1 q}^s(\Gamma) \\
{\scriptstyle id^j}\Big\downarrow & & \Big\downarrow{\scriptstyle id_\Gamma} \\
2^{-j\frac{d}{p_2}}\ell_{p_2}^{M_j} & \xleftarrow{\quad B \quad} & L_{p_2}(\Gamma)
\end{array} \tag{20.57}$$

given by

$$A \{a_r : r = 1, \ldots, M_j\} = \sum_{r=1}^{M_j} a_r \, \varphi \left(2^j (x - x^{j,r})\right) |\Gamma \qquad (20.58)$$

and

$$B f = \left\{ c_{j,r} \, 2^{jd} \int_\Gamma f(\gamma) \, \psi \left(2^j (\gamma - x^{j,r})\right) \mu(d\gamma) : r = 1, \ldots, M_j \right\}, \qquad (20.59)$$

respectively. Furthermore, id_Γ and id^j are the embeddings indicated. We interpret the right-hand side of (20.58) as an atomic decomposition in $B^\sigma_{p_1 q}(\mathbb{R}^n)$ with $\sigma = s + \frac{n-d}{p_1}$ according to Theorem 13.8 where the needed normalizing factor for the atom is $2^{-j(\sigma - \frac{n}{p_1})} = 2^{-j(s - \frac{d}{p_1})}$. Hence we obtain

$$\|A\{a_r\} \mid \mathbb{B}^s_{p_1 q}(\Gamma)\| \leq \| \sum_r a_r \, \varphi \left(2^j (\cdot - x^{j,r})\right) \mid B^\sigma_{p_1 q}(\mathbb{R}^n)\|$$

$$\qquad (20.60)$$

$$\leq c \, 2^{j(s - \frac{d}{p_1})} \, \|\{a_r\} \mid \ell^{M_j}_{p_1}\|,$$

where c is independent of j. Let b^j_r be the coefficients in the brackets in (20.59). By Hölder's inequality with $\frac{1}{p_2} + \frac{1}{p'_2} = 1$, the above disjoint balls $B_{j,r}$, and (3.1) we obtain

$$|b_{j,r}|^{p_2} \leq c \, 2^{jd p_2} \int_{B_{j,r} \cap \Gamma} |f(\gamma)|^{p_2} \, \mu(d\gamma) \, 2^{-jd \frac{p_2}{p'_2}}$$

$$\qquad (20.61)$$

$$= c \, 2^{jd} \int_{B_{j,r} \cap \Gamma} |f(\gamma)|^{p_2} \, \mu(d\gamma)$$

and

$$\|B f \mid \ell^{M_j}_{p_2}\| = \left(\sum_{r=1}^{M_j} |b^j_r|^{p_2} \right)^{\frac{1}{p_2}} \leq c \, 2^{j \frac{d}{p_2}} \, \|f \mid L_{p_2}(\Gamma)\| \qquad (20.62)$$

(usual modification if $p_2 = \infty$), where again c is independent of j. In other words, both A and B are linear and bounded operators where the corresponding norms can be estimated independently of j. By (20.55) we have

$$id^j = B \circ id_\Gamma \circ A \qquad (20.63)$$

and consequently

$$e_k(id^j) \leq c \, e_k(id_\Gamma), \quad k \in \mathbb{N}, \qquad (20.64)$$

where c is independent of j. By Proposition 7.2 with $k = 2M_j \sim c2^{jd}$ we obtain

$$e_{c \, 2^{jd}}(id^j) = 2^{-j(s - \frac{d}{p_1})} \, 2^{-j \frac{d}{p_2}} \, e_{c \, 2^{jd}} \left(id : \ell^{M_j}_{p_1} \to \ell^{M_j}_{p_2}\right)$$

$$\qquad (20.65)$$

$$\geq c' 2^{-js},$$

where $c > 0$ and $c' > 0$ are independent of j. We insert (20.65) in (20.64) and obtain (20.54) with $k = c \, 2^{jd}$. But this is sufficient to prove (20.54) for all $k \in \mathbb{N}$.

20.7 Non-compact embeddings

To prepare our considerations in Section 21 we discuss some limiting embeddings. Let again Γ be a compact d-set in $\mathbb{R}^n$ with $0 < d < n$ and let

$$0 < s_2 < s_1 < \infty, \quad 0 < p_1 < p_2 \leq \infty, \tag{20.66}$$

and

$$s_1 - \frac{d}{p_1} = s_2 - \frac{d}{p_2}. \tag{20.67}$$

These assumptions correspond to $\delta = \delta_+ = 0$ in Theorem 20.6. By Definition 20.2, (20.23), (20.24), and [Tri83], 2.7.1, p. 129, it follows

$$\mathbb{B}_{p_1 q}^{s_1}(\Gamma) \subset \mathbb{B}_{p_2 q}^{s_2}(\Gamma), \quad 0 < q \leq \infty. \tag{20.68}$$

We complement this limiting embedding by

$$\mathbb{B}_{p_1 \overline{p_2}}^{s}(\Gamma) \subset L_{p_2}(\Gamma), \quad s > 0, \ 0 < p_1 < p_2 \leq \infty, \tag{20.69}$$

with $\overline{p_2} = \min(1, p_2)$, and

$$s - \frac{d}{p_1} = -\frac{d}{p_2}, \tag{20.70}$$

which follows from (20.1), and again (20.23), (20.24), and [Tri83], 2.7.1, p. 129. However in contrast to Theorem 20.6 these limiting embeddings cannot be compact. We prove this claim in case of (20.69), (20.70). As in Step 2 of the proof of Theorem 20.6 we choose a non-trivial C^∞ function $\varphi(x)$ in $\mathbb{R}^n$ with support in the unit ball. Let $x^{j,r} \in \Gamma$ be the same points as there and let

$$\Phi_j(x) = \sum_{r=1}^{M_j} a_r \, \varphi\left(2^j (x - x^{j,r})\right), \quad x \in \mathbb{R}^n, \tag{20.71}$$

as in (20.58). We have

$$\|\Phi_j \mid L_{p_2}(\Gamma)\| \sim 2^{-j\frac{d}{p_2}} \left(\sum_{r=1}^{M_j} |a_r|^{p_2}\right)^{\frac{1}{p_2}} \tag{20.72}$$

where «$\sim$» is independent of $j \in \mathbb{N}$. On the other hand assuming that φ satisfies the necessary moment conditions it follows by Definition 20.2 and Theorem 13.8

$$\left\|\Phi_j \mid \mathbb{B}_{p_1 \overline{p_2}}^{s}(\Gamma)\right\| \leq \left\|\Phi_j \mid B_{p_1 \overline{p_2}}^{s+\frac{n-d}{p_1}}(\mathbb{R}^n)\right\|$$

$$\leq c \, 2^{j(s-\frac{d}{p_1})} \left(\sum_{r=1}^{M_j} |a_r|^{p_1}\right)^{\frac{1}{p_1}}, \tag{20.73}$$

where again c is independent of $j \in \mathbb{N}$. Hence, by (20.70), (20.72), (20.73) and the analogue of (20.63) we have

$$e_k \left(id : \ell_{p_1}^{M_j} \to \ell_{p_2}^{M_j} \right) \leq c\, e_k \left(id : \mathbb{B}_{p_1 \overline{p_2}}^s(\Gamma) \to L_{p_2}(\Gamma) \right), \quad k \in \mathbb{N}, \quad (20.74)$$

where c is independent of j, and where e_k are the respective entropy numbers. Assuming that the embedding (20.69) is compact one obtains by (20.74) a contradiction. Hence we arrive at the assertion:

The embedding (20.69), (20.70) is continuous but not compact. (20.75)

20.8 Frames
Subatomic decompositions of function spaces on $\mathbb{R}^n$ as treated in Section 14 can be rephrased in the fashionable term of *frames*, adopting here a slightly more general point of view than usual. As for frames connected with wavelets and preferably in Hilbert spaces we refer to [Mey92] and [Dau92]. By Definition 20.2 this can be extended to spaces on d-sets Γ in $\mathbb{R}^n$. We give a brief description restricting ourselves to the simplest case. As in Definition 14.2 we assume that $\psi \in S(\mathbb{R}^n)$ has a compact support (near the origin) and

$$\sum_{m \in \mathbb{Z}^n} \psi(x - m) = 1 \quad \text{for} \quad x \in \mathbb{R}^n. \quad (20.76)$$

Let $\psi_\beta(x) = x^\beta\, \psi(x)$ where $\beta \in \mathbb{N}_0^n$ is a multi-index. Then

$$(\beta qu)_{vm}(x) = 2^{-v\left(s - \frac{n}{p}\right)} \psi^\beta(2^v x - m) \quad (20.77)$$

with $v \in \mathbb{N}_0$, $\beta \in \mathbb{N}_0^n$, $s \in \mathbb{R}$, $0 < p \leq \infty$, is called an $(s, p) - \beta$-quark (more precisely $(s, p)_{-1} - \beta$-quark). Let, in addition, $\frac{n}{n+1} < p \leq \infty$, $0 < q \leq \infty$, and $s > \sigma_p$, then by Corollary 14.7, the above $(s, p) - \beta$-quarks

$$\{(\beta qu)_{vm} : \beta \in \mathbb{N}_0^n, v \in \mathbb{N}_0, m \in \mathbb{Z}^n\} \quad (20.78)$$

may be considered as a *frame* in $B_{pq}^s(\mathbb{R}^n)$: An element $f \in S'(\mathbb{R}^n)$ belongs to $B_{pq}^s(\mathbb{R}^n)$ if, and only if, it can be represented as

$$f = \sum_{\beta \in \mathbb{N}_0^n} \sum_{v=0}^{\infty} \sum_{m \in \mathbb{Z}^n} \lambda_{vm}^\beta\, (\beta qu)_{vm}(x) \quad (20.79)$$

with

$$\sup_\beta 2^{\varrho|\beta|} \|\lambda^\beta \mid b_{pq}\| < \infty, \quad (20.80)$$

where for fixed ϱ the infimum of (20.80) over all admissible representations (20.79) is an equivalent quasi-norm in $B^s_{pq}(\mathbb{R}^n)$. Here any $\varrho > 0$ sufficiently large is admitted and

$$\lambda^\beta = \left\{ \lambda^\beta_{\nu m} \in \mathbb{C} : \nu \in \mathbb{N}_0, \ m \in \mathbb{Z}^n \right\}.$$

Recall that the sequence space b_{pq} had been introduced in Definition 13.5. By Definition 20.2 the $\mathbb{R}^n$-frame (20.78) based on the quarks (20.77) has a counterpart for corresponding spaces on d-sets Γ in $\mathbb{R}^n$. Although we used this implicitly in the proof of Proposition 20.5 it might be useful to have an explicit formulation. We restrict ourselves again to the simplest case. Let Γ be a compact d-set in $\mathbb{R}^n$ with $0 < d < n$ according to Definition 3.1 and let

$$1 \le p \le \infty, \quad 0 < q \le \infty, \quad s \ge 0. \tag{20.81}$$

Then $\mathbb{B}^s_{pq}(\Gamma)$ are the spaces introduced in Definition 20.2, notationally complemented by $\mathbb{B}^0_{pq}(\Gamma) = L_p(\Gamma)$. The $(s, p) - \beta$-quarks on Γ are now given by

$$(\beta qu)_{\nu m}(\gamma) = 2^{-\nu(s-\frac{d}{p})} \psi^\beta (2^\nu \gamma - m), \quad \gamma \in \Gamma. \tag{20.82}$$

Let $\nu \in \mathbb{N}_0$ and

$$\mathbb{Z}^{n,\nu,\Gamma} = \{ m \in \mathbb{Z}^n : C Q_{\nu m} \cap \Gamma \ne \emptyset \} \tag{20.83}$$

in agreement with (20.28). Finally we modify b_{pq} introduced in Definition 13.5 by

$$
\begin{aligned}
b^\Gamma_{pq} = \Bigg\{ \lambda : \quad & \|\lambda \mid b^\Gamma_{pq}\| \\
& = \left(\sum_{\nu=0}^{\infty} \left(\sum_{m \in \mathbb{Z}^{n,\nu,\Gamma}} |\lambda_{\nu m}|^p \right)^{\frac{q}{p}} \right)^{\frac{1}{q}} < \infty \Bigg\}
\end{aligned}
\tag{20.84}
$$

(with the usual modification if $p = \infty$ and/or $q = \infty$), where

$$\lambda = \left\{ \lambda_{\nu m} \in \mathbb{C} : \nu \in \mathbb{N}_0, \ m \in \mathbb{Z}^{n,\nu,\Gamma} \right\}.$$

20.9 Proposition *Let Γ be a compact d-set in $\mathbb{R}^n$ with $0 < d < n$ according to Definition 3.1 and let p, q, s be given by (20.81). Let $\mathbb{B}^s_{pq}(\Gamma)$ be the spaces introduced in Definition 20.2 with the notational understanding that $\mathbb{B}^0_{pq}(\Gamma) = L_p(\Gamma)$ for any q. Then*

$$\left\{ (\beta qu)_{\nu m} : \quad \beta \in \mathbb{N}^n_0, \ \nu \in \mathbb{N}_0, \ m \in \mathbb{Z}^{n,\nu,\Gamma} \right\}, \tag{20.85}$$

given by (20.82) *and* (20.83), *is a frame in* $\mathbb{B}^s_{pq}(\Gamma)$ *in the following sense:* $f \in L_1(\Gamma)$ *belongs to* $\mathbb{B}^s_{pq}(\Gamma)$ *if, and only if, it can be represented as*

$$f = \sum_{\beta \in \mathbb{N}^n_0} \sum_{\nu=0}^{\infty} \sum_{m \in \mathbb{Z}^{n,\nu,\Gamma}} \lambda^\beta_{\nu m} (\beta q u)_{\nu m}(\gamma) \tag{20.86}$$

(convergence in $L_1(\Gamma)$*) with*

$$\sup_{\beta} 2^{\varrho|\beta|} \|\lambda^\beta | b^\Gamma_{pq}\| < \infty, \tag{20.87}$$

where for fixed ϱ *the infimum in* (20.87) *over all admissible representations* (20.86) *is an equivalent quasi-norm in* $\mathbb{B}^s_{pq}(\Gamma)$. *Here any* $\varrho > 0$ *sufficiently large is admitted,*

$$\lambda^\beta = \left\{ \lambda^\beta_{\nu m} \in \mathbb{C} : \quad \nu \in \mathbb{N}_0, \ m \in \mathbb{Z}^{n,\nu,\Gamma} \right\}$$

and $q = 1$ *in case of* $s = 0$.

Proof. By Definition 20.2 and Theorem 18.6 the assertion can be reduced to the traces of the spaces $B^{s+\frac{n-d}{p}}_{pq}(\mathbb{R}^n)$. However under the above restrictions we can apply the frame (20.78) and the representation (20.79), (20.80). Restriction to Γ yields the desired result. This includes the claimed L_1-convergence of (20.86) since the cardinal numbers M_ν of $\mathbb{Z}^{n,\nu,\Gamma}$ can be estimated by (20.29).

20.10 Remark The considerations in 20.8 and 20.9 can be extended to all other cases covered according to Section 14 and Definition 20.2, complemented by Theorem 18.6.

20.11 Diluted d-sets and weighted spaces
Let $0 < p_1 \le p_2 \le \infty$, $0 < q \le \infty$, and

$$s - d \left(\frac{1}{p_1} - \frac{1}{p_2} \right) > 0. \tag{20.88}$$

Then (20.44) with $s_2 = 0$ can be reformulated as

$$e_k \left(id : \quad B^{s+\frac{n-d}{p_1}}_{p_1 q}(\mathbb{R}^n) \to L_{p_2}(\Gamma) \right) \sim k^{-\frac{s}{d}}, \quad k \in \mathbb{N}. \tag{20.89}$$

Later on, in Section 22, in connection with anisotropic and nonisotropic d-sets, we concentrate on assertions of that type (estimates from above). Here we indicate some other possibilities. As mentioned in 14.14 and 14.13 there is a good chance to extend the subatomic decompositions in Theorem 14.4 and in the Corollaries 14.7 and 14.11 from the unweighted spaces $B^s_{pq}(\mathbb{R}^n)$ and $F^s_{pq}(\mathbb{R}^n)$ to some weighted spaces $B^s_{pq}(\mathbb{R}^n, w(\cdot))$ and $F^s_{pq}(\mathbb{R}^n, w(\cdot))$ considered in [HaT94a,b] and [ET96],

Ch. 4. There we calculated the entropy numbers for compact embeddings between these spaces. Combining the techniques used there with the results and the methods developed here it seems to be possible to estimate the entropy numbers for the embeddings of type

$$id: \quad B_{p_1 q}^{s+\frac{n-d}{p_1}} (\mathbb{R}^n, w(\cdot)) \to L_{p_2}(\Gamma), \tag{20.90}$$

where now Γ is a non-compact d-set according to Definition 3.1 typically with $\mathcal{H}^d(\Gamma) = \infty$. Also the replacement of $L_{p_2}(\Gamma)$ by some weighted versions $L_{p_2}(\Gamma, v(\cdot))$ makes sense. If one accepts that nature produces fractals then it is reasonable to have a closer look at non-compact fractals Γ which are locally d-sets but getting thinner and thinner if tending to infinity. We modify Definition 3.1:

Let $0 < d < n$. A closed set Γ in $\mathbb{R}^n$ is called a diluted d-set if there are two functions

$$0 < c_1(\gamma) \leq c_2(\gamma) \text{ for } \gamma \in \Gamma \text{ with } c_2(\gamma) \to 0 \text{ if } |\gamma| \to \infty$$

such that for all $\gamma \in \Gamma$ and all r with $0 < r < 1$

$$c_1(\gamma) r^d \leq \mathcal{H}^d \left(B(\gamma, r) \cap \Gamma \right) \leq c_2(\gamma) r^d. \tag{20.91}$$

Of course $\mathcal{H}^d$ stands for the Hausdorff measure. Now one can ask for the entropy numbers of possible embeddings

$$id: \quad B_{p_1 q}^{s+\frac{n-d}{p_1}} (\mathbb{R}^n) \to L_{p_2}(\Gamma) \tag{20.92}$$

and their weighted counterparts. Any result of this type can be used for spectral assertions of fractal (weighted) pseudodifferential operators following the methods developed in Chapter V below. But nothing has been done so far.

21 Fractal limiting embeddings

21.1 Preliminaries
Let Γ be a compact d-set in $\mathbb{R}^n$ according to Definition 3.1 with $0 < d < n$. Let $0 < p \leq \infty$ and $s > 0$. Then we put

$$\frac{1}{p^s} = \frac{1}{p} + \frac{s}{d} \tag{21.1}$$

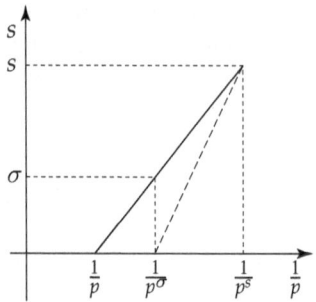

Fig. 21.1

analogously to (11.25). Hence, $\frac{1}{p}$ is the «foot point» of the line of slope d in Fig. 21.1. We can reformulate (20.75) as follows:

> *Let $0 < p \leq \infty$ and $s > 0$. Then the embedding*

$$id: \quad \mathbb{B}^s_{p^s q}(\Gamma) \to L_p(\Gamma), \quad q \leq \min(1, p), \tag{21.2}$$

is continuous but not compact.

On the other hand, let $0 < \sigma < s$, then we have by (20.43) with $\delta_+ = \delta$,

$$\delta = s - \frac{d}{p^s} + \frac{d}{p^\sigma} = \sigma > 0, \tag{21.3}$$

and hence by Theorem 20.6

$$e_k \left(id: \quad \mathbb{B}^s_{p^s q}(\Gamma) \to L_{p^\sigma}(\Gamma) \right) \sim k^{-\frac{s}{d}}, \quad 0 < q \leq \infty, \ k \in \mathbb{N}. \tag{21.4}$$

If p, q and s are fixed than the equivalence constants in (21.4) depend on σ. In particular one would expect that the corresponding constant $c(\sigma)$ in the estimate from above in (21.4) tends to infinity if $\sigma \to 0$. As we shall see we can estimate the dependence of $c(\sigma)$ on σ. This gives us the possibility to deal with the following problem:

Are there natural spaces larger than $L_p(\Gamma)$ but smaller than any of the spaces $L_{p^\sigma}(\Gamma)$ with $\sigma > 0$ such that the embedding of $\mathbb{B}^s_{p^s q}(\Gamma)$ into these spaces is compact, including a control about the related entropy numbers?

First candidates are the spaces $L_p(\log L)_a(\Gamma)$ with $a < 0$, and the related mappings are called *fractal limiting embeddings*. As we mentioned above some parts of this book might be considered as the *fractal twin of ET* (in the version [ET96]). This applies also to Section 21. We dealt in [ET96], 2.6, 3.4, and also later on in that book with the spaces $L_p(\log L)_a(\Omega)$ where Ω is a domain in $\mathbb{R}^n$ with $|\Omega| < \infty$ and their use in spectral theory in some detail. We rely on the techniques developed there and indicate the necessary modifications.

21.2 Definition *Let Γ be a compact d-set in $\mathbb{R}^n$ according to Definition 3.1 with $0 < d < n$.*

(i) *Let $0 < p < \infty$ and $a \in \mathbb{R}$. Then $L_p(\log L)_a(\Gamma)$ is the set of all μ-measurable complex-valued functions f such that*

$$\int_\Gamma |f(\gamma)|^p \, \log^{ap}(2 + |f(\gamma)|) \, \mu(d\gamma) < \infty. \tag{21.5}$$

(ii) *Let $a < 0$. Then $L_\infty(\log L)_a(\Gamma)$ is the set of all μ-measurable complex-valued functions f for which there exists a constant $\lambda > 0$ such that*

$$\int_\Gamma \exp\left\{ (\lambda|f(\gamma)|)^{-\frac{1}{a}} \right\} \mu(d\gamma) < \infty. \tag{21.6}$$

21.3 Comments and properties
This is the direct counterpart of Definition 2.6.1 in [ET96], p. 66. There one finds also the necessary references to these *logarithmic spaces*. We do not go into detail. Further assertions may be found in [BeS88]. The spaces $L_\infty(\log L)_a(\Gamma)$ might be better known as $L_{\exp,-a}(\Gamma)$. But our notation is justified by the proposition below. All these spaces are quasi-Banach spaces, equipped with suitable quasi-norms with the help of rearrangement functions. We again do not go into detail and refer to [ET96], pp. 66–67, and the references and comments given there. Analogously to Proposition 1 in [ET96], p. 67, we obtain immediately by the above definition:

(i) *Let $0 < \varepsilon < p < \infty$ and $-\infty < a_2 < a_1 < \infty$. Then*

$$L_{p+\varepsilon}(\Gamma) \subset L_p(\log L)_{a_1}(\Gamma) \subset L_p(\log L)_{a_2}(\Gamma) \subset L_{p-\varepsilon}(\Gamma) \tag{21.7}$$

and

$$L_p(\log L)_\varepsilon(\Gamma) \subset L_p(\Gamma) \subset L_p(\log L)_{-\varepsilon}(\Gamma). \tag{21.8}$$

(ii) *Let $-\infty < b_1 < b_2 < 0$. Then*

$$L_\infty(\Gamma) \subset L_\infty(\log L)_{b_2}(\Gamma) \subset L_\infty(\log L)_{b_1}(\Gamma). \tag{21.9}$$

In particular, as desired, the spaces $L_p(\log L)_a(\Gamma)$ with $a < 0$ are larger than $L_p(\Gamma)$ and smaller than any of the spaces $L_{p^\sigma}(\Gamma)$ with $\sigma > 0$. In [ET96], 2.6.2, pp. 69–75, we proved two basic theorems for the spaces $L_p(\log L)_a(\Omega)$, where Ω is a domain in $\mathbb{R}^n$ with $|\Omega| < \infty$, for $a < 0$ and $a > 0$, respectively. Both theorems have immediate counterparts for the respective spaces on Γ in place of Ω. We need here only the spaces $L_p(\log L)_a(\Gamma)$ with $a < 0$ and so we restrict the formulation below to that case. As for a proof and also references to the relevant literature we refer to [ET96], 2.6.2, pp. 69–75, and Remark 5 on p. 81. In agreement with (21.1) we set

$$\frac{1}{p^{\sigma_j}} = \frac{1}{p} + \frac{\sigma_j}{d} \quad \text{where} \quad 0 < p \le \infty \quad \text{and} \quad \sigma_j = 2^{-j}, \; j \in \mathbb{N}. \tag{21.10}$$

21.4 Proposition *Let Γ be a compact d-set in $\mathbb{R}^n$ according to Definition 3.1 with $0 < d < n$. Let $0 < p \le \infty$ and $a < 0$. Then $L_p(\log L)_a(\Gamma)$ is the set of all μ-measurable complex-valued functions f such that*

$$\left(\sum_{j=1}^{\infty} 2^{jap} \|f \mid L_{p^{\sigma_j}}(\Gamma)\|^p\right)^{\frac{1}{p}} < \infty \tag{21.11}$$

(with the usual modification if $p = \infty$), and (21.11) defines an equivalent quasi-norm on $L_p(\log L)_a(\Gamma)$.

21.5 Remark Analogously to [ET96], 2.6.2, there is a continuous version of (21.11) and also a counterpart for the spaces $L_p(\log L)_a(\Gamma)$ with $1 \le p < \infty$ and $a > 0$. By (21.11) it is also clear that $L_p(\log L)_a(\Gamma)$ with $1 < p \le \infty$ and $a < 0$ is a Banach space (equivalent norming with $\sum\limits_{j=J}^{\infty}$ in place of $\sum\limits_{j=1}^{\infty}$, where $J \in \mathbb{N}$ is chosen sufficiently large).

21.6 Embeddings
Let Γ be again a compact d-set in $\mathbb{R}^n$ with $0 < d < n$. By (21.2), (21.4), and (21.8) it makes sense to ask for which $a < 0$ the embeddings

$$id : \quad \mathbb{B}_{p^s q}^{s}(\Gamma) \to L_p(\log L)_a(\Gamma) \tag{21.12}$$

exist and whether they are compact. Assuming that we can control the constant $c(\sigma)$ mentioned after (21.4) then Proposition 21.4 paves the way to deal with this problem. Although not really necessary we assume in addition $1 < p \le \infty$: this is the case of interest in connection with the spectral theory in the next chapter. If $\sigma > 0$ is small such that $p^\sigma > 1$ then by Theorem 18.6

$$tr_\Gamma \, B_{p^\sigma,1}^{\frac{n-d}{p^\sigma}}(\mathbb{R}^n) = L_{p^\sigma}(\Gamma). \tag{21.13}$$

As mentioned after (20.36) the estimate from above in (21.4) is covered by (20.22) with $s_2 = 0$, $q_2 = 1$ and the necessary notational changes. The proof of (20.22) was based on the subatomic decompositions in Theorem 14.4 and Corollary 14.7 on the one hand and Theorem 9.2 on the other hand. However in our case we can replace Theorem 9.2 by Corollary 9.4 and obtain

$$e_k \left(id : \mathbb{B}_{p^s q}^{s}(\Gamma) \to L_{p^\sigma}(\Gamma)\right) \le c\, \delta^{-1-2(\frac{1}{p^s} - \frac{1}{p^\sigma})}\, k^{-\frac{\delta}{d} + \frac{1}{p^\sigma} - \frac{1}{p^s}}$$
$$\le c\, \sigma^{-1-2\frac{s-\sigma}{d}}\, k^{-\frac{s}{d}} \le c'\, \sigma^{-1-2\frac{s}{d}}\, k^{-\frac{s}{d}}, \quad k \in \mathbb{N}, \tag{21.14}$$

where in the second inequality we used (21.1) and (21.3). The constants c and c' are independent of $\sigma > 0$.

21.7 Theorem *Let* Γ *be a compact d-set in* $\mathbb{R}^n$ *with* $0 < d < n$ *according to Definition 3.1. Let*

$$1 < p \le \infty, \ s > 0, \ 0 < q \le \infty, \quad and \quad a < -1 - \frac{2s}{d}. \tag{21.15}$$

Let p^s *be given by* (21.1) *and let* $\mathbb{B}^s_{p^s q}(\Gamma)$ *and* $L_p(\log L)_a(\Gamma)$ *be the spaces introduced in the Definitions 20.2 and 21.2, respectively. Then the embedding of* $\mathbb{B}^s_{p^s q}(\Gamma)$ *into* $L_p(\log L)_a(\Gamma)$ *is compact and for the related entropy numbers we have*

$$e_k\left(id: \ \mathbb{B}^s_{p^s q}(\Gamma) \to L_p(\log L)_a(\Gamma)\right) \sim k^{-\frac{s}{d}}, \quad k \in \mathbb{N}. \tag{21.16}$$

Proof.
Step 1. We prove the estimate from above. Given $J \in \mathbb{N}$ and $a < b < -1 - \frac{2s}{d}$; let

$$k_j \sim 2^{\frac{d}{s}(j-J)(b+1+\frac{2s}{d})} \quad \text{if} \quad j = 1, \dots, J, \tag{21.17}$$

where $\sim$ means that k_j is a nearby natural number. By (21.14) with $\sigma_j = 2^{-j}$ we have with obvious notation

$$e_{k_j}\left(id: \ \mathbb{B}^s_{p^s q}(\Gamma) \to 2^{jb} L_{p^{\sigma_j}}(\Gamma)\right)$$
$$\le c \, 2^{jb} \, 2^{j(1+2\frac{s}{d})} \, 2^{(J-j)(b+1+2\frac{s}{d})} = c \, 2^{J(b+1+2\frac{s}{d})}. \tag{21.18}$$

Cover the unit ball U of $\mathbb{B}^s_{p^s q}(\Gamma)$ by 2^{k_1} balls in $2^b L_{p^{\sigma_1}}(\Gamma)$ of radius $C \, e_{k_1}$, each ball having centre in U, and $C > 1$ is an appropriate number. Let U_1 be one of these balls, and cover $U \cap U_1$ by 2^{k_2} balls in $2^{2b} L_{p^{\sigma_2}}(\Gamma)$ of radius $C \, e_{k_2}$, where we may assume that the centres of these balls are in $U \cap U_1$. By iteration, we obtain a covering of $U \cap U_1 \cap \cdots \cap U_{J-1}$ by 2^{k_J} balls in $2^{Jb} L_{p^{\sigma_J}}(\Gamma)$ of radius $C \, e_{k_J}$ where the centres of these balls are in $U \cap U_1 \cap \cdots \cap U_{J-1}$. Let these centres be g_l with $l = 1, \dots, L$ where

$$L = 2^{k_1 + \cdots + k_J} \quad \text{and} \quad \sum_{j=1}^{J} k_j \sim 2^{\frac{d}{s}J|b+1+2\frac{s}{d}|}. \tag{21.19}$$

Thus when $f \in U$ is given, there is one of these centres g_l such that

$$2^{jb} \|f - g_l \,|\, L_{p^{\sigma_j}}(\Gamma)\| \le c \, e_{k_j} \le c' \, 2^{J(b+1+2\frac{s}{d})} \tag{21.20}$$

if $j = 1, \dots, J$. Let $j > J$. Then it follows by (21.14) with $k = 1$,

$$2^{jb} \|f - g_l \,|\, L_{p^{\sigma_j}}(\Gamma)\| \le c \, 2^{j(b+1+2\frac{s}{d})} \le c \, 2^{J(b+1+2\frac{s}{d})}. \tag{21.21}$$

With $k = \sum_{j=1}^{J} k_j$ we obtain by (21.19)–(21.21), Proposition 21.4, and $a < b$,

$$e_k \left(id \; : \; \mathbb{B}^s_{p^s q}(\Gamma) \to L_p(\log L)_a(\Gamma) \right) \leq c \, 2^{J(b+1+2\frac{s}{d})} \leq c' \, k^{-\frac{s}{d}}. \tag{21.22}$$

In view of the monotonicity the estimate from above in (21.16) follows.

Step 2. By (21.4) and (21.7) we have

$$
\begin{aligned}
c_1 \, k^{-\frac{s}{d}} &\leq e_k \left(id \; : \; \mathbb{B}^s_{p^s q}(\Gamma) \to L_{p^\sigma}(\Gamma) \right) \\
&\leq c_2 \, e_k \left(id \; : \; \mathbb{B}^s_{p^s q}(\Gamma) \to L_p(\log L)_a(\Gamma) \right)
\end{aligned} \tag{21.23}
$$

for some positive c_1 and c_2 and all $k \in \mathbb{N}$. This proves the estimate from below in (21.16).

21.8 Remark We followed closely the arguments in [ET96], pp. 136–7. On the other hand we shift a comparison of the above results with those ones obtained in [ET96] to Section 23.

21.9 The case $0 > a \geq -\frac{d+2s}{d}$
In connection with Theorem 21.7 the question arises what happens if a is larger than or equal to $-\frac{d+2s}{d}$. It might well happen that the answer depends on q. In case of $q \leq 1$ one obtains at least an estimate of the corresponding entropy numbers.

21.10 Corollary *Let Γ be a compact d-set in $\mathbb{R}^n$ with $0 < d < n$ according to Definition 3.1. Let*

$$1 < p \leq \infty, \; s > 0, \; 0 < q \leq 1 \quad and \quad -\frac{d+2s}{d} \leq a < 0. \tag{21.24}$$

Let p^s be given by (21.1) and let $\mathbb{B}^s_{p^s q}(\Gamma)$ and $L_p(\log L)_a(\Gamma)$ be the spaces introduced in the Definitions 20.2 and 21.2, respectively. Then the embedding of $\mathbb{B}^s_{p^s q}(\Gamma)$ into $L_p(\log L)_a(\Gamma)$ is compact and for any $\varepsilon > 0$ there is a constant c_ε such that for the related entropy numbers we have

$$e_k \left(id \; : \; \mathbb{B}^s_{p^s q}(\Gamma) \to L_p(\log L)_a(\Gamma) \right) \leq c_\varepsilon \, k^{\frac{s}{d+2s}a+\varepsilon}, \quad k \in \mathbb{N}. \tag{21.25}$$

Proof. We apply Proposition 21.4. Let $b < -\frac{d+2s}{d}$ and $a = \Theta b$. Then we have $0 < \Theta < 1$. By Hölder's inequality we have

$$
\begin{aligned}
2^{ja} \|f \,|\, L_{p^{\sigma_j}}(\Gamma)\| &\leq c \, \|f \,|\, L_p(\Gamma)\|^{1-\Theta} \left(2^{jb} \, \|f \,|\, L_{p^{\sigma_j}}(\Gamma)\| \right)^{\Theta} \\
&\leq c \, 2^{-j\eta\Theta} \, \|f \,|\, L_p(\Gamma)\|^{1-\Theta} \, \|f \,|\, L_p(\log L)_{b+\eta}(\Gamma)\|^{\Theta}
\end{aligned} \tag{21.26}
$$

for any small $\eta > 0$. We obtain

$$\|f \mid L_p(\log L)_a(\Gamma)\|$$
$$\leq c \|f \mid L_p(\Gamma)\|^{1-\Theta} \|f \mid L_p(\log L)_{b+\eta}(\Gamma)\|^{\Theta}. \tag{21.27}$$

We assume $b + \eta < -\frac{d+2s}{d}$. Then it follows by (21.2) and Theorem 21.7

$$id : \quad \mathbb{B}^s_{p^s q}(\Gamma) \to L_p(\Gamma) \quad \text{is continuous,} \tag{21.28}$$

and

$$e_k \left(id : \mathbb{B}^s_{p^s q}(\Gamma) \to L_p(\log L)_{b+\eta}(\Gamma) \right) \sim k^{-\frac{s}{d}}, \quad k \in \mathbb{N}. \tag{21.29}$$

However (21.27)–(21.29) allow to apply the interpolation property for entropy numbers where we gave some references in 6.7. In particular, by [ET96], Theorem 1.3.2, p. 13, it follows

$$e_k \left(id : \mathbb{B}^s_{p^s q}(\Gamma) \to L_p(\log L)_a(\Gamma) \right)$$
$$\leq c\, e_k \left(id : \mathbb{B}^s_{p^s q}(\Gamma) \to L_p(\log L)_{b+\eta}(\Gamma) \right)^{\Theta} \leq c'\, k^{-\frac{s}{d}\Theta}. \tag{21.30}$$

Any Θ with

$$\Theta < \frac{|a|d}{d + 2s} \tag{21.31}$$

is admitted. Inserting (21.31) in (21.30) we obtain (21.25) for any given $\varepsilon > 0$.

22 Nonisotropic embeddings

22.1 Preliminaries and speculations

In case of an (isotropic) d-set Γ in $\mathbb{R}^n$ we introduced in Definition 20.2 the spaces $\mathbb{B}^s_{pq}(\Gamma)$ where $s > 0$. This was justified by Theorem 18.6 and (20.1). Furthermore we obtained in Theorem 20.6 rather satisfactory results for entropy numbers of compact embeddings between these spaces. If the fractal Γ is not an (isotropic) d-set the situation is less favourable. In the Theorems 18.15 and 18.17 we obtained reasonable counterparts of Theorem 18.6 and (20.1) where Γ is a compact anisotropic or nonisotropic d-set in $\mathbb{R}^2$ according to the Definitions 5.2 or 5.3, respectively. Besides d with $0 < d < 2$ also the so-called deviation a with $0 \leq a \leq 1$ occurs in (18.56) and (18.62). However to take these embeddings as starting points to introduce corresponding spaces on Γ analogously to Definition 20.2 is not reasonable. For example, our discussion on anisotropic curves in in 4.21–4.25 shows that it may happen that some fractals can be represented in many ways as anisotropic d-sets with different values of d (and a). Neither fractals in general, nor the more special anisotropic and nonisotropic fractals introduced in Definitions 5.2 and 5.3, respectively, can be characterized by one or two numbers, such as d and a. This applies also to the constructive and more special PXT and OF fractals according to Proposition 5.6 with references to 4.17 and 4.19. On the other hand, PXT and OF sets are special fractals generated by *iterated function systems*. In Section 19 we did a first step to find a natural link between fractals generated by iterated function systems and characterizing *designer distributions* and measures. But as we speculated in 19.2 it might well be the case that this is the right way to introduce natural *designer function spaces* connected with those fractals. Afterwards one could try to study entropy numbers between these function spaces which, in turn, pave the way to a related spectral theory as developed in Chapter V. Our aim in this Section 22 is more modest. We avoid any discussion about function spaces on anisotropic or nonisotropic fractals. Instead of the operator *id* in (20.44) we look at the trace operator tr_Γ from some spaces $B^\sigma_{p_1 q}(\mathbb{R}^2)$ into $L_{p_2}(\Gamma)$. The general background for anisotropic and nonisotropic fractals has been described in 18.14 which will not be repeated here.

22.2 Theorem *Let Γ be a compact anisotropic d-set in $\mathbb{R}^2$ with $0 < d < 2$ and the deviation $0 \le a \le 1$ according to Definition 5.2. Let $0 < p_1 \le p_2 \le \infty$, $0 < q \le \infty$,*

$$s(p_1, p_2) = \frac{2}{p_1} - \frac{d}{(1+a)p_2}, \quad \text{and} \quad s > 0. \tag{22.1}$$

Then the trace operator

$$tr_\Gamma : \quad B^{s(p_1,p_2)+s}_{p_1 q}(\mathbb{R}^2) \to L_{p_2}(\Gamma) \tag{22.2}$$

is compact and for the related entropy numbers we have

$$e_k\left(tr_\Gamma : B^{s(p_1,p_2)+s}_{p_1 q}(\mathbb{R}^2) \to L_{p_2}(\Gamma)\right) \le c\, k^{-\varrho} \tag{22.3}$$

with

$$\varrho = \frac{1+a}{d+2a} s + \frac{1}{p_1} - \frac{1}{p_2}$$

for some $c > 0$ and all $k \in \mathbb{N}$.

Proof. We follow Step 1 of the proof of Proposition 20.5. Let $\sigma = s(p_1, p_2) + s$ and let $g \in B^\sigma_{p_1 q}(\mathbb{R}^2)$, assuming that g is zero outside of a fixed neighbourhood of Γ. We expand g according to the subatomic representation Theorem 14.4(i) and Corollary 14.11(i). By the same reasoning as between the formulas (20.25) and (20.26) we suppose, without restriction of generality, that Corollary 14.7 can be applied and that we have obvious counterparts of (20.26), (20.27) adapted to our situation. Let again $\psi(x)$ be given by (14.11), (14.12) and $\psi^\beta(x) = x^\beta \psi(x)$ now in $\mathbb{R}^2$. We replace 2^ν in (20.26) by $2^{\nu(1+a)}$ where a is the anisotropic deviation. Then we have analogously to (20.26) and (20.27)

$$g = \sum_{\beta \in \mathbb{N}_0^2} \sum_{\nu=0}^{\infty} \sum_{m \in \mathbb{Z}^2} \lambda^\beta_{\nu m}\, 2^{-\nu(1+a)(\sigma - \frac{2}{p_1})}\, \psi^\beta(2^{\nu(1+a)}x - m) \tag{22.4}$$

and

$$\sup_{\beta} 2^{\varrho|\beta|} \, \|\lambda^{\beta} \, | \, b_{p_1 q}\| \sim \|g \, | \, B^{\sigma}_{p_1 q}(\mathbb{R}^2)\| \tag{22.5}$$

with $\varrho > 0$ large. Again let

$$\lambda^{\beta} = \left(\lambda^{\beta}_{\nu m} : \nu \in \mathbb{N}_0 \quad \text{and} \quad m \in \mathbb{Z}^2 \right).$$

In modification of the cubes $Q_{\nu m}$ and $c \, Q_{\nu m}$ in connection with (20.26) we have now the cubes $Q^a_{\nu m}$; centred at $2^{-\nu(1+a)} m$ and with side-length $2^{-\nu(1+a)}$ where $\nu \in \mathbb{N}_0$ and $m \in \mathbb{Z}^2$; and $cQ^a_{\nu m}$. Here a is the above anisotropic deviation. We modify (20.28) by

$$\lambda^{\beta,\Gamma} = \left(\lambda^{\beta}_{\nu m} : \nu \in \mathbb{N}_0, \, m \in \mathbb{Z}^2, \, C \, Q^a_{\nu m} \cap \Gamma \neq \emptyset \right), \tag{22.6}$$

where again we assume that $C > 1$ is chosen sufficiently large. Next we have to find the counterpart of (20.29), where M_{ν} is now the number of cubes $Q^a_{\nu m}$ such that

$$C \, Q^a_{\nu m} \cap \Gamma \neq \emptyset.$$

For fixed $\nu \in \mathbb{N}_0$ it follows by (5.6) that there are at most $c2^{\nu d}$ rectangles R^{ν}_l according to 5.1 and 5.2, where $c > 0$ is independent of ν. By a volume argument a given rectangle R^{ν}_l has a non-empty intersection with at most

$$c \, 2^{-2\nu} \, 2^{2(1+a)\nu} = c \, 2^{2a\nu} \tag{22.7}$$

cubes $C \, Q^a_{\nu m}$. Hence we have

$$M_{\nu} \leq c \, 2^{\nu(d+2a)}, \quad \nu \in \mathbb{N}_0, \tag{22.8}$$

for some $c > 0$ which is independent of ν. Now we follow the construction of the operators S, T, id, and tr_{Γ} as in (20.30)–(20.36) and indicate the necessary changes. Based on (22.4), (22.5) we introduce the (non-linear) operator S,

$$S \, : \, B^{\sigma}_{p_1 q}(\mathbb{R}^2) \to \ell_{\infty} \left[2^{\varrho_1|\beta|} \, \ell_q \left(2^{\nu\delta} \, \ell^{M_{\nu}}_{p_1} \right) \right], \tag{22.9}$$

where S is given by (20.31), (20.32) with $n = 2$ and $Q^a_{\nu m}$ in place of $Q_{\nu m}$. The value of $\delta > 0$ will be determined later on. By (22.4), (22.5) it follows that S is a (non-linear) bounded map, where all the sequence spaces have the previous meaning. The counterpart of the linear operator T, given by (20.33), (20.34), reads as follows,

$$T \, : \, \ell_{\infty} \left[2^{\varrho_2|\beta|} \, \ell_{\overline{p_2}} \left(\ell^{M_{\nu}}_{p_2} \right) \right] \to L_{p_2}(\Gamma), \tag{22.10}$$

where $\overline{p_2} = \min(1, p_2)$, $\varrho_1 > \varrho_2 > 0$ sufficiently large, and

$$T\varkappa = \sum_{\beta \in \mathbb{N}_0^2} \sum_{\nu=0}^{\infty} \sum_{m} \varkappa^{\beta}_{\nu m} \, 2^{\frac{d\nu}{p_2}} \, \psi^{\beta} \left(2^{\nu(1+a)}\gamma - m \right), \tag{22.11}$$

$\gamma \in \Gamma$, where $\varkappa = (\varkappa^{\beta,\Gamma} : \beta \in \mathbb{N}_0^2)$ and the sum over m in (22.11) is taken according to (22.6) with $\varkappa_{\nu m}^\beta$ in place of $\lambda_{\nu m}^\beta$. By Theorem 5.5 we have for fixed $\beta \in \mathbb{N}_0^2$,

$$\left\| \sum_{\nu=0}^{\infty} \sum_m \varkappa_{\nu m}^\beta \, 2^{\frac{d\nu}{p_2}} \, \psi^\beta \left(2^{\nu(1+a)}\gamma - m \right) \mid L_{p_2}(\Gamma) \right\|^{\overline{p_2}}$$

$$\leq \sum_{\nu=0}^{\infty} \left\| \sum_m \varkappa_{\nu m}^\beta \, 2^{\frac{d\nu}{p_2}} \, \psi^\beta \left(2^{\nu(1+a)}\gamma - m \right) \mid L_{p_2}(\Gamma) \right\|^{\overline{p_2}}$$

$$\leq c^{|\beta|} \sum_{\nu=0}^{\infty} \left(\sum_m |\varkappa_{\nu m}^\beta|^{p_2} \right)^{\frac{\overline{p_2}}{p_2}} \leq c^{|\beta|} \left\| \varkappa^{\beta,\Gamma} \mid \ell_{\overline{p_2}} \left(\ell_{p_2}^{M_\nu} \right) \right\|^{\overline{p_2}}, \quad (22.12)$$

where $c \geq 1$ is a constant which is independent of β. Since ϱ_1 in (22.9), and hence also ϱ_2 in (22.10), is at our disposal, we obtain (22.10) for large values of ϱ_2 as a consequence of (22.12). As in (20.35) we complement S and T by

$$id : \ell_\infty \left[2^{\varrho_1 |\beta|} \, \ell_q \left(2^{\nu\delta} \, \ell_{p_1}^{M_\nu} \right) \right] \to \ell_\infty \left[2^{\varrho_2 |\beta|} \, \ell_{\overline{p_2}} \left(\ell_{p_2}^{M_\nu} \right) \right] \qquad (22.13)$$

with $\varrho_1 > \varrho_2 > 0$ sufficiently large. We wish to determine δ in such a way that

$$tr_\Gamma = T \circ id \circ S. \qquad (22.14)$$

By (22.4), (20.31), (20.32), and (22.11) this is the case if we choose δ such that

$$-(1+a)\left(\sigma - \frac{2}{p_1}\right) = -\delta + \frac{d}{p_2}. \qquad (22.15)$$

Since

$$\sigma - \frac{2}{p_1} = s(p_1, p_2) + s - \frac{2}{p_1} = -\frac{d}{(1+a)p_2} + s \qquad (22.16)$$

we have

$$\delta = (1+a)s. \qquad (22.17)$$

By (22.8) and (22.17) the counterpart of (20.39) is given by

$$e_k(id) \leq c \, k^{-\varrho}, \quad k \in \mathbb{N}, \qquad (22.18)$$

with

$$\varrho = \frac{1+a}{d+2a} s + \frac{1}{p_1} - \frac{1}{p_2}.$$

Now we obtain (22.3) by the same reasoning as in Step 1 of the proof of Proposition 20.5.

22.3 Discussion

By 22.1 one cannot expect in general that (22.3) is an equivalence, in contrast to Theorem 20.6. We compare the Theorems 20.6 and 22.2. Let the deviation $a = 0$. Then by Theorem 5.5 the fractal Γ becomes an (isotropic) d-set. In this case we have by (22.3) and Definition 20.2

$$e_k \left(id : \mathbb{B}^\sigma_{p_1 q}(\Gamma) \to L_{p_2}(\Gamma) \right) \leq c\, k^{-\frac{s}{d} + \frac{1}{p_2} - \frac{1}{p_1}} \tag{22.19}$$

with

$$\sigma + \frac{2 - d}{p_1} = s(p_1, p_2) + s = \frac{2}{p_1} - \frac{d}{p_2} + s \tag{22.20}$$

and hence

$$\frac{s}{d} + \frac{1}{p_1} - \frac{1}{p_2} = \frac{\sigma}{d}. \tag{22.21}$$

This is the same exponent as in (20.44). Furthermore, by (22.21)

$$\delta = \sigma - d\left(\frac{1}{p_1} - \frac{1}{p_2} \right) > 0 \tag{22.22}$$

in (20.43) coincides with $s > 0$ as required in the above theorem. Hence the assumptions in the above theorem are at least natural. Furthermore if $p_1 = p_2 = p$ and $0 \leq a \leq 1$ then

$$s(p, p) = \frac{1}{p}\left(2 - \frac{d}{1 + a} \right) \tag{22.23}$$

what is just the exponent in (18.56). Hence, by the above arguments one obtains the following generalization of (18.56).

22.4 Corollary *Let Γ be a compact anisotropic d-set in $\mathbb{R}^2$ with $0 < d < 2$ and the deviation $0 \leq a \leq 1$ according to Definition 5.2. Let $0 < p \leq \infty$ and $0 < q \leq \min(1, p)$. Then*

$$tr_\Gamma\, B^{\frac{1}{p}(2 - \frac{d}{1+a})}_{pq}(\mathbb{R}^2) \subset L_p(\Gamma). \tag{22.24}$$

Proof. We follow the proof of Theorem 22.2 now with

$$s = 0 \quad \text{and} \quad s(p_1, p_2) = s(p, p) = \frac{1}{p}\left(2 - \frac{d}{1 + a} \right)$$

in (22.1). By (22.17) we have also $\delta = 0$ This shows that id in (22.13) with $p_1 = p_2$ and $q \leq \bar{p} = \min(1, p)$ is continuous. The rest is the same as in the proof of Theorem 22.2.

22.5 Theorem *Let Γ be a compact nonisotropic d-set in $\mathbb{R}^2$ with $0 < d < 2$ and the deviation $0 \le a \le 1$ according to Definition 5.3. Let $0 < p_1 \le p_2 \le \infty$, $0 < q \le \infty$,*

$$s(p_1, p_2) = \frac{2}{p_1} - \frac{d}{(1+a)p_2}, \quad \text{and} \quad s > \frac{1}{2p_2}. \tag{22.25}$$

Then the trace operator

$$\mathrm{tr}_\Gamma \;:\; B_{p_1 q}^{s(p_1, p_2) + s}(\mathbb{R}^2) \to L_{p_2}(\Gamma) \tag{22.26}$$

is compact and for the related entropy numbers we have

$$e_k\left(\mathrm{tr}_\Gamma \;:\; B_{p_1 q}^{s(p_1, p_2) + s}(\mathbb{R}^2) \to L_{p_2}(\Gamma)\right) \le c\, k^{-\varrho} \tag{22.27}$$

with

$$\varrho = \frac{2(s - \frac{1}{2p_2})(1+a)}{2d + 1 + 5a} + \frac{1}{p_1} - \frac{1}{p_2}$$

for some $c > 0$ and all $k \in \mathbb{N}$.

Proof. We follow Step 1 of the proof of Proposition 20.5 and the proof of Theorem 22.2 and indicate the necessary changes. Let again $\sigma = s(p_1, p_2) + s$ and let $g \in B_{p_1 q}^{\sigma}(\mathbb{R}^2)$ as in 22.2 be expanded by (22.4) and (22.5). We have also (22.6). Next we have to find the counterpart of (22.8) where M_ν again is the number of cubes $Q_{\nu m}^a$ such that $C\, Q_{\nu m}^a \cap \Gamma \neq \emptyset$. Instead of the rectangles R_l^ν in the proof of Theorem 22.2 we have now to consider the ellipses E_l^ν according to Definition 5.3. For fixed $\nu \in \mathbb{N}_0$, by (5.6) there are at most $2^{\nu d}$ ellipses E_l^ν, where $c > 0$ is independent of ν. Now we must modify the volume argument in 22.2 which resulted in the rectangular case in the estimate (22.8). We have an additional defect. By (18.63) a given ellipse E_l^ν has a non-empty intersection with at most

$$c\, 2^{-2\nu}\, 2^{\frac{5}{2}(1+a)\nu} = c\, 2^{\frac{5a+1}{2}\nu} \tag{22.28}$$

cubes $C\, Q_{\nu m}^a$. Hence, instead of (22.8) we have now

$$M_\nu \le c\, 2^{\frac{5a+1+2d}{2}\nu}, \quad \nu \in \mathbb{N}_0, \tag{22.29}$$

for some $c > 0$ which is independent of ν. The operators S and T have the same meaning as in (22.9) and (22.11), respectively. But (22.12) must be modified. By our arguments after (18.63) a given cube $Q_{\nu m}^a$ has now at most $c\, 2^{\frac{1+a}{2}\nu}$ non-empty intersections with ellipses E_l^ν. Hence the left-hand side of (22.12) can now be estimated from above by

$$c^{|\beta|} \left\| \varkappa^{\beta, \Gamma} \mid \ell_{\overline{p_2}}\left(2^{\frac{1+a}{2p_2}\nu}\, \varrho M_\nu\right) \right\|^{\overline{p_2}}. \tag{22.30}$$

We must change (22.13) and (22.10) accordingly, whereas (22.14) and (22.17) are the same as before. The counterpart of (22.18) follows now from Theorem 9.2 applied to

$$
id : \ell_\infty \left[2^{\varrho_1 |\beta|} \, \ell_q \left(2^{\nu\delta} \, \ell_{p_1}^{M_\nu} \right) \right] \rightarrow \ell_\infty \left[2^{\varrho_2 |\beta|} \, \ell_{\overline{p_2}} \left(2^{\frac{1+a}{2p_2} \, \nu} \, \ell_{p_2}^{M_\nu} \right) \right] , \tag{22.31}
$$

where δ and d on the right-hand side of (9.5) must be replaced by

$$
(1+a)\, s - \frac{1+a}{2p_2} = (1+a)\,(s - \frac{1}{2p_2}) > 0 \tag{22.32}
$$

and $\frac{1}{2}(5a + 1 + 2d)$, respectively. This is the exponent ϱ in (22.27).

22.6 Remark In the discussion 22.3 and in Corollary 22.4 we proved in the anisotropic case a generalization of (18.56). By the same reasoning we obtain by the above theorem the following generalization of (18.62).

22.7 Corollary *Let Γ be a compact nonisotropic d-set in $\mathbb{R}^2$ with $0 < d < 2$ and the deviation $0 \le a \le 1$ according to Definition 5.3. Let $0 < p \le \infty$ and $0 < q \le \min(1, p)$. Then*

$$
tr_\Gamma \; B_{pq}^{\frac{1}{p}\left(\frac{5}{2} - \frac{d}{1+a}\right)}(\mathbb{R}^2) \subset L_p(\Gamma). \tag{22.33}
$$

22.8 Remark We refer to our discussion in 18.18 which makes clear that the results for the nonisotropic case have some shortcomings. On the other hand in Theorem 30.7 below we shall play the anisotropic drum. The rather satisfactory assertions obtained there are based on Theorem 22.2.

23 The spaces $B_{pq}^s(\Omega)$

23.1 Introduction
Let $s \in \mathbb{R}$, $0 < p \le \infty$ ($p < \infty$ for the F-spaces), and $0 < q \le \infty$. In Definition 10.3 we introduced the spaces $B_{pq}^s(\mathbb{R}^n)$ and $F_{pq}^s(\mathbb{R}^n)$. Let Ω be an arbitrary bounded domain in $\mathbb{R}^n$. For all admitted values of s, p, q then the space $B_{pq}^s(\Omega)$ is defined as the restriction of $B_{pq}^s(\mathbb{R}^n)$ to Ω, which means

$$
B_{pq}^s(\Omega) = \{ f \in D'(\Omega) : \text{ there is a } g \in B_{pq}^s(\mathbb{R}^n) \text{ with } g|\Omega = f \} , \tag{23.1}
$$

$$
\| f \mid B_{pq}^s(\Omega) \| = \inf \| g \mid B_{pq}^s(\mathbb{R}^n) \| , \tag{23.2}
$$

where the infimum is taken over all $g \in B_{pq}^s(\mathbb{R}^n)$ such that its restriction to Ω, denoted by $g|\Omega$, coincides in $D'(\Omega)$ with f. In the same way one can introduce the spaces $F_{pq}^s(\Omega)$ with $H_p^s(\Omega) = F_{p,2}^s(\Omega)$ as special cases. In 20.4 we collected some

results for compact embeddings between the spaces $B_{pq}^s(\Omega)$ in order to provide a
better understanding for corresponding results for the spaces $\mathbb{B}_{pq}^s(\Gamma)$ on fractals
Γ as described in Theorem 20.6 and in Section 21. We return to these comments
and references later on. By (11.15) and the above definitions we have

$$B_{pu}^s(\Omega) \subset F_{pq}^s(\Omega) \subset B_{pv}^s(\Omega), \quad s \in \mathbb{R}, \ 0 < p < \infty, \tag{23.3}$$

with

$$0 < q \leq \infty, \quad u = \min(p,q) \quad \text{and} \quad v = \max(p,q).$$

In other words, any assertion about entropy numbers for B_{pq}^s-spaces where the
q-indices do not play any role applies also to the related F_{pq}^s-spaces. This is the
case in all non-limiting situations. We shall not stress this point in the sequel.
To facilitate the comparison with the corresponding results in [ET96] we use the
same notation as there. Recall the meaning of «~» as explained in (20.19). Again
we put $a_+ = \max(a,0)$ if $a \in \mathbb{R}$.

23.2 Theorem *Let Ω be a bounded domain in $\mathbb{R}^n$. Let*

$$-\infty < s_2 < s_1 < \infty,$$

$$0 < p_1 \leq \infty, \ 0 < p_2 \leq \infty, \ 0 < q_1 \leq \infty, \ 0 < q_2 \leq \infty, \tag{23.4}$$

and

$$\delta_+ = s_1 - s_2 - n \left(\frac{1}{p_1} - \frac{1}{p_2} \right)_+ > 0. \tag{23.5}$$

*Then the embedding of $B_{p_1q_1}^{s_1}(\Omega)$ into $B_{p_2q_2}^{s_2}(\Omega)$ is compact and for the related
entropy numbers we have*

$$e_k \left(id : B_{p_1q_1}^{s_1}(\Omega) \to B_{p_2q_2}^{s_2}(\Omega) \right) \sim k^{-\frac{s_1-s_2}{n}}, \quad k \in \mathbb{N}. \tag{23.6}$$

Proof.

Step 1. The estimate of e_k from above by $c\, k^{-\frac{s_1-s_2}{n}}$ is covered by the above
definitions and the proof of Proposition 20.5. It may happen that s_1 (and then also
s_2) or s_2 are negative. But this does not matter since the subatomic representation
Theorem 14.4(i) and Corollary 14.11(i) apply to all B_{pq}^s-spaces.

Step 2. To prove the estimate from below we have to show that there is a constant
$c > 0$ such that

$$e_k \left(id : B_{p_1q_1}^{s_1}(\Omega) \to B_{p_2q_2}^{s_2}(\Omega) \right) k^{\frac{s_1-s_2}{n}} \geq c, \quad k \in \mathbb{N}. \tag{23.7}$$

Assume that there is no such $c > 0$. Then we find a sequence $k_j \to \infty$ such that

$$e_{k_j} \left(id : B_{p_1q_1}^{s_1}(\Omega) \to B_{p_2q_2}^{s_2}(\Omega) \right) k_j^{\frac{s_1-s_2}{n}} \to 0 \quad \text{if } j \to \infty. \tag{23.8}$$

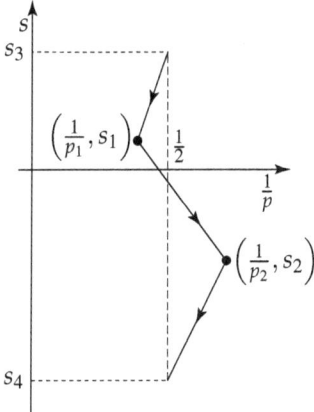

Fig. 23.1

Whatever the location of $(\frac{1}{p_1}, s_1)$ and $(\frac{1}{p_2}, s_2)$ in Fig. 23.1 might be there are always numbers $s_3 > s_1$ and $s_4 < s_2$ such that by Step 1

$$e_k \left(id \; : \; B^{s_3}_{2,2}(\Omega) \to B^{s_1}_{p_1 q_1}(\Omega) \right) \leq c\, k^{-\frac{s_3 - s_1}{n}}, \quad k \in \mathbb{N}, \tag{23.9}$$

and

$$e_k \left(id \; : \; B^{s_2}_{p_2 q_2}(\Omega) \to B^{s_4}_{2,2}(\Omega) \right) \leq c\, k^{-\frac{s_4 - s_2}{n}}, \quad k \in \mathbb{N}. \tag{23.10}$$

By (23.8)–(23.10) and the multiplication property (6.8) it follows

$$e_{3k_j} \left(id \; : \; B^{s_3}_{2,2}(\Omega) \to B^{s_4}_{2,2}(\Omega) \right) k_j^{\frac{s_3 - s_4}{n}} \to 0 \quad \text{if} \quad j \to \infty. \tag{23.11}$$

We may assume $s_4 < 0 < s_3$. Recall $L_2(\Omega) = B^0_{2,2}(\Omega)$ and

$$\|f \,|\, L_2(\Omega)\| \leq c \left\| f \,|\, B^{s_3}_{2,2}(\Omega) \right\|^{1-\Theta} \left\| f \,|\, B^{s_4}_{2,2}(\Omega) \right\|^{\Theta} \tag{23.12}$$

with

$$(1 - \Theta)s_3 + \Theta s_4 = 0. \tag{23.13}$$

This is a well-known property which also follows from (23.1), (23.2) and (10.6). We apply the interpolation property for entropy numbers as discussed in [ET96], 1.3.2, p. 13, and obtain by (23.11)–(23.13)

$$k_j^{\frac{s_3}{n}} e_{3k_j} \left(id \; : \; B^{s_3}_{2,2}(\Omega) \to L_2(\Omega) \right)$$

$$\leq c \left[k_j^{\frac{s_3 - s_4}{n}} e_{3k_j} \left(id \; : \; B^{s_3}_{2,2}(\Omega) \to B^{s_4}_{2,2}(\Omega) \right) \right]^{\Theta} \to 0 \tag{23.14}$$

if $j \to \infty$. This is impossible as follows by the same arguments as in Step 2 of the proof of Theorem 20.6.

23.3 Remark The theorem is not new. We proved it in [ET96], 3.3.3, p. 118, and 3.5, p. 151. There one finds also the necessary references to the original papers on which our considerations in [ET96] are based. As far as non-smooth bounded domains are concerned, assertions of the above type may also be found in [TrW96a], shortly mentioned in [ET96], 3.5. Entropy numbers for compact embeddings between function spaces in domains have been studied since some thirty years. Historical comments and related references may be found in [ET96], 3.3.5, pp. 126–128.

23.4 Comparison
In [ET96] the proof of (23.6) under the assumptions (23.4), (23.5) was essentially based on the original Fourier analytical Definition 10.3 of the underlying spaces. The estimates (23.6) for entropy numbers were reduced to corresponding estimates for entropy numbers between ℓ_p^M-spaces according to Theorem 7.3. In contrast to this approach we used here systematically atomic and subatomic decompositions of the underlying spaces $B_{pq}^s(\Omega)$ as presented in Section 14. On this way we shifted most of the problems from the spaces B_{pq}^s to weighted ℓ_p^M-spaces and developed the corresponding theory separately in Sections 8 and 9. We hope that the results in Sections 8, 9, and, in particular, 14, are of self-contained interest. Taking together all these efforts and the specific arguments in this Section 23, which, in turn, are based on the preceding sections, in particular Section 20, the proof of (23.6) under the assumptions (23.4), (23.5), presented here might not be shorter than the corresponding proof in [ET96]. But the proofs given here seem to be more transparent, in particular if one takes the results obtained in the Sections 8, 9, and 14 for granted. Furthermore the new method developed in this book so far applies also to spaces on fractals and on compact embeddings between function spaces related to fractals. As far as we can see this is not possible on the basis of the technique in [ET96] and the underlying papers. Of course, just the fractals and their relations to function spaces are one of the main topics of this book.

23.5 Limiting embeddings
By obvious reasoning also the fractal limiting embedding Theorem 21.7 has an immediate counterpart when the fractal Γ is replaced by a bounded (smooth or non-smooth) domain Ω in $\mathbb{R}^n$. Hence,

if

$$1 < p \leq \infty,\ s > 0,\ 0 < q \leq \infty, \quad \text{and} \quad a < -1 - \frac{2s}{n}, \qquad (23.15)$$

and

$$\frac{1}{p^s} = \frac{1}{p} + \frac{s}{n}, \qquad (23.16)$$

then the embedding of $B_{p^s q}^s(\Omega)$ into $L_p(\log L)_a(\Omega)$ is compact and for the related entropy numbers we have

$$e_k\left(id\ :\ B_{p^s q}^s(\Omega) \to L_p(\log L)_a(\Omega)\right) \sim k^{-\frac{s}{n}}, \quad k \in \mathbb{N}. \qquad (23.17)$$

We have the same limiting situation as in Fig. 21.1, where the slope of the line with the «foot point» $\frac{1}{p}$ is now n. There is a corresponding counterpart of Corollary 21.10 with Ω and n in place of Γ and d, respectively, in (21.24) and (21.25). In [ET96], 3.4, we dealt extensively with limiting embeddings of the indicated type, sometimes with more specific spaces in place of $B_{pq}^s(\Omega)$ where we started from. The results in [ET96], 3.4, and obtained here, are comparable, but not identical; they complement each other. For instance, in case of $p = \infty$, the assertion (23.15)–(23.17) is slightly better than [ET96], Theorem 3.4.2(ii), p. 130. On the other hand, if $1 < p < \infty$, then [ET96], Theorem 3.4.3/1, p. 144, is not covered by the above method.

23.6 Weighted spaces
Based on [HaT94a,b] we estimated in [ET96], 4.3.2, pp. 170–171, entropy numbers of compact embeddings for weighted function spaces on $\mathbb{R}^n$. The technique developed in the preceding sections might also be useful in connection with problems of this type. We refer also to our remarks in 20.11.

24 Quadratic forms and approximation numbers

24.1 Introduction
We misused Chapter II which deals with ℓ_p-spaces, to introduce in Section 6 entropy numbers in abstract quasi-Banach spaces. Now we do something similar with this Chapter IV which deals with concrete function spaces, to introduce on an abstract level quadratic forms in Hilbert spaces and approximation numbers in quasi-Banach spaces. We collect known assertions without proofs but with the necessary references. A comment about our intentions seems to be in order. In the following Chapter V we deal with the distribution of eigenvalues of fractal differential and pseudodifferential operators. Estimates from above for the related eigenvalues rely on 6.9 and 6.10 and on concrete assertions for the entropy numbers of compact embeddings between function spaces obtained in the preceding sections of this chapter. The *root systems*, that means the collection of the eigenfunctions and associated eigenfunctions, of the considered fractal pseudodifferential operators are independent of the basic space in which they are considered (under a few appropriate assumptions). In other words, one may look for optimal basic spaces to get estimates for the eigenvalues from above and from below. As far as estimates from above are concerned this aspect is not of great interest (maybe besides some limiting cases) since the related inequalities for entropy numbers combined with 6.9 and 6.10 yield always the same results. As for estimates of eigenvalues from below we first remark that 6.8–6.10 cannot be applied. However in some interesting cases it comes out that Hilbert spaces are admitted in which the operator considered is self-adjoint and compact. It generates a non-negative quadratic form or, adopting the opposite point of view, a given non-negative quadratic form generates a desirable (compact) non-negative self-adjoint operator. The related

eigenvalues coincide with the *approximation numbers* of corresponding embeddings. These remarks make clear that, at least in principle, approximation numbers (in Hilbert spaces) pave the way to obtain estimates from below for some fractal pseudodifferential operators. In this Section 24 we describe the necessary abstract background. In the following Section 25 we give a first rather simple application: quadratic forms on fractals and in domains generating suitable elliptic operators. Chapter V contains more sophisticated applications.

24.2 Quadratic forms

Let H be a complex Hilbert space with the scalar product $(\cdot,\cdot)_H$ and let $a(\cdot,\cdot)$ be a *non-negative quadratic form* defined on $D \times D$, where D is linear and dense in H. Recall that, by definition,

$$a(\lambda_1 f_1 + \lambda_2 f_2, g) = \lambda_1 a(f_1, g) + \lambda_2 a(f_2, g), \tag{24.1}$$

where $\lambda_1 \in \mathbb{C}$, $\lambda_2 \in \mathbb{C}$, $f_1 \in D$, $f_2 \in D$, $g \in D$,

$$a(f,g) = \overline{a(g,f)} \in \mathbb{C}, \quad f \in D, \quad g \in D, \tag{24.2}$$

and

$$a(f,f) \geq 0, \quad f \in D. \tag{24.3}$$

The *non-negative quadratic form a* is called *closed* if D, equipped with the scalar product

$$a(f,g) + (f,g)_H, \quad f \in D, \quad g \in D, \tag{24.4}$$

becomes a Hilbert space. Let A be a non-negative self-adjoint operator in H. Then both its domain of definition $dom(A)$ and also $dom\,(A^{\frac{1}{2}})$ equipped with the scalar products

$$(Af, Ag)_H + (f,g)_H, \quad f \in dom\,(A), \quad g \in dom\,(A), \tag{24.5}$$

and

$$\left(A^{\frac{1}{2}}f, A^{\frac{1}{2}}g\right)_H + (f,g)_H, \quad f \in dom\,(A^{\frac{1}{2}}), \quad g \in dom\,(A^{\frac{1}{2}}), \tag{24.6}$$

respectively, become Hilbert spaces. In particular,

$$a(f,g) = \left(A^{\frac{1}{2}}f, A^{\frac{1}{2}}g\right)_H, \quad D = dom\,(A^{\frac{1}{2}}), \tag{24.7}$$

is a non-negative closed quadratic form. Recall that a non-negative self-adjoint operator A in H can be recovered uniquely from its quadratic form $a(f,g)$ in (24.7) as follows:

$f \in H$ belongs to $dom\,(A)$ if, and only if, $f \in dom\,(A^{\frac{1}{2}})$ and there is an $h \in H$ such that

$$a(f,g) = (h,g)_H \quad \text{for all } g \in dom\,(A^{\frac{1}{2}}). \tag{24.8}$$

In that case $Af = h$.

It is one of the basic assertions of the theory of quadratic forms in Hilbert spaces that these observations have the following converse:

Let a be a non-negative closed quadratic form in H, defined on $D \times D$, where D is linear and dense in H. Then there is a (uniquely determined) non-negative self-adjoint operator A with (24.7), (24.8).

$$(24.9)$$

A direct short proof may be found in [Dav95], 4.4, pp. 81–84. We refer also to [EEv87], pp. 168–178, where corresponding assertions are given in the more general context of sectorial forms based on the Lax-Milgram theorem. One should be aware that equivalent non-negative closed quadratic forms on a given dense subset D of H may result in rather different operators by the above procedure. But this causes no trouble for us in the sequel: With exception of the following Section 25, which serves merely as an illustration of our technique, we follow in Chapter V the scheme:

Given a fractal pseudodifferential operator we find in some distinguished cases suitable Hilbert spaces in which it generates non-negative closed quadratic forms. Then we try to estimate first the related approximation numbers, and secondly the related eigenvalues.

$$(24.10)$$

In this sense we turn now to approximation numbers.

24.3 Definition *Let A and B be complex quasi-Banach spaces and let $T \in L(A,B)$. Then*

$$a_k(T) = \inf\{\|T - L\| : L \in L(A,B), \ \text{rank} \, L < k\}, \quad k \in \mathbb{N}, \qquad (24.11)$$

is the k th approximation number of T, where rank T is the dimension of the range of L.

24.4 Properties All the general notation have the same meaning as in 6.1. This definition is the counterpart of the corresponding Definition 6.2 for the entropy numbers and we have analogously to Proposition 6.4:

Let A, B, C be quasi-Banach spaces, let $S, T \in L(A,B)$ and $R \in L(B,C)$. Then

$$\|T\| = a_1(T) \geq a_2(T) \geq \cdots, \qquad (24.12)$$

$$a_{k+l-1}(R \circ S) \leq a_k(R) \, a_l(S), \quad k \in \mathbb{N}, \, l \in \mathbb{N}, \qquad (24.13)$$

and, if in addition B is a p-Banach space with $0 < p \leq 1$,

$$a_{k+l-1}^p(S + T) \leq a_k^p(S) + a_l^p(T), \quad k \in \mathbb{N}, \, l \in \mathbb{N}. \qquad (24.14)$$

These formulations coincide essentially with [ET96], Lemma 2 on p. 11, where one finds also a short proof. Further properties, comments and references to the literature may be found in [ET96], pp. 11–12, 15–18. We restrict ourselves to those assertions which we need later on.

24.5 Proposition (i) *Let A and B be Banach spaces and let $T \in L(A, B)$. Then $a_k(T) = 0$ if, and only if, rank $T < k$.*

(ii) *Let A be a Banach space with* $\dim A \geq n$ *and let* id $: A \to A$ *be the identity map; then* $a_k(T) = 1$ *for* $k = 1, \ldots, n$.

(iii) *Let H be a Hilbert space and let* $T \in L(H)$ *be a compact self-adjoint operator. Let* $\{\lambda_k\}_{k \in \mathbb{N}}$ *be the sequence of all eigenvalues of T, repeated according to their geometric multiplicity and ordered so that*

$$|\lambda_1(T)| \geq |\lambda_2(T)| \geq \cdots. \tag{24.15}$$

Then holds

$$|\lambda_k(T)| = a_k(T), \quad k \in \mathbb{N}. \tag{24.16}$$

24.6 Remark Proofs of these well-known assertions may be found in [EEv87], Proposition 2.3, p. 54, as for (i) and (ii), and Theorem 5.10, p. 91, as for (iii). The results proved there are even more general. We refer also to [ET96], pp. 11–12, 21–22, for further information.

24.7 Approximation numbers in function spaces
Our approach in this book is almost exclusively based on entropy numbers for compact embeddings between function spaces as developed in the preceding sections of this chapter, in particular in the Sections 20–23. On the other hand, for compact embeddings between spaces $B_{pq}^s(\Omega)$ and $F_{pq}^s(\Omega)$ on bounded smooth domains Ω in $\mathbb{R}^n$ approximation numbers have been studied in great detail since some thirty years. Detailed references may be found in [Tri78], 4.10.2, pp. 349–354, [EEv87], pp. 242, 299, and [ET96], pp. 126–128. In [ET96], 3.3.4, pp. 119–126, we studied in detail approximation numbers of compact embeddings between the spaces $B_{pq}^s(\Omega)$ and $F_{pq}^s(\Omega)$, and extended these considerations to limiting cases in [ET96], 3.4.2, p. 129. We always used the fact that for bounded smooth domains Ω there are linear and bounded extension operators from, say, $B_{pq}^s(\Omega)$ into $B_{pq}^s(\mathbb{R}^n)$. The existence of such linear extension operators is not clear if the bounded domain Ω is non-smooth, and there is little hope of corresponding assertions for function spaces on and of fractals, at least on the basis of the methods developed so far in [ET96]. Complementing 23.6 we remark that approximation numbers for compact embeddings between weighted function spaces on $\mathbb{R}^n$ have been treated briefly in [ET96], 4.3.3, pp. 179–183, and in detail in [Har95].

25   Quadratic forms on fractals and in domains

25.1 The spaces $H^s(\Gamma)$
In this Section 25 we always assume that Γ is a *compact d-set* in $\mathbb{R}^n$ according to Definition 3.1 *with* $0 < d < n$ and the finite Hausdorff measure μ. In Definition 20.2 we introduced the spaces

$$\mathbb{B}_{pq}^s(\Gamma) \quad \text{with} \quad s > 0, \, 0 < p \leq \infty, \, 0 < q \leq \infty.$$

We put for brevity

$$H^s(\Gamma) = \mathbb{B}_{2,2}^s(\Gamma), \quad s > 0. \tag{25.1}$$

There is no need to write $\mathbb{H}^s(\Gamma)$ instead of $H^s(\Gamma)$ since the possible confusion as described in (20.10), (20.11) cannot happen if $p = q = 2$. In 10.5(iii) we introduced the Sobolev spaces $H_p^\sigma(\mathbb{R}^n)$. If $p = 2$ then we now put

$$H^\sigma(\mathbb{R}^n) = H_2^\sigma(\mathbb{R}^n) = B_{2,2}^\sigma(\mathbb{R}^n) = F_{2,2}^\sigma(\mathbb{R}^n), \quad \sigma \in \mathbb{R}. \tag{25.2}$$

By (10.13) the spaces $H^\sigma(\mathbb{R}^n)$ can be equipped with the scalar product

$$(f,g)_{H^\sigma(\mathbb{R}^n)} = \int_{\mathbb{R}^n} \left(\langle \xi \rangle^\sigma \hat{f} \right)^{\vee}(x) \, \overline{\left(\langle \xi \rangle^\sigma \hat{g} \right)^{\vee}(x)} \, dx. \tag{25.3}$$

Normed in this way $H^\sigma(\mathbb{R}^n)$ becomes a Hilbert space. According to Definition 20.2 we specify (25.1) by

$$H^s(\Gamma) = tr_\Gamma \, H^{s + \frac{n-d}{2}}(\mathbb{R}^n), \quad s > 0, \tag{25.4}$$

equipped with the norm

$$\|f \mid H^s(\Gamma)\| = \inf \left\| g \mid H^{s + \frac{n-d}{2}}(\mathbb{R}^n) \right\|, \tag{25.5}$$

where the infimum is taken over all

$$g \in H^{s + \frac{n-d}{2}}(\mathbb{R}^n) \quad \text{with} \quad tr_\Gamma \, g = f.$$

By standard arguments and with obvious notation we have

$$H^{s + \frac{n-d}{2}}(\mathbb{R}^n) = \left\{ g \in H^{s + \frac{n-d}{2}}(\mathbb{R}^n) : tr_\Gamma \, g = 0 \right\} \oplus H^s(\Gamma). \tag{25.6}$$

In particular, $H^s(\Gamma)$, normed by (25.5), is a Hilbert space densely embedded in $L_2(\Gamma)$. It is well-known that the corresponding scalar product in $H^s(\Gamma)$ is given by

$$\begin{aligned} 4(f,g)_{H^s(\Gamma)} = &\|f + g \mid H^s(\Gamma)\|^2 - \|f - g \mid H^s(\Gamma)\|^2 \\ &+ i\|f + ig \mid H^s(\Gamma)\|^2 - i\|f - ig \mid H^s(\Gamma)\|^2, \end{aligned} \tag{25.7}$$

see e. g. [Tri92*], p. 95. Hence, according to 24.2,

$$a(f,g) = (f,g)_{H^s(\Gamma)}, \quad D = H^s(\Gamma), \tag{25.8}$$

is a closed quadratic form in the Hilbert space $L_2(\Gamma)$. By (25.4) and Theorem 18.6 we have

$$\|f \mid L_2(\Gamma)\| \le c \, \|f \mid H^s(\Gamma)\|, \quad f \in H^s(\Gamma), \tag{25.9}$$

for some $c > 0$. Let A_s be the related self-adjoint operator according to (24.9). By (25.9) this operator is positive-definite and we have

$$\left\| A_s^{\frac{1}{2}} f \mid L_2(\Gamma) \right\| = \| f \mid H^s(\Gamma) \|, \quad dom\left(A_s^{\frac{1}{2}} \right) = H^s(\Gamma). \tag{25.10}$$

By our remarks after (24.9) the operators A_s depend decisively on the given quadratic form: equivalent quadratic forms may create rather different operators. In other words, to introduce distinguished operators on Γ, such as a Γ-*Laplacian* or so, one needs surely additional specific considerations, see also 25.6.

Recall that a self-adjoint operator is said to be an operator with *pure point spectrum* if its continuous spectrum is empty. If A is a positive-definite operator with pure point spectrum then its eigenvalues μ_k, repeated according to their geometric multiplicities, can be ordered such that

$$0 < \mu_1 \le \mu_2 \le \cdots, \quad \mu_k \to \infty \quad \text{if} \quad k \to \infty. \tag{25.11}$$

25.2 Theorem *Let Γ be a compact d-set in $\mathbb{R}^n$ according to Definition 3.1 with $0 < d < n$. Let $s > 0$ and let A_s be the operator introduced in 25.1, in particular*

$$(f,g)_{H^s(\Gamma)} = (A_s f, g)_{L_2(\Gamma)}, \quad f \in dom(A_s), \ g \in H^s(\Gamma). \tag{25.12}$$

Then A_s is a positive-definite self-adjoint operator in $L_2(\Gamma)$ with pure point spectrum, and there are two numbers $0 < c_1 \le c_2 < \infty$ with

$$c_1 k^{\frac{2s}{d}} \le \mu_k \le c_2 k^{\frac{2s}{d}}, \quad k \in \mathbb{N}, \tag{25.13}$$

where μ_k are the eigenvalues of A_s ordered by (25.11).

Proof.

Step 1. The eigenvalues of the non-negative compact self-adjoint operator $A_s^{-\frac{1}{2}}$ in $L_2(\Gamma)$ are $\varrho_k = \mu_k^{-\frac{1}{2}}$. Furthermore, $A_s^{-\frac{1}{2}}$ is also an isomorphic map from $L_2(\Gamma)$ onto $H^s(\Gamma)$. Hence, by

$$A_s^{-\frac{1}{2}} (L_2(\Gamma) \to L_2(\Gamma)) = id\,(H^s(\Gamma) \to L_2(\Gamma)) \circ A_s^{-\frac{1}{2}} (L_2(\Gamma) \to H^s(\Gamma)), \tag{25.14}$$

Corollary 6.10, and Theorem 20.6 we have

$$\varrho_k \le c\,e_k\,(id : H^s(\Gamma) \to L_2(\Gamma)) \le c'\,k^{-\frac{s}{d}}, \quad k \in \mathbb{N}. \tag{25.15}$$

This proves the left-hand side of (25.13).

Step 2. We prove the converse assertion. We use the construction in Step 2 of the proof of Theorem 20.6, in particular (20.57) with $p_1 = p_2 = 2$ and $H^s(\Gamma)$ in place of $\mathbb{B}^s_{p,q}(\Gamma)$. Let a_k be the approximation numbers introduced in Definition 24.3. By (24.13) and (20.57) with the indicated modifications we have

$$a_k \left(id \ : \ 2^{js} \, \ell_2^{M_j} \to \ell_2^{M_j} \right) \le c \, a_k \left(id \ : \ H^s(\Gamma) \to L_2(\Gamma) \right), \quad k \in \mathbb{N}. \qquad (25.16)$$

Recall $M_j \sim 2^{jd}$. By (25.14) and Proposition 24.5(iii) it follows

$$a_k \left(id \ : \ 2^{js} \, \ell_2^{M_j} \to \ell_2^{M_j} \right) \le c \, \varrho_k, \quad k \in \mathbb{N}. \qquad (25.17)$$

By Proposition 24.5(ii) and $k = M_j - 1 \sim 2^{jd}$ we obtain

$$2^{-js} \le c' \varrho_{c2^{jd}}, \quad j \in \mathbb{N}, \qquad (25.18)$$

and finally

$$\varrho_k \ge c \, k^{-\frac{s}{d}}, \quad k \in \mathbb{N}. \qquad (25.19)$$

But this is the right-hand side of (25.13).

25.3 Quadratic forms on domains

To shed more light on our discussion at the end of 25.1 we add a few remarks on quadratic forms on domains. We restrict ourselves to the simplest case. Let Ω be a bounded (not necessarily smooth) domain in $\mathbb{R}^n$. Recall that the classical Sobolev space $\overset{\circ}{W}{}^1_2(\Omega) = \overset{\circ}{H}{}^1(\Omega)$ can be defined as the completion of $C_0^\infty(\Omega) = D(\Omega)$ in the norm

$$\left(\| \, |\nabla f| \, | L_2(\Omega) \|^2 + \| f \, | L_2(\Omega) \|^2 \right)^{\frac{1}{2}}$$

$$= \left(\int_\Omega \left(\sum_{j=1}^{n} \left| \frac{\partial f}{\partial x_j}(x) \right|^2 + |f(x)|^2 \right) dx \right)^{\frac{1}{2}}. \qquad (25.20)$$

This coincides with (25.3) and (25.4) with Ω in place of Γ. By Friedrichs' inequality, see e. g. [Tri92*], p. 357, we have

$$c \, \| f \, | L_2(\Omega) \| \le \| \, |\nabla f| \, | L_2(\Omega) \|, \quad f \in \overset{\circ}{H}{}^1(\Omega), \qquad (25.21)$$

for some $c > 0$. Hence $\| \, |\nabla f| \, | L_2(\Omega) \|$ is an equivalent norm on $\overset{\circ}{H}{}^1(\Omega)$. Let $a_{jk}(x) = a_{kj}(x)$ with $j = 1, \ldots, n$ and $k = 1, \ldots, n$ be real L_∞-functions in Ω and let

$$\sum_{j,k=1}^{n} a_{jk}(x)\xi_j \overline{\xi_k} \ge c \, |\xi|^2, \quad \xi = (\xi_1, \ldots, \xi_n) \in \mathbb{C}^n, \qquad (25.22)$$

for some $c > 0$ and almost all $x \in \Omega$ (*ellipticity condition*). Hence

$$a(f,g) = \int_\Omega \sum_{j,k=1}^{n} a_{jk}(x) \frac{\partial f}{\partial x_j}(x) \overline{\frac{\partial g}{\partial x_k}}(x) \, dx \qquad (25.23)$$

with $f \in \overset{\circ}{H}{}^1(\Omega)$ and $g \in \overset{\circ}{H}{}^1(\Omega)$ is a quadratic form where $L_2(\Omega)$ is the underlying Hilbert space. By (25.22) and (25.21) it follows that $a(f,f)^{\frac{1}{2}}$ is an equivalent norm in $\overset{\circ}{H}{}^1(\Omega)$. In particular $a(f,g)$ is a closed quadratic form. Let A be the related self-adjoint positive-definite operator according to (24.9). We have

$$\left\| A^{\frac{1}{2}} f \mid L_2(\Omega) \right\| = a(f,f)^{\frac{1}{2}} \sim \| f \mid H^1(\Omega) \|, \quad dom\,(A^{\frac{1}{2}}) = \overset{\circ}{H}{}^1(\Omega) \tag{25.24}$$

analogously to (25.10). Although one cannot say very much about $dom\,(A)$ in general we have for $f \in dom\,(A)$, and $g \in D(\Omega)$ analogously to (25.12)

$$(Af,g)_{L_2(\Omega)} = \sum_{j,k=1}^{n} \int_{\Omega} a_{jk}(x) \frac{\partial f}{\partial x_j}(x) \overline{\frac{\partial g}{\partial x_k}(x)}\, dx$$

$$= \int_{\Omega} - \sum_{j,k=1}^{n} \frac{\partial}{\partial x_k} \left(a_{jk}(x) \frac{\partial f}{\partial x_j}(x) \right) \overline{g(x)}\, dx \tag{25.25}$$

and hence

$$Af = - \sum_{j,k=1}^{n} \frac{\partial}{\partial x_k} \left(a_{jk}(x) \frac{\partial f}{\partial x_j}(x) \right), \quad f \in dom\,(A) \subset \overset{\circ}{H}{}^1(\Omega), \tag{25.26}$$

interpreted in $D'(\Omega)$. More details may be found in [Tri92*], 6.2.2, pp. 361–365.

25.4 Theorem *Let A be the positive-definite self-adjoint operator in $L_2(\Omega)$ introduced in 25.3. Then A is an operator with pure point spectrum, and there are two numbers $0 < c_1 \le c_2 < \infty$ with*

$$c_1 k^{\frac{2}{n}} \le \mu_k \le c_2 k^{\frac{2}{n}}, \quad k \in \mathbb{N}, \tag{25.27}$$

where μ_k are the eigenvalues of A ordered by (25.11).

Proof. The proof is the same as the proof of Theorem 25.2. Instead of Theorem 20.6 we rely on Theorem 23.2.

25.5 Remark If $a_{jk}(x)$ are sufficiently smooth in $\overline{\Omega}$ then

$$dom\,(A) = H^2(\Omega) \cap \overset{\circ}{H}{}^1(\Omega) \tag{25.28}$$

and (25.26) has an obvious meaning. It is quite clear that these considerations can be extended to higher order elliptic equations.

25.6 Spectral theory
The distribution of eigenvalues of (elliptic) operators has attracted much attention since H. Weyl's seminal contributions in 1911: a sharper version of (25.27) in case of the Laplacian $-\Delta$ in bounded (smooth) domains, see also 26.1 below. References to some older papers may be found in [Tri78], 5.6.2, pp. 395–397. More recent publications, especially related to fractals or to elliptic differential operators in bounded domains with «non-Weylian» or even fractal boundaries are listed in [Lap91], [Lap93], [LeV96], and [SaV96]. We return to these questions later on. There is only one point which we wish to discuss in connection with Theorem 25.2 now. As we mentioned briefly after (24.9) and at the end of 25.1 the operator A_s depends in a sensitive way on the corresponding quadratic form. Equivalent quadratic forms may generate rather different operators. Hence it is somewhat doubtful whether one finds in that way a distinguished quadratic form generating an operator which might serve as the Laplacian on Γ. Just on the contrary there are several papers in recent times aiming to find out what should be called a *Laplacian* and to study spectral properties of type (25.13). The underlying fractal is specialized with a preference of the *Sierpinski triangle* depicted in Fig. 19.1. We do not go into detail and refer to [Kig93], [Shi93], and especially to the survey [Kig95].

Chapter V
Spectra of fractal pseudodifferential operators

26 Introduction and preliminaries

26.1 The Weyl problem

Let Ω be a bounded smooth domain in $\mathbb{R}^n$ and let $-\Delta$ be the Dirichlet Laplacian considered in the Hilbert space $L_2(\Omega)$ with its domain of definition

$$dom(-\Delta) = H^2(\Omega) \cap \overset{\circ}{H}{}^1(\Omega) \tag{26.1}$$

according to (25.28). Recall that $H^2(\Omega)$ is the restriction of

$$H^2(\mathbb{R}^n) = B^2_{2,2}(\mathbb{R}^n)$$

on Ω as introduced in (23.1) and that $\overset{\circ}{H}{}^1(\Omega)$ is the completion of $C_0^\infty(\Omega) = D(\Omega)$ in the norm given by (25.20). Let

$$0 < \lambda_1 \le \lambda_2 \le \cdots \le \lambda_j \le \cdots, \quad \lambda_j \to \infty \quad \text{if} \quad j \to \infty, \tag{26.2}$$

be the ordered eigenvalues with respect to their multiplicities. The counting function $N(\lambda)$ is given by

$$N(\lambda) = \#\{\lambda_j < \lambda\}, \quad \lambda > 0, \tag{26.3}$$

the number of eigenvalues less than λ. Under some additional geometric assumptions one has

$$N(\lambda) = (2\pi)^{-n} \omega_n \, vol(\Omega) \, \lambda^{\frac{n}{2}} - \varkappa_n \, vol(\partial\Omega) \, \lambda^{\frac{n-1}{2}} (1 + o(1)) \tag{26.4}$$

as $\lambda \to \infty$, where ω_n is the volume of the unit ball in $\mathbb{R}^n$, and $\varkappa_n$ is some positive number depending only on n. The equivalence (25.27) with λ_k in place μ_k is an immediate consequence of (26.4). Historical remarks and references may be found in the papers mentioned at the beginning of 25.6. We do not go into detail. It is well-known that this problem goes back to H. Weyl in 1911/12 and has attracted much attention up to our time. If $n = 2$ then $\sqrt{\lambda_j}$ can be interpreted as the eigenfrequencies of a *drum* with the vibrating membrane Ω.

H. Triebel, *Fractals and Spectra*, Modern Birkhäuser Classics,
DOI 10.1007/978-3-0348-0034-1_5, © Birkhäuser Verlag 1997

26.2 Fractal drums

One may ask what happens with the counting function $N(\lambda)$ or its weaker version

$$\lambda_k \sim k^{\frac{2}{n}}, \quad k \in \mathbb{N}, \tag{26.5}$$

if the smooth bounded domain Ω in $\mathbb{R}^n$ is replaced by something bounded, but non-smooth. If $n = 2$ or also $n = 3$ this may be related as above to the eigenfrequencies of a suitable membrane. There are two versions in literature and we add a third one.

(i) Let Ω be again a bounded domain in $\mathbb{R}^n$ and let $\partial\Omega$ be a fractal with the Hausdorff dimension d, where $n - 1 < d < n$. Then $-\Delta$ makes sense at least according to 25.3 and 25.4. M. V. Berry conjectured that one has again (26.4) with the same main term on the right-hand side and $\mathcal{H}^d(\partial\Omega) \lambda^{\frac{d}{2}}$ in place of $vol(\partial\Omega) \lambda^{\frac{n-1}{2}}$. This bold conjecture cannot be correct in this generality. However if $\partial\Omega$ is a compact d-set according to Definition 3.1 then a partial affirmative answer was given in [Lap91], where one finds a proof of (26.4) with unchanged main term and the remainder term $O(\lambda^{\frac{d}{2}})$ on the right-hand side. We refer to this paper and also to [Lap93], [LeV96], [Vas91], [FlV93], and [FLV95] for a thorough discussion of these problems.

(ii) Even bolder is the assumption that both Ω and $\partial\Omega$ are fractals, describing, say, the vibrating earth with its fractal core and its fractal surface (whatever this means). But then one has to say what is meant by the Laplacian on Ω. This is just the problem we addressed in 25.6 and the references given there reflect the state of art.

(iii) Our intention here is different from that one in (i) or in (ii). Let $(-\Delta)^{-1}$ be the inverse of the Dirichlet Laplacian $-\Delta$ in a smooth bounded domain Ω in $\mathbb{R}^n$ considered in (i). For its eigenvalues $\mu_k = \lambda_k^{-1}$ we have

$$\mu_k \sim k^{-\frac{2}{n}}, \quad k \in \mathbb{N}. \tag{26.6}$$

Recall that «$\sim$» means that there are two constants $0 < c_1 \leq c_2 < \infty$ such that

$$c_1 k^{-\frac{2}{n}} \leq \mu_k \leq c_2 k^{-\frac{2}{n}}, \quad k \in \mathbb{N}. \tag{26.7}$$

If $n = 2$ then $(-\Delta)^{-1}$ represents in the above-described way a vibrating membrane where the mass is evenly distributed, that means the mass density $m(x)$, $x \in \Omega$, is constant. If this is not so, then it follows by the usual physical reasoning that $(-\Delta)^{-1}$ must be replaced by

$$B : f \mapsto (-\Delta)^{-1} m(x) f. \tag{26.8}$$

Assuming that the whole mass is concentrated on a compact fractal Γ with $\Gamma \subset \Omega$. Say, Γ is a compact d-set according to Definition 3.1 with the measure μ. Then B is given by, say,

$$B = (-\Delta)^{-1} b(\gamma) \mu, \quad b(\gamma) \in L_p(\Gamma), \tag{26.9}$$

for some $1 \leq p \leq \infty$. First one has to give (26.9) a precise meaning. This will be done with the help of the results obtained in Section 20. It turns out that under some circumstances, for example $b(\gamma) = 1$ and $n - 2 < d < n$, the operator B is self-adjoint in a suitable Hilbert space, non-negative and compact. For the positive eigenvalues μ_k one has now

$$\mu_k \sim k^{-\frac{2-n+d}{d}}, \quad k \in \mathbb{N}, \tag{26.10}$$

instead of (26.6). In contrast to what had been said in (i) now the *main term* in (26.4) depends on the fractal dimension d. Problems of that type will be discussed in detail in Section 30.

26.3 Degenerate pseudodifferential operators
Almost nothing special is used to handle operators of type (26.8) or their fractal counterpart (26.9). One can replace $(-\Delta)^{-1}$ in (26.8) by reasonable other operators such as *fractional powers of elliptic differential operators* or by *pseudodifferential operators*. In the context of $\mathbb{R}^n$ instead of Ω, one arrives at operators of the type

$$B = b_2(x)\, b(x, D)\, b_1(x) \tag{26.11}$$

with

$$b_1(x) \in L_{r_1}(\mathbb{R}^n), \quad b_2(x) \in L_{r_2}(\mathbb{R}^n), \quad \text{supp}\, b_2 \quad \text{compact}, \tag{26.12}$$

and

$$b(x, D) \in \Psi_{1,\varrho}^{-\varkappa}(\mathbb{R}^n), \quad \varkappa > 0, \quad 0 \leq \varrho \leq 1, \tag{26.13}$$

where the *Hörmander class* of pseudodifferential operators $\Psi_{1,\varrho}^{-\varkappa} = \Psi_{1,\varrho}^{-\varkappa}(\mathbb{R}^n)$ will be explained later on in 26.4. We studied spectral properties of such *degenerate pseudodifferential operators* in [ET96], Ch. 5, in detail. We formulate a typical assertion which gives us the possibility to compare our later fractal-minded results with degenerate ones obtained in [ET96]:

Suppose that

$$1 \leq r_1 \leq \infty,\ 1 \leq r_2 \leq \infty,\ 1 \leq p \leq \infty,\ \varkappa > 0,\ 0 \leq \varrho \leq 1, \tag{26.14}$$

where

$$\frac{1}{r_2} < \frac{1}{p} < 1 - \frac{1}{r_1} \quad \text{and} \quad \frac{\varkappa}{n} > \frac{1}{r_1} + \frac{1}{r_2}. \tag{26.15}$$

Let $b_1(x)$, $b_2(x)$ and $b(x, D)$ be given by (26.12) and (26.13), respectively. Then the degenerate pseudodifferential operator B defined by (26.11) is compact in $L_p(\mathbb{R}^n)$ and its non-vanishing eigenvalues μ_k, repeated according to algebraic multiplicity, and ordered so that

$$|\mu_1| \geq |\mu_2| \geq \cdots \tag{26.16}$$

satisfy

$$|\mu_k| \le c \, \|b_1 \,|\, L_{r_1}(\mathbb{R}^n)\| \, \|b_2 \,|\, L_{r_2}(\mathbb{R}^n)\| \, k^{-\frac{\varkappa}{n}}, \quad k \in \mathbb{N}, \tag{26.17}$$

where c is independent of b_1, b_2 and k, but depends on $\text{supp} \, b_2$.

This assertion is essentially covered by [ET96], 5.2.6, pp. 205–206. We changed the formulation given there slightly to adapt it to our later needs. As far as *spectral theory for compact operators in (quasi-)Banach spaces* is concerned we gave the necessary information at the end of 6.1, in 6.6 and 6.8. As for pseudodifferential operators we collect what we need now.

26.4 Pseudodifferential operators

First we explain what is meant by (26.13). We follow [ET96], 5.2.3, pp. 189–190. Let $\varkappa \in \mathbb{R}$ and $0 \le \varrho \le 1$, then the Hörmander class $S_{1,\varrho}^{\varkappa}$ consists of all complex-valued C^{∞} functions $(x, \xi) \to b(x, \xi)$ on $\mathbb{R}^n \times \mathbb{R}^n$ such that for all multi-indices α, β there is a positive constant $c_{\alpha\beta}$ with

$$|D_\xi^\alpha D_x^\beta b(x, \xi)| \le c_{\alpha\beta} \langle \xi \rangle^{\varkappa - |\alpha| + \varrho|\beta|} \tag{26.18}$$

for all $x \in \mathbb{R}^n$, $\xi \in \mathbb{R}^n$. Here D_ξ^α and D_x^β refer to derivatives with respect to ξ and x, respectively; also $\langle \xi \rangle = \left(1 + |\xi|^2\right)^{\frac{1}{2}}$. Given any such function (or *symbol*) b, the corresponding pseudodifferential operator $b(x, D)$ is defined by

$$
\begin{aligned}
b(x, D) f(x) &= \int_{\mathbb{R}^n} e^{ix\xi} b(x, \xi) \hat{f}(\xi) d\xi \\
&= (2\pi)^{-\frac{n}{2}} \int_{\mathbb{R}^{2n}} e^{i(x-y)\xi} b(x, \xi) f(y) \, dy \, d\xi \\
&= \int_{\mathbb{R}^n} K(x, y) f(y) \, dy, \quad f \in S'(\mathbb{R}^n),
\end{aligned}
\tag{26.19}
$$

where the final equality is the *kernel representation* in the sense of Schwartz, with $K \in D'(\mathbb{R}^n \times \mathbb{R}^n)$. The class of all such pseudodifferential operators will be denoted by $\Psi_{1,\varrho}^{\varkappa}$ including the *exotic* subclass $\Psi_{1,1}^{\varkappa}$. For the necessary background material we refer to [Tay81], [Tay91] and [Hör85]. Next we recall mapping properties of the above pseudodifferential operators in the spaces $B_{pq}^s(\mathbb{R}^n)$ and $F_{pq}^s(\mathbb{R}^n)$ as introduced in Definition 10.3.

   Let

$$b(x, D) \in \Psi_{1,\varrho}^{\varkappa} \quad \text{for some} \quad \varkappa \in \mathbb{R} \quad \text{and} \quad 0 \le \varrho \le 1. \tag{26.20}$$

(i) *Let $0 < p \le \infty$, $0 < q \le \infty$, $s \in \mathbb{R}$ (with $s - \varkappa > n\left(\frac{1}{p} - 1\right)_+$ in the exotic case $\varrho = 1$). Then $b(x, D)$ is a linear and bounded map from*

$$B_{pq}^s(\mathbb{R}^n) \quad \text{into} \quad B_{pq}^{s-\varkappa}(\mathbb{R}^n). \tag{26.21}$$

(ii) *Let $0 < p < \infty$, $0 < q \le \infty$, $s \in \mathbb{R}$ (with $s - \varkappa > n \left(\frac{1}{\min(p,q)} - 1 \right)_+$ in the exotic case $\varrho = 1$). Then $b(x, D)$ is a linear and bounded map from*

$$F_{pq}^s(\mathbb{R}^n) \quad into \quad F_{pq}^{s-\varkappa}(\mathbb{R}^n). \tag{26.22}$$

Proofs may be found in [Päi83], [Run85], [Torr90], [Torr91], [Tri92], 6.2.2, p. 258. An extension to weighted spaces is given in [HaT94b]. In particular, the above mapping properties apply to the *Zygmund spaces*

$$\mathcal{C}^s(\mathbb{R}^n) = B_{\infty,\infty}^s(\mathbb{R}^n), \quad s \in \mathbb{R},$$

given by (10.16) and to the *Sobolev spaces*

$$H_p^s(\mathbb{R}^n) = F_{p,2}^s(\mathbb{R}^n), \quad s \in \mathbb{R}, \ 1 < p < \infty, \tag{26.23}$$

where details may be found in 10.5.

26.5 Fractal pseudodifferential operators

Besides our intention to play the fractal drum tuned by 26.2(iii) we are looking for the fractal counterpart of what had been said in 26.3 about the spectral theory of degenerate pseudodifferential operators. As in case of what might be called a fractal drum there are several possibilities to extend (26.11) from degenerate to fractal. We list three of them.

(i) **Type 1.** *Operators on fractals.* Let, say, Γ be a compact d-set in $\mathbb{R}^n$ according to Definition 3.1 with $0 < d < n$, then $A_s^{-\frac{1}{2}}$, where A_s is the operator introduced in 25.1 and 25.2, may serve as a substitute of $b(x, D)$ in (26.11). The estimate from above for the related eigenvalues is based on (25.15). By Theorem 20.6 and the technique developed in [ET96] it is quite clear that then one can replace $A_s^{-\frac{1}{2}}$ by

$$B = b_2(\gamma) \, A_s^{-\frac{1}{2}} \, b_1(\gamma), \quad b_j \in L_{r_j}(\Gamma), \tag{26.24}$$

with appropriate restrictions for r_1, r_2. On this way one obtains the counterpart of (26.17) with $\frac{s}{d}$ in place of $\frac{\varkappa}{n}$. But we do not go into detail. Firstly the method is rather clear if one follows the technique developed in [ET96] and which also will be used later on in this book. Secondly, the discussion in 25.6 makes clear that a substantial theory of (elliptic) operators on Γ depends presumably on the intrinsic geometry of Γ. But this is not well reflected by the constructions in 25.1 and 25.2.

(ii) **Type 2.** *Pseudodifferential operators with fractal coefficients.* More interesting in our context is the question whether the coefficients $b_1(x)$ and $b_2(x)$ in (26.11) can be singular distributions. Let, for instance, $b(x, D)$ be given by (26.13) and let

$$B = b(x) \, b(x, D), \quad b(x) \in B_{rq}^s(\mathbb{R}^n), \quad supp \, b \quad compact, \tag{26.25}$$

where $s < 0$ and r, q are chosen appropriately. In that case and related ones, we speak about *pseudodifferential operators with fractal coefficients*. After giving B in (26.25) a precise meaning in a suitable basic space one can ask for a related spectral theory, in particular for a counterpart of (26.17). It turns out that under suitable circumstances one has estimates of type (26.17) with

$$k^{-\frac{\varkappa - 2|s|}{n}} \quad \text{in place of} \quad k^{-\frac{\varkappa}{n}}. \tag{26.26}$$

Hence there is a *fractal defect*.

(iii) **Type 3.** *Fractal pseudodifferential operators.* Let again Γ be a compact d-set in $\mathbb{R}^n$ according to Definition 3.1 with $0 < d < n$. As before let $b(x, D)$ be given by (26.13) and let

$$B = b_2(\gamma)\, b(x, D)\, b_1(\gamma) \tag{26.27}$$

with

$$b_1(\gamma) \in L_{r_1}(\Gamma) \quad \text{and} \quad b_2(\gamma) \in L_{r_2}(\Gamma), \tag{26.28}$$

where the numbers $\varkappa$, r_1, r_2 will be chosen later on similarly as in (26.14), (26.15). By Theorem 18.2 it is quite clear that B in (26.27) specializes the above type 2 operators with fractal coefficients. Hence we have more information and the outcome improves (26.26) by

$$|\mu_k| \leq c\, k^{-\frac{\varkappa - (n-d)}{d}}, \quad k \in \mathbb{N}. \tag{26.29}$$

In this case we speak about *fractal pseudodifferential operators*. We are mainly interested in this type of operators including modifications such as the fractal drum (26.9). As in (26.10) where $\varkappa = 2$, the exponent in (26.29) is sharp.

26.6 Plan of the chapter

Pseudodifferential operators with fractal coefficients of type (26.25) will be considered in Section 27, whereas fractal pseudodifferential operators as roughly defined in (26.27) and (26.28) are treated in the Sections 28 and 29. Here Section 28 is the fractal counterpart of 26.3. In Section 29 we are interested in some limiting cases where spaces of type $L_p(\log L)_a$ are involved. In 29.1 we seize the opportunity of looking backwards of what has been achieved so far, and forwards to further topics complementing of what has been said here. In Section 30 we play the fractal drum in the version (26.9). If Γ is an (isotropic) d-set we obtain rather final results. But we are also interested in more general fractals, especially in anisotropic and nonisotropic d-sets according to Sections 4 and 5. In quantum mechanics one is interested in the so-called «negative spectrum» of operators of the type

$$Hf = -\Delta f + \beta\, V(x)\, f, \tag{26.30}$$

in $\mathbb{R}^n$ where Δ is the Laplacian, $V(x)$ is a potential function and β is the coupling constant. Of interest is the behaviour of the negative spectrum if $\beta \to \infty$. This comes from the observation that $\sqrt{\beta}$ is, roughly speaking, the inverse of the Planck constant $\hbar$, and that this admittedly tiny constant has the miraculous property of tending to zero in the semi-classical limit. We dealt in [ET96] with problems of this type. In Section 31 we study the fractal counterpart, where $V(x)$ is replaced by, say, $V(\gamma)\,\mu$, where μ is the Hausdorff measure on a compact d-set and $V(\gamma) \in L_r(\Gamma)$. Finally in Section 32 we deal briefly with elliptic operators having fractal nonlinearities.

27 Pseudodifferential operators with fractal coefficients

27.1 Prerequisites

In this section we develop a spectral theory for operators of type (26.25). Somewhat in contrast to the rather systematic study in [ET96], Ch. 5, we restrict ourselves to typical model situations. Consulting [ET96] it will be quite clear that the results obtained can be generalized in many respects. We always assume that

$$b(x,D) \in \Psi_{1,\varrho}^{-\varkappa}, \quad 0 \leq \varrho \leq 1, \tag{27.1}$$

with $\varkappa > 0$, is a pseudodifferential operator in $\mathbb{R}^n$ according 26.4. Recall our notation introduced in (11.25) and slightly extended in 11.8:

$$\frac{1}{p^s} = \frac{1}{p} + \frac{s}{n} \geq 0, \quad 1 \leq p \leq \infty, \quad s \in \mathbb{R}, \tag{27.2}$$

in particular $\infty^s = \frac{n}{s}$, see also Fig. 11.1 with the strip G given by (11.24). Finally as in 11.8 we abbreviate $B_{pp}^s(\mathbb{R}^n)$ by $B_p^s(\mathbb{R}^n)$, whereas $H_p^s(\mathbb{R}^n)$ are the Sobolev spaces according to (11.32) with the classical special cases listed in 10.5(ii). On an abstract level we collected in 6.1 and 6.8 the necessary assertions about the spectrum of a compact operator B in a complex quasi-Banach space. In particular the sequence $\{\mu_k\}$ of all non-zero eigenvalues of B, repeated according to algebraic multiplicity can be ordered so that

$$|\mu_1| \geq |\mu_2| \geq \cdots \quad ; \quad \mu_j \to 0. \tag{27.3}$$

27.2 Theorem *Let*

$$1 < p < \infty, \quad s < 0, \quad and \quad \frac{1}{p^s} = \frac{1}{p} + \frac{s}{n} \geq 0. \tag{27.4}$$

Let $b(x,D)$ be the pseudodifferential operator given by (27.1) with

$$\varkappa > 2|s| \quad and \quad \varkappa > \frac{n}{p}, \tag{27.5}$$

and let

$$b(x) \in B^s_{p^s}(\mathbb{R}^n) \quad \text{with} \quad \text{supp } b \quad \text{compact.} \tag{27.6}$$

Then

$$B = b(x) b(x, D) \tag{27.7}$$

is compact in $B^s_{p^s}(\mathbb{R}^n)$ and

$$|\mu_k| \le c \, \|b \, | \, B^s_{p^s}(\mathbb{R}^n)\| \, k^{-\frac{\varkappa - 2|s|}{n}}, \quad k \in \mathbb{N}, \tag{27.8}$$

where $\mu_k \in \mathbb{C}$ are the eigenvalues of B according to (27.3) and c is independent of b and k.

Proof. We extend the technique developed in [ET96], 5.2.4 and 5.2.5, to the cases considered here. In particular by the arguments below it will be clear that the right-hand side of (27.7) makes sense in $S'(\mathbb{R}^n)$. Let χ be a C^∞ function in $\mathbb{R}^n$ with compact support and which is identically 1 in a neighbourhood of *supp b*. Let K be an (open) ball in $\mathbb{R}^n$ with *supp* $\chi \subset K$. We factorize B as indicated in

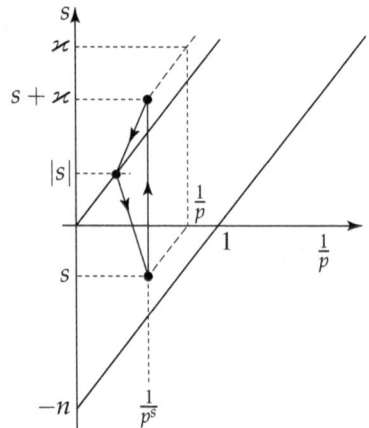

Fig. 27.1

Fig. 27.1 by

$$B = b \circ id \circ \chi b(\cdot, D) \tag{27.9}$$

with

$$\chi b(\cdot, D) \; : \; B^s_{p^s}(\mathbb{R}^n) \to B^{s+\varkappa}_{p^s}(K)$$

$$id \; : \; B^{s+\varkappa}_{p^s}(K) \to B^{|s|}_{\infty|s|,1}(K) \tag{27.10}$$

$$b \; : \; B^{|s|}_{\infty|s|,1}(K) \to B^s_{p^s}(\mathbb{R}^n).$$

The first map is continuous by 26.4(i), (26.21). The identity *id* is compact and we have by Theorem 23.2 for the corresponding entropy numbers

$$e_k(id) \sim k^{-\frac{\varkappa - 2|s|}{n}}, \quad k \in \mathbb{N}. \tag{27.11}$$

Finally the last line in (27.10) is a consequence of (11.35). By the multiplication property (6.8) of the entropy numbers we obtain that

$$e_k(B) \leq c \, \|b \mid B_{p^s}^s(\mathbb{R}^n)\| \, k^{-\frac{\varkappa - 2|s|}{n}}, \quad k \in \mathbb{N}, \tag{27.12}$$

where c is independent of b and k. Finally, (27.8) follows from (27.12) and Corollary 6.10.

27.3 Discussion
The proof is surprisingly short. However each of the three lines in (27.10) combined with (27.11) are rather substantial assertions. There arise several questions. First, as we shall see later on in 27.9, one can change the order order of $b(x)$ and $b(x, D)$ in (27.7). Secondly, more interesting is the question about the conditions (27.5) and the exponent $\frac{\varkappa - 2|s|}{n}$ in (27.8), also in comparison with the «Weyl exponent» $\frac{\varkappa}{n}$ in (26.17). We do not know whether $\frac{\varkappa - 2|s|}{n}$ is sharp. But in Theorem 27.15 we construct (a rather complicated) example of an operator of the above type where the estimate (27.8) from above is complemented by an estimate from below with the exponent $\frac{\varkappa - |s|}{n}$ in place of $\frac{\varkappa - 2|s|}{n}$, see (27.65) with $\varkappa$ instead of $2\varkappa$. Hence, in any case,

the deviation from the Weyl exponent $\frac{\varkappa}{n}$ caused by the «fractal coefficient» $b(x)$ is not an artefact which originates from the method used.

The restriction $\varkappa > \frac{n}{p}$ in (27.5) is technical, and the next theorem deals with the case $\varkappa \leq \frac{n}{p}$. Finally it is of some interest that $p^s = \infty$ is admitted in the above theorem. Let, as usual,

$$\mathscr{C}^s(\mathbb{R}^n) = B_\infty^s(\mathbb{R}^n) = B_{\infty,\infty}^s(\mathbb{R}^n), \quad s \in \mathbb{R}, \tag{27.13}$$

be the (extended) Zygmund scale. Then

$$B = b(x) \, (id - \Delta)^{-\frac{\varkappa}{2}}, \quad b \in \mathscr{C}^s(\mathbb{R}^n), \quad \text{supp } b \quad \text{compact}, \tag{27.14}$$

is a rather typical example covered by the theorem under the indicated restrictions for the involved parameters $s < 0$ and $\varkappa$. Of course, Δ stands for the Laplacian.

27.4 Theorem *Let*

$$1 < p < \infty, \quad s < 0, \quad and \quad \frac{1}{p^s} = \frac{1}{p} + \frac{s}{n} > 0. \tag{27.15}$$

Let $b(x, D)$ be the pseudodifferential operator given by (27.1) and let

$$b(x) \in H^s_{q^s}(\mathbb{R}^n) \quad with \quad supp\, b \quad compact, \tag{27.16}$$

and

$$\frac{|s|}{n} < \frac{1}{q} < \frac{\varkappa}{n} \le \frac{1}{p}, \quad \varkappa > 2|s|. \tag{27.17}$$

Then

$$B = b(x)\, b(x, D) \tag{27.18}$$

is compact in $H^s_{p^s}(\mathbb{R}^n)$ and

$$|\mu_k| \le c \, \|b \,|\, H^s_{q^s}(\mathbb{R}^n)\| \, k^{-\frac{\varkappa - 2|s|}{n}}, \quad k \in \mathbb{N}, \tag{27.19}$$

where $\mu_k \in \mathbb{C}$ are the eigenvalues of B according to (27.3) and c is independent of b and k.

Proof. We describe the modifications compared with the proof of Theorem 27.2 as indicated in Fig. 27.2. The function χ has the same meaning as there. We factorize

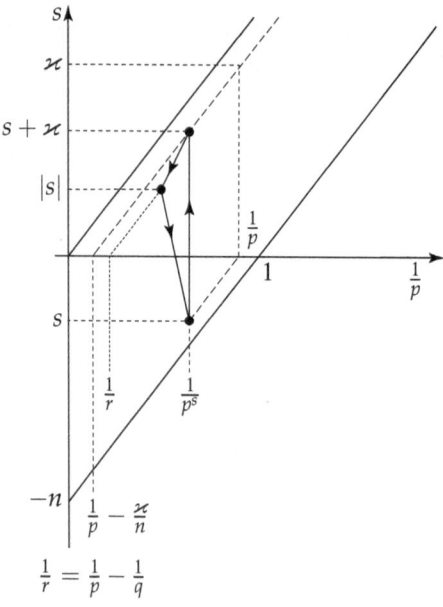

Fig. 27.2 $\frac{1}{r} = \frac{1}{p} - \frac{1}{q}$

B by

$$B = b \circ id \circ \chi\, b(\cdot, D) \tag{27.20}$$

now with

$$\chi\, b(\cdot, D) \;:\; H^s_{p^s}(\mathbb{R}^n) \to H^{s+\varkappa}_{p^s}(K)$$

$$id \;:\; H^{s+\varkappa}_{p^s}(K) \to H^{|s|}_{r|s|}(K) \tag{27.21}$$

$$b \;:\; H^{|s|}_{r|s|}(K) \to H^s_{p^s}(\mathbb{R}^n).$$

The continuity of the first map follows from 26.4(ii) and (26.23). By (27.17) we have

$$1 > \frac{1}{r} = \frac{1}{p} - \frac{1}{q} > \frac{1}{p} - \frac{\varkappa}{n} \ge 0. \tag{27.22}$$

Hence id is compact and by Theorem 23.2 and (23.3) we have again (27.11). The last line in (27.21) is a consequence of (27.22),

$$\frac{1}{q^s} = \frac{1}{q} - \frac{|s|}{n} > 0,$$

and (11.33). The rest is now the same as in the proof of Theorem 27.2 with an obvious counterpart of (27.12).

27.5 Remark Maybe the best way to understand the conditions in the last theorem is to look at Fig. 27.2 and afterwards at (27.17). Furthermore we recall

$$H^s_{q^s_1}(K) \subset H^s_{q^s_2}(K) \quad \text{if} \quad q_2 \le q_1 \tag{27.23}$$

and hence $q^s_2 \le q^s_1$, provided that $0 < q^s_2 \le q^s_1 < \infty$. This follows from 11.2(i) where we described the possibility to find equivalent (quasi-)norms based on local means. Hence q in (27.17) should be chosen as small as possible and $q = \frac{n}{\varkappa}$ is a (so far excluded) limiting case.

27.6 Limiting cases
By Theorems 27.2 and 27.4, and by the last remark

$$(i) \quad q = \frac{n}{\varkappa} \quad \text{and / or} \quad (ii) \quad \varkappa = 2|s| \tag{27.24}$$

are limiting cases. We add a discussion and assume in addition $\varkappa < \frac{n}{p}$, what corresponds to Fig. 27.2.

(i) Let $q = \frac{n}{\varkappa}$ and $\varkappa \ge 2|s|$ then $\frac{1}{r} = \frac{1}{p} - \frac{\varkappa}{n}$ in (27.22) and the map id in (27.21) is continuous (but not compact) according to (11.20). This applies in particular to $\varkappa = 2|s|$, where $r_{|s|} = p_s$. We obtain

$$\|B\| \le c \,\|b \mid H^s_{q^s}(\mathbb{R}^n)\|, \tag{27.25}$$

where c is independent of b. We approximate b in $H_{q^s}^s(\mathbb{R}^n)$ by $b_j(x) \in C_0^\infty(\mathbb{R}^n)$. The related operators B_j are compact and by (27.25) we have

$$\|B - B_j\| \le c \|b - b_j \mid H_{q^s}^s(\mathbb{R}^n)\|. \tag{27.26}$$

Hence, B is also compact. But we have no information of type (27.19) about the distribution of eigenvalues.

(ii) Let $\varkappa = 2|s|$. The worst case is $q = \frac{n}{\varkappa}$ which is covered by the above arguments. Hence in any case B is compact, but we cannot say more.

27.7 Corollary *Let $b(x)$ be a complex Radon measure in $\mathbb{R}^n$ with a compact support. Let $b(\cdot, D)$ be the pseudodifferential operator given by (27.1) with $\varkappa > n$. Then*

$$B = b(x) b(x, D) \tag{27.27}$$

is compact in $B_{1,\infty}^0(\mathbb{R}^n)$ and

$$|\mu_k| \le c \|b \mid B_{1,\infty}^0(\mathbb{R}^n)\| k^{-\frac{\varkappa}{n}}, \quad k \in \mathbb{N}, \tag{27.28}$$

where $\mu_k \in \mathbb{C}$ are the eigenvalues of B according to (27.3) and c is independent of b and k.

Proof. By (18.19) the measure b belongs to $B_{1,\infty}^0(\mathbb{R}^n)$. The function χ has the same meaning as in the proof of Theorem 27.2 Again we have the factorization (27.9) where (27.10) must be replaced by

$$\chi b(\cdot, D) : B_{1,\infty}^0(\mathbb{R}^n) \to B_{1,\infty}^\varkappa(K)$$

$$id : B_{1,\infty}^\varkappa(K) \to B_{\infty,1}^0(K) \tag{27.29}$$

$$b : B_{\infty,1}^0(K) \to B_{1,\infty}^0(\mathbb{R}^n),$$

as indicated in Fig. 27.3. The last line in (27.29) follows from the fact that

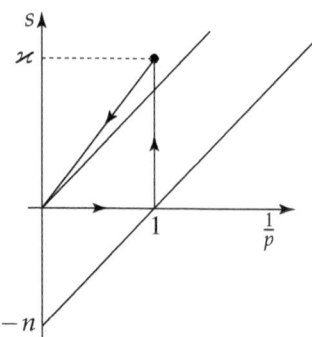

Fig. 27.3

$B_{\infty,1}^0(\mathbb{R}^n)$ is continuously embedded in $C(\mathbb{R}^n)$, the space of complex-valued bounded and uniformly continuous functions in $\mathbb{R}^n$. We refer to [Tri83], p. 130, formula (11). Otherwise the references after (27.10) can be applied.

27.8 Comparison
It is clear that the above corollary applies to any compactly supported $b \in B^0_{1,\infty}(\mathbb{R}^n)$. On the other hand, as described in 26.5 we modify b in Section 28 by, say, $b \in L_p(\Gamma)$, where $1 < p < \infty$ and Γ is a compact d-set in $\mathbb{R}^n$ according to Definition 3.1 with $0 < d < n$. Under some restrictions we obtain estimates of type (26.29). If, say, $b = \mathcal{H}^d|\Gamma$ is the corresponding Hausdorff measure restricted to Γ and $\varkappa > n$ then we have also (27.28). Hence, one has to compare the exponents in (26.29) and (27.28) under the assumption $\varkappa > n$. By

$$\frac{\varkappa - (n - d)}{d} = \frac{\varkappa - n}{d} + 1 > \frac{\varkappa}{n} \tag{27.30}$$

it follows that the exponent in (26.29) is better than the Weyl exponent in (27.28), as it should be.

27.9 Theorem *Let*

$$1 < p < \infty, \quad 1 < q < \infty, \quad \frac{1}{p} + \frac{1}{q} < 1, \tag{27.31}$$

and let

$$\varkappa > 2s > 0, \quad \frac{s}{n} < \frac{1}{q} < \frac{\varkappa}{n}. \tag{27.32}$$

Let $b(x, D)$ be the pseudodifferential operator given by (27.1) and let

$$b(x) \in H^{-s}_{q}(\mathbb{R}^n) \quad \text{with} \quad \text{supp } b \quad \text{compact.} \tag{27.33}$$

Then

$$B = b(x, D)\, b(x) \tag{27.34}$$

is compact in $H^s_{p^s}(\mathbb{R}^n)$ and

$$|\mu_k| \le c \,\|b\,|\, H^{-s}_{q}(\mathbb{R}^n)\|\, k^{-\frac{\varkappa - 2s}{n}}, \quad k \in \mathbb{N}, \tag{27.35}$$

where $\mu_k \in \mathbb{C}$ are the eigenvalues of B according to (27.3) and c is independent of b and k.

Proof. We modify the proofs of the Theorems 27.2 and 27.4. Let K be the same ball as there. As indicated in Fig 27.4 we factorize B by

$$B = b(\cdot, D) \circ id \circ b \tag{27.36}$$

with

$$b \;:\; H^s_{p^s}(\mathbb{R}^n) \to H^{-s}_{r-s}(K), \quad \frac{1}{r} = \frac{1}{p} + \frac{1}{q}$$

$$id \;:\; H^{-s}_{r-s}(K) \to H^{-\varkappa+s}_{p^s}(\mathbb{R}^n) \tag{27.37}$$

$$b(\cdot, D) \;:\; H^{-\varkappa+s}_{p^s}(\mathbb{R}^n) \to H^s_{p^s}(\mathbb{R}^n).$$

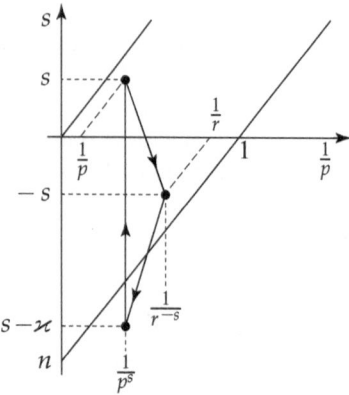

Fig. 27.4

The continuity of the first map follows from (11.33) and $\frac{1}{r} = \frac{1}{p} + \frac{1}{q} < 1$. As for *id* we first remark that $-s > -\varkappa + s$. Furthermore we have

$$
\begin{aligned}
-\varkappa + s - \frac{n}{p^s} &= -\varkappa + s - n\left(\frac{1}{p} + \frac{s}{n}\right) = -\varkappa - \frac{n}{p} \\
&< -n\left(\frac{1}{p} + \frac{1}{q}\right) = -\frac{n}{r} \\
&= -s - n\left(\frac{1}{r} - \frac{s}{n}\right) = -s - \frac{n}{r^{-s}}.
\end{aligned} \tag{27.38}
$$

Now it follows that *id* is a compact embedding and we obtain by Theorem 23.2 and (23.3)

$$
e_k(id) \sim k^{-\frac{\varkappa - 2s}{n}}, \quad k \in \mathbb{N}. \tag{27.39}
$$

The continuity of the last map in (27.37) follows again form 26.4(ii) and (26.23). Now we are in the same position as in the proofs of the Theorems 27.2 and 27.4 which leads us to (27.35).

27.10 Remark and outlook The above theorem is the counterpart of Theorem 27.4. If one replaces $H^s_{p^s}(\mathbb{R}^n)$ in the above theorem by $B^s_{\infty^s,1}(\mathbb{R}^n)$ then one finds a corresponding counterpart of Theorem 27.2. But we do not go into detail. Furthermore we refer to [ET96], Ch. 5, where we developed for degenerate pseudodifferential operators, as described in 26.3, a *smoothness theory* and where we studied the *independence of the root system of the underlying spaces*. All this can also be done for the fractal pseudodifferential operators. But we shall not develop such a theory here. Compared with [ET96] more interesting for us is the new problem about the exponents $\frac{\varkappa - 2|s|}{n}$, respectively $\frac{\varkappa - 2s}{n}$, in the Theorems 27.2, 27.4, and 27.9 and estimates from below for the absolute values of the eigenvalues μ_k. For this reason we modify the general background somewhat and prepare in that way also our later considerations in connection with fractal drums. Roughly speaking we replace pseudodifferential operators $b(x, D)$ by fractional powers of regular elliptic differential operators. We begin with the collection of the necessary basic facts.

27.11 Fractional powers of elliptic operators and spaces on domains
So far we concentrated on pseudodifferential operators with fractal coefficients in $\mathbb{R}^n$ of type (26.25) where the desired compactness of B originates from the assumption that b has compact support. As for fractal pseudodifferential operators of type (26.27) we shall be in Section 28 in a similar position, where the compactness of B comes from the compactness of the d-set Γ. Of perhaps equal importance in view of our later intentions to play the fractal drum, is the replacement of the pseudodifferential operator $b(x, D)$ on $\mathbb{R}^n$ by *fractional powers of regular elliptic differential operators in bounded smooth domains in* $\mathbb{R}^n$. We give a brief description following closely [Tri78], in particular 4.3.1 and 4.9.2, pp. 333–336, where more details and further information may be found.

Let Ω be a bounded domain in $\mathbb{R}^n$ with C^∞ boundary $\partial\Omega$. Then $B^s_{pq}(\Omega)$ and $H^s_p(\Omega)$ have the same meaning as in 23.1, based on 10.3 and 10.5. For our purpose and in agreement with [Tri78] we restrict now the parameters s, p, q by

$$s \in \mathbb{R}, \quad 1 < p < \infty, \quad 1 \leq q \leq \infty. \tag{27.40}$$

In particular, as in (10.9), (10.10), (10.15), if $s \in \mathbb{N}_0$, then

$$W^s_p(\Omega) = H^s_p(\Omega) = \{f \in L_p(\Omega) \; : \; D^\alpha f \in L_p(\Omega), \; |\alpha| \leq s\} \tag{27.41}$$

are the classical Sobolev spaces in smooth domains. This follows from the well-known existence of a linear extension operator, see e. g. [Tri78], 4.2, pp. 309–316. Following [Tri78], 4.3.2, pp. 317–318, we introduce for the above values of the parameters s, p, q,

$$\widetilde{B}^s_{pq}(\Omega) = \{f \in B^s_{pq}(\mathbb{R}^n) \; : \; supp\, f \subset \overline{\Omega}\} \tag{27.42}$$

and

$$\widetilde{H}^s_p(\Omega) = \{f \in H^s_p(\mathbb{R}^n) \; : \; supp\, f \subset \overline{\Omega}\} . \tag{27.43}$$

To shed some light on these spaces we repeat some assertions which may be found in [Tri78], 4.3.2, pp. 317–319. Let $\overset{\circ}{B}{}^s_{pq}(\Omega)$ and $\overset{\circ}{H}{}^s_p(\Omega)$ be the completion of $C^\infty_0(\Omega)$ in $B^s_{pq}(\Omega)$ and $H^s_p(\Omega)$, respectively. It holds:

(i) Let $1 < p < \infty$, $1 \leq q < \infty$, and $-\infty < s < \frac{1}{p}$; then

$$B^s_{pq}(\Omega) = \overset{\circ}{B}{}^s_{pq}(\Omega) \quad \text{and} \quad H^s_p(\Omega) = \overset{\circ}{H}{}^s_p(\Omega); \tag{27.44}$$

(ii) Let $1 < p < \infty$, $1 \le q < \infty$, $\frac{1}{p} - 1 < s < \infty$, and $s - \frac{1}{p} \ne \mathbb{N}_0$; then

$$\widetilde{B}^s_{pq}(\Omega) = \overset{\circ}{B}{}^s_{pq}(\Omega) \quad \text{and} \quad \widetilde{H}^s_p(\Omega) = \overset{\circ}{H}{}^s_p(\Omega). \tag{27.45}$$

As for the tricky cases $s = \frac{1}{p} + \mathbb{N}_0$ we refer to [Tri78], 4.3.2, p. 319. There is some overlapping of the diverse types of spaces if $\frac{1}{p} - 1 < s < \frac{1}{p}$, which provides a better understanding for introducing the following types of scales of spaces:

(i) *Let* $1 < p < \infty$*; then*

$$\overline{H}^s_p(\Omega) = H^s_p(\Omega) \text{ if } s < 0, \quad \text{and} \quad \overline{H}^s_p(\Omega) = \widetilde{H}^s_p(\Omega) \text{ if } s \ge 0. \tag{27.46}$$

(ii) *Let* $1 < p < \infty$ *and* $1 \le q \le \infty$*; then*

$$\overline{B}^s_{pq}(\Omega) = B^s_{pq}(\Omega) \text{ if } s < 0, \quad \text{and} \quad \overline{B}^s_{pq}(\Omega) = \widetilde{B}^s_{pq}(\Omega) \text{ if } s \ge 0. \tag{27.47}$$

Having the necessary spaces we turn to *regular elliptic differential operators* (of order $2m$). For sake of simplicity we restrict ourselves to the prototype of these operators, this is A,

$$A u = (-\Delta)^m u, \tag{27.48}$$

considered as an unbounded operator in $L_p(\Omega)$, where $1 < p < \infty$, and with the domain of definition

$$dom(A) = \left\{ g \in H^{2m}_p(\Omega) \; : \; D^\gamma g | \partial\Omega = 0 \quad \text{if} \quad 0 \le |\gamma| \le m - 1 \right\}. \tag{27.49}$$

Here $\Delta = \sum\limits_{j=1}^{n} \dfrac{\partial^2}{\partial x_j^2}$ stands for the Laplacian. For these operators we developed in [Tri78], 4.9.1, 4.9.2, pp. 333–336, a theory of the corresponding fractional powers A^τ, $\tau \in \mathbb{R}$, within the scales (27.46) and (27.47). As usual we shall not indicate by additional indices the spaces where we start from or the target spaces. We have the following assertions:

(i) *Let* $1 < p < \infty$, $\varkappa \in \mathbb{R}$, $s_1 \in \mathbb{R}$, $s_2 = s_1 - \varkappa$ *with*

$$-m - 1 < s_1 - \frac{1}{p} < m, \quad -m - 1 < s_2 - \frac{1}{p} < m. \tag{27.50}$$

Then $A^{\frac{\varkappa}{2m}}$ *maps*

$$\overline{H}^{s_1}_p(\Omega) \quad \text{\textit{isomorphically onto}} \quad \overline{H}^{s_2}_p(\Omega). \tag{27.51}$$

(ii) *Let* $1 < p < \infty$, $1 \le q \le \infty$, $\varkappa \in \mathbb{R}$, $s_1 \in \mathbb{R}$, $s_2 = s_1 - \varkappa$ *with* (27.50). *Then* $A^{\frac{\varkappa}{2m}}$ *maps*

$$\overline{B}^{s_1}_{pq}(\Omega) \quad \text{\textit{isomorphically onto}} \quad \overline{B}^{s_2}_{pq}(\Omega). \tag{27.52}$$

A proof may be found in [Tri78], 4.9.2, pp. 335–336.

27.12 Comments First we wish to emphasize that A given by (27.48), (27.49) with its fractional powers A^τ, surfing freely through the indicated scales of spaces, is a prototype representing a much larger class of *regular elliptic differential operators* with complementing boundary operators which are not necessarily of Dirichlet type. If one wishes to have assertions of the above type one needs some additional conditions which go back to S. Agmon and R. Seeley. As for a description including the necessary references we refer again to [Tri78], 4.9.1, 4.9.2, 5.2.1, pp. 333–336, 361–364.

27.13 Fractal elliptic operators
In 26.5 we described the two types (26.25) and (26.27) of fractal pseudodifferential operators we are interested in. By the mapping properties described at the end of 27.11 we are now in the position outlined at the beginning of 27.11. In other words, after the necessary modifications one can replace $b(\cdot, D)$ in (26.25), (26.27), Theorems 27.2, 27.4, 27.9, and in the related assertions in Section 28 by fractional powers $A^{-\frac{\varkappa}{2m}}$ of regular elliptic differential operators. We will not do this here in a systematic way, somewhat in contrast to [ET96], Ch. 5, with respect to the degenerate case. We restrict ourselves here to the above prototype (27.48), (27.49) in order to study the problem of exponents according to the outlook in 27.10 and to play diverse fractal drums.

27.14 Orthogonal wavelets
Instead of pseudodifferential operators given by (27.34) of order $-\varkappa$ and with the fractal coefficients $b(x)$, we have now a closer look at (fractional) elliptic operators

$$B = A^{-\frac{\varkappa}{m}} \circ b \tag{27.53}$$

of order $-2\varkappa$ and

$$b \in \mathscr{C}^s(\Omega) = B^s_{\infty,\infty}(\Omega), \quad s < 0, \tag{27.54}$$

where A is given by (27.48), (27.49) and the fractional powers of A have the meaning explained in 27.11. It is our aim to find suitable coefficients b where one can complement (27.35) by two-sided estimates. For that purpose we rely on $L_2(\Omega)$-orthogonal wavelets which have been constructed by I. Daubechies, S. Mallat and Y. Meyer, see [Dau92] and [Mey92]. We follow here [Mey92], 3.9, p. 108:

Let $r \in \mathbb{N}$ be given. There exists an r times differentiable real function $\varphi(x)$ in $\mathbb{R}^n$ with a support near the origin,

$$\int_{\mathbb{R}^n} \varphi^2(x)\, dx = 1 \tag{27.55}$$

and

$$\int_{\mathbb{R}^n} x^\gamma\, \varphi(x)\, dx = 0, \quad 0 \le |\gamma| \le r, \tag{27.56}$$

such that

$$\{\varphi(2^\nu x - k) \; : \; \nu \in \mathbb{N}_0, \; k \in \mathbb{Z}^n\} \tag{27.57}$$

is orthogonal in $L_2(\mathbb{R}^n)$.

One may assume that for fixed ν the functions $\varphi(2^\nu x - k)$ with $k \in \mathbb{Z}^n$ have disjoint supports. Hence the crucial point is to compare functions with different ν's. Recall that Ω is a bounded domain in $\mathbb{R}^n$ with C^∞ boundary $\partial\Omega$. Let

$$b(x) = \sum_{\nu \in \mathbb{N}_0} \sum_{k \in \mathbb{Z}^n}^{\nu, \Omega} 2^{-\nu s} \, \varphi \, (2^\nu x - k), \tag{27.58}$$

where $\displaystyle\sum_{k \in \mathbb{Z}^n}^{\nu, \Omega}$ means that for fixed $\nu \in \mathbb{N}_0$ the summation over k is restricted to those k with

$$\operatorname{supp} \varphi \, (2^\nu \cdot -k) \subset \Omega. \tag{27.59}$$

If $s < 0$ is given and $r \in \mathbb{N}$ is chosen sufficiently large then (27.58) might be considered as an atomic decomposition and by Theorem 13.8 we have

$$b \in \mathscr{C}^s(\Omega) = B^s_{\infty,\infty}(\Omega). \tag{27.60}$$

Using equivalent quasi-norms based on local means, see 11.2, it follows that b belongs also to any of the spaces $B^s_{pq}(\Omega)$ with the same s and $0 < p \leq \infty$, $0 < q \leq \infty$.

To prepare what follows we put

$$\widetilde{H}^\varkappa(\Omega) = \widetilde{H}^\varkappa_2(\Omega). \tag{27.61}$$

Assume $0 < \varkappa < m + \frac{1}{2}$ and let A be the regular elliptic operator of order $2m$ given by (27.48), (27.49). By (27.51) and (27.46) we may equip the Hilbert space $\widetilde{H}^\varkappa(\Omega)$ with the norm

$$\left\| A^{\frac{\varkappa}{2m}} u \,|\, L_2(\Omega) \right\|$$

and a corresponding scalar product. In the theorem below we assume in addition $\varkappa > \frac{n}{2}$. This corresponds to Theorem 27.2. But this is unimportant and one can also find a counterpart of Theorem 27.4. Finally we refer to Section 24 where we discussed operators generated by quadratic forms.

27.15 Theorem *Let Ω be a bounded domain in $\mathbb{R}^n$ with C^∞ boundary. Let the regular elliptic differential operator A of order $2m$ and its fractional powers be given by (27.48), (27.49), and as explained in 27.11. Let*

$$m + \frac{1}{2} > \varkappa > \frac{n}{2} \tag{27.62}$$

and let $b \in \mathscr{C}^s(\Omega)$ with $-\frac{n}{2} < s < 0$ be the singular distribution (27.58). Then

$$B = A^{-\frac{\varkappa}{m}} \circ b \tag{27.63}$$

is a self-adjoint compact operator in the Hilbert space $\widetilde{H}^{\varkappa}(\Omega)$, generated by the (appropriately interpreted) quadratic form

$$\int_{\Omega} b(x)\, g(x)\, \overline{h(x)}\, dx = (Bg, h)_{\widetilde{H}^{\varkappa}}, \quad h \in C_0^{\infty}(\Omega). \tag{27.64}$$

Let μ_k be the non-vanishing eigenvalues of B, ordered by their magnitude and counted according to their multiplicities. Then there are two positive constants c_1 and c_2 such that

$$c_1\, k^{-\frac{2\varkappa - |s|}{n}} \le |\mu_k| \le c_2\, k^{-\frac{2\varkappa - 2|s|}{n}}, \quad k \in \mathbb{N}. \tag{27.65}$$

Proof.
Step 1. By [Tri78], 4.8.1, p. 332, we have the duality assertion

$$\left(\widetilde{H}^{|s|}(\Omega)\right)' = H^s(\Omega), \tag{27.66}$$

interpreted within the framework of the dual pairing $(D(\Omega), D'(\Omega))$. Since $|s| < \varkappa$ we obtain in particular

$$\left|\int_{\Omega} b(x)\, g(x)\, \overline{h(x)}\, dx\right| \le c\, \|b\, g\,|\, H^s(\Omega)\|\, \|h\,|\, \widetilde{H}^{|s|}(\Omega)\|, \tag{27.67}$$

where b has the above meaning, $g \in \widetilde{H}^{\varkappa}(\Omega)$, and $h \in \widetilde{H}^{\varkappa}(\Omega)$. We apply (11.35) with p as indicated in Fig. 27.5 such that $p^s = 2$. With $B_2^s = H^s$ and by the remark

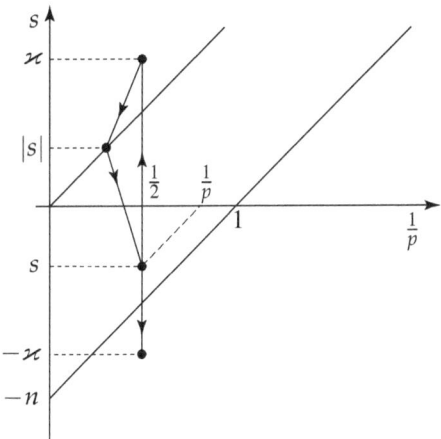

Fig. 27.5

after (27.60) we have

$$\|bg \mid H^s(\Omega)\| \le c \, \|g \mid B_{\infty|s|,1}^{|s|}(\Omega)\| \, \|b \mid H^s(\Omega)\|$$
$$\le c' \, \|g \mid \widetilde{H}^{\varkappa}(\Omega)\| \, \|b \mid \mathcal{C}^s(\Omega)\|, \tag{27.68}$$

where we used in addition that $\widetilde{H}^{\varkappa}(\Omega)$ is continuously (and compactly) embedded in $B_{\infty|s|,1}^{|s|}(\Omega)$, see Fig. 27.5. Inserting (27.68) in (27.67) we obtain

$$\left| \int_\Omega b(x) \, g(x) \, \overline{h(x)} \, dx \right| \tag{27.69}$$
$$\le c \, \|b \mid \mathcal{C}^s(\Omega)\| \, \|g \mid \widetilde{H}^{\varkappa}(\Omega)\| \, \|h \mid \widetilde{H}^{\varkappa}(\Omega)\|.$$

Hence the left-hand side of (27.64) is a bounded bilinear form in the Hilbert space $\widetilde{H}^{\varkappa}(\Omega)$. Then there is a uniquely determined bounded operator B in $\widetilde{H}^{\varkappa}(\Omega)$ with (27.64). Since b is real by construction, the bilinear form with $h = g$ is also real, and hence B is selfadjoint. It remains to prove that B defined by (27.64) coincides with $A^{-\frac{\varkappa}{m}} \circ b$ according to (27.63). Let $h \in C_0^\infty(\Omega)$ and $g \in \widetilde{H}^{\varkappa}(\Omega)$, then we have at least formally

$$\int_\Omega b(x) \, g(x) \, \overline{h(x)} \, dx = (Bg, h)_{\widetilde{H}^{\varkappa}}$$
$$= \left(A^{\frac{\varkappa}{2m}} Bg, A^{\frac{\varkappa}{2m}} h \right)_{L_2} \tag{27.70}$$
$$= \left(A^{\frac{\varkappa}{m}} Bg, h \right)_{L_2}.$$

Recall that we fixed the norm in $\widetilde{H}^{\varkappa}(\Omega)$ by $\left\| A^{\frac{\varkappa}{2m}} f \mid L_2(\Omega) \right\|$. This justifies the middle equality in (27.70). The last equality in (27.70) is based on the observation

$$\left(A^{\frac{\varkappa}{2m}} v, A^{-\frac{\varkappa}{2m}} w \right)_{L_2} = (v, w)_{L_2}, \tag{27.71}$$

interpreted in the framework of dual pairing $(D(\Omega), D'(\Omega))$, where $v = A^{\frac{\varkappa}{2m}} Bg \in L_2(\Omega)$, $w = A^{\frac{\varkappa}{2m}} h \in L_2(\Omega)$, and $|\varkappa| < m + \frac{1}{2}$ as required; which, in turn, follows from (27.50), (27.51), and the underlying duality arguments in [Tri78], 4.9.2, p. 336. Hence we have at least in $D'(\Omega)$

$$A^{\frac{\varkappa}{m}} Bg = bg, \quad g \in \widetilde{H}^{\varkappa}(\Omega). \tag{27.72}$$

Again by the indicated lifting properties we arrive at (27.63).

Step 2. We prove the estimate from above in (27.65) which is a modification of the proof of (27.8), cf. Fig.'s 27.1 and 27.5. Much as there we factorize B by

$$B = A^{-\frac{\varkappa}{m}} \circ id_2 \circ b \circ id_1 \tag{27.73}$$

with

$$id_1 \; : \; \widetilde{H}^{\varkappa}(\Omega) \to B^{|s|}_{\infty|s|,1}(\Omega)$$

$$b \; : \; B^{|s|}_{\infty|s|,1}(\Omega) \to H^{s}(\Omega)$$

$$id_2 \; : \; H^{s}(\Omega) \to H^{-\varkappa}(\Omega) \tag{27.74}$$

$$A^{-\frac{\varkappa}{m}} \; : \; H^{-\varkappa}(\Omega) \to \widetilde{H}^{\varkappa}(\Omega).$$

The embedding id_1 is compact and we have by Theorem 23.2 for the corresponding entropy numbers

$$e_k(id_1) \sim k^{-\frac{\varkappa - |s|}{n}}, \quad k \in \mathbb{N}. \tag{27.75}$$

The continuity of the map b follows from (27.68). The embedding id_2 is also compact and we have again

$$e_k(id_2) \sim k^{-\frac{\varkappa - |s|}{n}}, \quad k \in \mathbb{N}. \tag{27.76}$$

The last line in (27.74) fits into the scheme (27.50), (27.51) and is an isomorphic map. By the multiplication property (6.8) for entropy numbers we obtain

$$e_k(B) \leq c \, \|b \,|\, \mathcal{C}^s(\Omega)\| \, k^{-\frac{2\varkappa - 2|s|}{n}}, \quad k \in \mathbb{N}. \tag{27.77}$$

Finally the right-hand side of (27.65) follows from (27.77) and Corollary 6.10.

Step 3. We prove the estimate from below. Since B is a self-adjoint compact operator in the Hilbert space $\widetilde{H}^{\varkappa}(\Omega)$ we may apply Proposition 24.5(iii) where $a_k(B)$ are the corresponding approximation numbers according to Definition 24.3. In other words it is sufficient to estimate $a_k(B)$ from below. This is the point where the explicit structure of $b(x)$ comes in. Let

$$\varphi_{\nu k}(x) = \varphi(2^{\nu}x - k), \quad \nu \in \mathbb{N}_0, \quad k \in \mathbb{Z}^n, \tag{27.78}$$

be the L_2-orthogonal wavelets introduced in 27.14. Recall that $b(x)$ is given by (27.58). Let $\varrho \in \mathbb{N}_0$ be fixed and let

$$g_{\varrho}(x) = \sum_{k \in \mathbb{Z}^n}^{\varrho,\Omega} d_k^{\varrho} \, 2^{-\varrho(\varkappa - \frac{n}{2})} \, \varphi_{\varrho k}(x), \tag{27.79}$$

where d_k^{ϱ} are complex numbers and the sum has the same meaning as in (27.58). By the localization property in [ET96], 2.3.2, pp. 35–36, we have

$$\|g_{\varrho} \,|\, \widetilde{H}^{\varkappa}(\Omega)\| \sim \left(\sum_k^{\varrho,\Omega} |d_k^{\varrho}|^2 \right)^{\frac{1}{2}}, \tag{27.80}$$

where «$\sim$» means that the corresponding equivalence constants may be chosen independently of ϱ (and, of course, of d_k^ϱ). We have

$$\|Bg_\varrho \,|\, \tilde{H}^\varkappa(\Omega)\| = \sup \left| (Bg_\varrho, f)_{\tilde{H}^\varkappa} \right|, \tag{27.81}$$

where the supremum is taken over all $f \in \tilde{H}^\varkappa(\Omega)$ with $\|f\,|\,\tilde{H}^\varkappa(\Omega)\| \le 1$. Let χ be a suitably chosen C^∞ function with compact support near the origin and $\chi(x) = 1$ if $x \in \text{supp}\,\varphi$. Let $\chi_{\varrho k}(x)$ be the obvious counterpart of (27.78). By (27.80), up to unimportant constants,

$$f_\varrho(x) = \left(\sum_l {}^{\varrho,\Omega} |d_l^\varrho|^2 \right)^{-\frac{1}{2}} \sum_k {}^{\varrho,\Omega} d_k^\varrho \, 2^{-\varrho(\varkappa - \frac{n}{2})} \chi_{\varrho k}(x) \tag{27.82}$$

is an admitted function f in (27.81). By (27.58) and (27.64) we have

$$(Bg_\varrho, f_\varrho)_{\tilde{H}^\varkappa} = \left(\sum_m {}^{\varrho,\Omega} |d_m^\varrho|^2 \right)^{-\frac{1}{2}}$$

$$\times \int_\Omega \sum_{\nu \in \mathbb{N}_0} \sum_k {}^{\nu,\Omega} 2^{-\nu s} \varphi_{\nu k}(x) \sum_l {}^{\varrho,\Omega} d_l^\varrho \, 2^{-\varrho(\varkappa - \frac{n}{2})} \varphi_{\varrho l}(x) \overline{d_l^\varrho} \, 2^{-\varrho(\varkappa - \frac{n}{2})} \, dx$$

$$= \left(\sum_m {}^{\varrho,\Omega} |d_m^\varrho|^2 \right)^{-\frac{1}{2}} 2^{-\varrho s - \varrho(2\varkappa - n)} \sum_l {}^{\varrho,\Omega} |d_l^\varrho|^2 \int_\Omega \varphi_{\varrho l}^2(x)\, dx, \tag{27.83}$$

where we used the orthogonality of the wavelets $\varphi_{\varrho l}(x)$. By (27.55) the last integral equals $2^{-\varrho n}$. Hence by (27.81) and (27.83) we obtain

$$\|Bg_\varrho \,|\, \tilde{H}^\varkappa(\Omega)\| \ge c \, 2^{-\varrho(2\varkappa - |s|)} \left(\sum_m {}^{\varrho,\Omega} |d_m^\varrho|^2 \right)^{\frac{1}{2}} \tag{27.84}$$

$$\ge c' \, 2^{-\varrho(2\varkappa - |s|)} \|g_\varrho \,|\, \tilde{H}^\varkappa(\Omega)\|,$$

where c and c' are positive constants which are independent of $\varrho \in \mathbb{N}$. We assume that the dimension of the span of the admitted functions g_ϱ in (27.79) is larger than $2^{\varrho n}$ (not to speak about constants). Let T be a linear operator in $\tilde{H}^\varkappa(\Omega)$ with rank of at most $2^{\varrho n}$. Then we find a function g_ϱ of type (27.79) with

$$\|g_\varrho \,|\, \tilde{H}^\varkappa(\Omega)\| = 1 \quad \text{and} \quad T(g_\varrho) = 0. \tag{27.85}$$

However then it follows by (24.11), (27.84) and (27.85)

$$a_{2^{\varrho n}}(B) \ge c \, 2^{-\varrho(2\varkappa - |s|)}, \quad \varrho \in \mathbb{N}, \tag{27.86}$$

where $c > 0$ is independent of ϱ. This is equivalent to

$$a_k(B) \ge c \, k^{-\frac{2\varkappa - |s|}{n}}, \quad k \in \mathbb{N}, \tag{27.87}$$

for some $c > 0$. By (24.16) we obtain the left-hand side of (27.65).

27.16 The fractal defect
We return to the discussion in 27.3. Obviously, B in (27.63) is an elliptic operator of order $-2\varkappa$ with the fractal coefficient b of smoothness $s < 0$. Hence the right-hand side of (27.65), now with $2\varkappa$ in place of $\varkappa$, coincides with (27.8) and the corresponding assertions in the Theorems 27.4 and 27.9. On the other hand, even in case of degenerate pseudodifferential operators as discussed in 26.3 we have always the Weyl exponent $\frac{\varkappa}{n}$, or better now $\frac{2\varkappa}{n}$ adapted to the situation in the above theorem. It turns out by the left-hand side of (27.65) that this deviation from the Weyl exponent $\frac{2\varkappa}{n}$ is not an artefact of the used method, but it is the price to pay if one steps from degenerate to fractal. On the other hand there remains a gap between the two exponents in (27.65) and it is not clear what is the correct exponent (if there is any) in the understanding of the worst what could happen. Since the technique is quite natural there might be a temptation to conjecture that $\frac{2\varkappa-2|s|}{n}$ is nearer to the truth than $\frac{2\varkappa-|s|}{n}$. In the next section we strengthen the assumptions about the fractal coefficients. In that case we shall obtain final answers as far as the exponents are concerned.

28 Fractal pseudodifferential operators

28.1 The set-up
So far we dealt with two types of elliptic operators connected with fractals. Firstly, the operators A_s in Theorem 25.2 are generated by the scalar product in $H^s(\Gamma)$, considered as a quadratic form. Here Γ is a compact d-set in $\mathbb{R}^n$ with $0 < d < n$. As discussed in 20.3 the surrounding $\mathbb{R}^n$ is completely unimportant. Hence A_s *is a genuine d-dimensional affair* on Γ. This is well-reflected by the behaviour of the eigenvalues as described in (25.13), where $\frac{2s}{d}$ is the expected Weyl exponent. As indicated in 26.5(i) and (26.24) these considerations can be extended to some degenerate operators of type $A_s^{-\frac{1}{2}}$. Secondly, as announced in 26.5(ii) we dealt in Section 27 in some detail with pseudodifferential operators with fractal coefficients of type (26.25). As described in (26.26) the typical exponent in the related distribution of eigenvalues is now $\frac{\varkappa-2|s|}{n}$, where n reminds of the dimension of $\mathbb{R}^n$ and $\varkappa - 2|s|$ in place of $\varkappa$ is the fractal defect. Despite this fractal influence *these operators live in $\mathbb{R}^n$*, or in bounded domains in $\mathbb{R}^n$. Now we come to the third (and most interesting) type of operators discussed in 26.5, characterized by (26.27), (26.28). It is quite clear that we are now in a mixed situation. On the one hand, these operators may *live on the compact d-set Γ*, but on the other hand *they communicate with the surrounding $\mathbb{R}^n$*. This is also well-reflected by the exponent $\frac{\varkappa-n+d}{d}$ in (26.29). Here d reminds of the dimension of Γ and the deviation $n - d$ from $\varkappa$ is just the difference of the two involved dimensions n and d. In contrast to the situation discussed in 27.16 we are now in the position to prove that this exponent is sharp.

We complement the above general remarks by some technicalities. Let Ω be a bounded domain in $\mathbb{R}^n$ with C^∞ boundary $\partial\Omega$, and let Γ be a compact d-set in $\mathbb{R}^n$ with $0 < d < n$ according to Definition 3.1 such that $\Gamma \subset \Omega$. Let, for example, A be given by (27.48), (27.49) and let A^τ be the related fractional powers. Then one type of fractal pseudodifferential operators of order $-2\varkappa < 0$ we are interested in is given by

$$B = A^{-\frac{\varkappa}{m}} \circ b(\gamma) \tag{28.1}$$

where $b(\gamma) \in L_r(\Gamma)$ is a typical assumption. This operator can be interpreted in two different ways:

(i) B is a (compact) operator in some space $L_p(\Gamma)$. In this case

$$f \in L_p(\Gamma) \mapsto bf \in L_q(\Gamma)$$

is Hölder's inequality. Then bf will be interpreted by (18.6), (18.7) as an element of some spaces $B_{q,\infty}^{-\lambda}(\Omega)$. Assume that $A^{-\frac{\varkappa}{m}}$ can be applied according to (27.52), then Bf belongs to the space $\widetilde{B}_{p,\infty}^{2\varkappa-\lambda}(\Omega)$. Finally one has to apply the trace theorem 18.6, or (20.3). If the parameters are suitably chosen one obtains in this way a compact operator B in $L_p(\Gamma)$.

(ii) B is a (compact) operator in some space $\widetilde{H}_p^\varkappa(\Omega)$ according to (27.46). Then first, if $f \in \widetilde{H}_p^\varkappa(\Omega)$ one has to ask for $tr_\Gamma f$ belonging to $L_p(\Gamma)$ or better to some $\mathbb{B}_{pq}^s(\Gamma)$ introduced in (20.3). After multiplication with b via Hölder's inequality and interpretation of the outcome as an element in some space $B_{q,\infty}^{-\lambda}(\Omega)$, one applies $A^{-\frac{\varkappa}{m}}$ and arrives by embedding in $\widetilde{H}_p^\varkappa(\Omega)$.

These two interpretations are typical for our approach. They are not so different as it might seem at first glance:

We travel through spaces in commutative diagrams.

Then the difference between (i) and (ii) is just the question from which space we are starting. We return to these interpretations in 28.8. *In the sequel we shall not distinguish between, say, $b(\gamma)$ as an element of some $L_p(\Gamma)$ and as the distribution belonging to some $B_{p,\infty}^{-\lambda}(\Omega)$ according to* (18.6). It will be clear from the context what is meant. To avoid any misunderstanding we emphasize that the trace operator has two different meanings which we distinguish by tr_Γ and tr^Γ if extra clarity is desirable. Let, for example, $1 < p < \infty$; then

$$tr_\Gamma \; : \; B_{p,1}^{\frac{n-d}{p}}(\Omega) \to L_p(\Gamma) \tag{28.2}$$

by (18.23) and

$$tr^\Gamma \; : \; B_{p,1}^{\frac{n-d}{p}}(\Omega) \to B_{p,\infty}^{-\frac{n-d}{p'}}(\Omega) \tag{28.3}$$

if one applies in addition (18.7). The latter can be rephrased asking for an optimal extension of tr^Γ considered as a mapping from $D(\Omega)$ into $D'(\Omega)$ given by

$$\left(tr^\Gamma \varphi \right)(\psi) = \int_\Gamma \varphi(\gamma)\,\psi(\gamma)\,\mu(d\gamma), \tag{28.4}$$

where $\varphi \in D(\Omega)$ and $\psi \in D(\Omega)$, and μ is the Hausdorff measure $\mathcal{H}^d$ restricted to Γ. We refer in this context also to the duality argument in 18.18.

First we look for the counterparts of 26.3 and 27.2. All notation has been explained in Section 26 and in 27.1.

28.2 Theorem *Let Γ be a compact d-set in $\mathbb{R}^n$ with $0 < d < n$ according to Definition 3.1. Suppose that*

$$1 \le r_1 \le \infty,\ 1 \le r_2 \le \infty,\ 1 \le p \le \infty,\ \varkappa > 0,\ 0 \le \varrho \le 1, \tag{28.5}$$

where

$$\frac{1}{r_2} < \frac{1}{p} < 1 - \frac{1}{r_1} \quad \text{and} \quad \varkappa - n + d > d\left(\frac{1}{r_1} + \frac{1}{r_2}\right). \tag{28.6}$$

Let

$$b_1 \in L_{r_1}(\Gamma), \quad b_2 \in L_{r_2}(\Gamma), \quad b(x,D) \in \Psi_{1,\varrho}^{-\varkappa}. \tag{28.7}$$

Then

$$B = b_2\,b(\cdot,D)\,b_1 \tag{28.8}$$

is compact in $L_p(\Gamma)$ and its non-vanishing eigenvalues μ_k, repeated according to algebraic multiplicity, and ordered so that

$$|\mu_1| \ge |\mu_2| \ge \cdots \tag{28.9}$$

satisfy

$$|\mu_k| \le c\,\|b_1 \,|\, L_{r_1}(\Gamma)\|\,\|b_2 \,|\, L_{r_2}(\Gamma)\|\ k^{-\frac{\varkappa-n+d}{d}}, \quad k \in \mathbb{N}, \tag{28.10}$$

where c is independent of b_1, b_2, and k.

Proof. By our explanations in 28.1 and by Theorem 18.2 the line between $-n+d$ and 1 in Fig. 28.1 represents $L_p(\Gamma)$. We factorize B as indicated in Fig. 28.1 by

$$B = b_2 \circ id_2 \circ tr_\Gamma \circ b(\cdot,D) \circ id_1 \circ b_1 \tag{28.11}$$

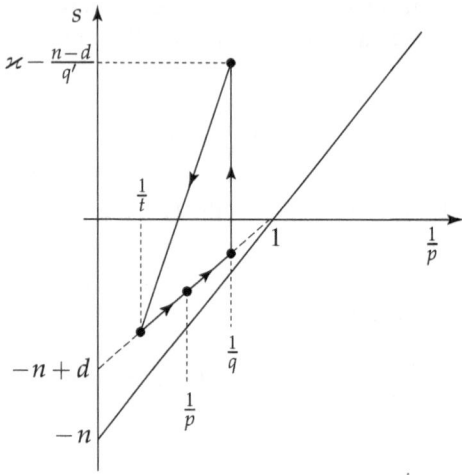

Fig. 28.1

with

$$b_1 \; : \; L_p(\Gamma) \to L_q(\Gamma) \quad \text{where} \quad \frac{1}{q} = \frac{1}{p} + \frac{1}{r_1}$$

$$id_1 \; : \; L_q(\Gamma) \to B_{q,\infty}^{-\frac{n-d}{q'}}(\mathbb{R}^n)$$

$$b(\cdot,D) \; : \; B_{q,\infty}^{-\frac{n-d}{q'}}(\mathbb{R}^n) \to B_{q,\infty}^{\varkappa-\frac{n-d}{q'}}(\mathbb{R}^n)$$

$$tr_\Gamma \; : \; B_{q,\infty}^{\varkappa-\frac{n-d}{q'}}(\mathbb{R}^n) \to \mathbb{B}_{q,\infty}^{\varkappa-n+d}(\Gamma)$$

$$id_2 \; : \; \mathbb{B}_{q,\infty}^{\varkappa-n+d}(\Gamma) \to L_t(\Gamma) \quad \text{where} \quad \frac{1}{t} = \frac{1}{p} - \frac{1}{r_2}$$

$$b_2 \; : \; L_t(\Gamma) \to L_p(\Gamma) \quad \text{since} \quad \frac{1}{p} = \frac{1}{t} + \frac{1}{r_2}.$$

(28.12)

The multiplications by b_1 and b_2 are covered by Hölder's inequality. Since $q > 1$ the continuity of id_1 comes from (18.7). Since $\varkappa > n - d$, the third line follows from (26.21), also in the exotic case. The continuity of tr_Γ is just the definition (20.3). Since

$$\frac{1}{q} - \frac{1}{t} = \frac{1}{r_1} + \frac{1}{r_2}$$

we have by (28.6)

$$\varkappa - n + d - \frac{d}{q} > -\frac{d}{t}.$$

(28.13)

Hence, by Theorem 20.6 the operator id_2 is compact and

$$e_k(id_2) \sim k^{-\frac{\varkappa-n+d}{d}}, \quad k \in \mathbb{N}.$$

(28.14)

This is the counterpart of (27.11), and the rest is the same as there.

28.3 Remark It is quite clear that $b(\cdot, D) \in \Psi_{1,\varrho}^{-\varkappa}$ in the above theorem can be replaced by any other operator having the mapping properties needed in the third line in (28.12). In particular, fractional powers of regular elliptic differential operators in bounded smooth domains are distinguished replacements of the above pseudodifferential operators. This was done in [ET96], 5.2, in connection with degenerate elliptic operators in a rather systematic way. One can do the same in case of the fractal elliptic operators now considered. But we restrict ourselves here to the direct counterpart of the above theorem and to the specific preparations done so far in 27.11. This causes the additional ugly restrictions (28.17) which can be removed as we are going to indicate in the remark after the next corollary. Otherwise we rely on the interpretation (i) in 28.1 after (28.1).

28.4 Corollary *Let Ω be a bounded domain in $\mathbb{R}^n$ with C^∞ boundary. Let Γ be a compact d-set in $\mathbb{R}^n$ according to Definition 3.1 such that $\Gamma \subset \Omega$ and $0 < d < n$. Let*

$$1 \le r_1 \le \infty, \; 1 \le r_2 \le \infty, \; 1 \le p \le \infty, \; \varkappa > 0, \; m \in \mathbb{N}, \tag{28.15}$$

where

$$\frac{1}{r_2} < \frac{1}{p} < 1 - \frac{1}{r_1}, \quad 2\varkappa - n + d > d\left(\frac{1}{r_1} + \frac{1}{r_2}\right) \tag{28.16}$$

and

$$m > (n - d - 1)\left(1 - \frac{1}{p} - \frac{1}{r_1}\right), \quad 2\varkappa < m + 1 + (n - d - 1)\left(1 - \frac{1}{p} - \frac{1}{r_1}\right). \tag{28.17}$$

Let

$$b_1 \in L_{r_1}(\Gamma), \quad b_2 \in L_{r_2}(\Gamma), \tag{28.18}$$

and let $A^{-\frac{\varkappa}{m}}$ be the fractional powers of A given by (27.48), (27.49), and as introduced in 27.11. Then

$$B = b_2 A^{-\frac{\varkappa}{m}} b_1 \tag{28.19}$$

is compact in $L_p(\Gamma)$ and its non-vanishing eigenvalues μ_k, repeated according to algebraic multiplicity, and ordered by (28.9) satisfy

$$|\mu_k| \le c \, \|b_1 \,|\, L_{r_1}(\Gamma)\| \, \|b_2 \,|\, L_{r_2}(\Gamma)\| \, k^{-\frac{2\varkappa - n + d}{d}}, \quad k \in \mathbb{N}, \tag{28.20}$$

where c is independent of b_1, b_2, and k.

Proof. With $\frac{1}{q} = \frac{1}{p} + \frac{1}{r_1}$ and $\frac{1}{q} + \frac{1}{q'} = 1$ the assumptions (28.17) can be rewritten as

$$m > \frac{n - d - 1}{q'}, \quad 2\varkappa < m + 1 + \frac{n - d - 1}{q'}. \tag{28.21}$$

Hence by (27.50) and (27.52) with $2\varkappa$ and q in place of $\varkappa$ and p, respectively, we have

$$A^{-\frac{\varkappa}{m}} \quad : \quad B_{q,\infty}^{-\frac{n-d}{q'}}(\Omega) \to \widetilde{B}_{q,\infty}^{2\varkappa - \frac{n-d}{q'}}(\Omega). \tag{28.22}$$

Having this in mind one can follow the proof of Theorem 28.2, always with $2\varkappa$ in place of $\varkappa$.

28.5 Remark Compared with Theorem 28.2 we have now the extra condition (28.17) to meet the requirements (27.50). But this is not really necessary. If $\varkappa$ is not restricted as in (28.17) or (28.21) then one has to replace the right-hand side of (28.22) by $dom\left(A^{\frac{\sigma}{2m}}\right)$ with $\sigma = 2\varkappa - \frac{n-d}{q'}$. The other arguments remain unchanged. We refer to [ET96], 5.2.2, pp. 187–189, for details. As for the first condition in (28.17) or (28.21) the situation seems to be more complicated since we needed this condition in 27.11 in connection with the underlying duality theory. However later on this does not play any role. Instead of the machinery (27.50)–(27.52) we rely on hand-made arguments. Finally as we said before, A given by (27.48), (27.49) is only a prototype of a regular elliptic differential operator of order $2m$ and can be replaced to a large extent by more general regular elliptic differential operators of order $2m$, as we did in [ET96].

This applies also to the following theorem where we replace b in Theorem 27.15 by the Hausdorff measure $\mathcal{H}^d$ restricted to the above compact d-set. We use the same notations as introduced there. In particular, the Hilbert space $\widetilde{H}^\varkappa(\Omega)$ given by (27.61), is normed by $\|A^{\frac{\varkappa}{2m}}u\,|\,L_2(\Omega)\|$ provided that $0 < \varkappa < m + \frac{1}{2}$, and $A^{\frac{\varkappa}{2m}}$ again are the fractional powers of the regular elliptic differential operator according to (27.48), (27.49).

28.6 Theorem *Let Ω be a bounded domain in $\mathbb{R}^n$ with C^∞ boundary. Let Γ be a compact d-set in $\mathbb{R}^n$ according to Definition 3.1 such that $\Gamma \subset \Omega$ and $0 < d < n$. Let $m \in \mathbb{N}$ and*

$$m + \frac{1}{2} > \varkappa > \frac{n-d}{2}. \tag{28.23}$$

Let $A^{-\frac{\varkappa}{m}}$ be the fractional powers of A, given by (27.48), (27.49), and as introduced in 27.11. Let tr^Γ be the trace operator in the interpretation of (28.3), (28.4). Then

$$B = A^{-\frac{\varkappa}{m}} \circ tr^\Gamma \tag{28.24}$$

is a non-negative compact self-adjoint operator of order $-2\varkappa$ in $\widetilde{H}^\varkappa(\Omega)$ with null-space

$$N(B) = \left\{ f \in \widetilde{H}^\varkappa(\Omega), \quad tr_\Gamma\, f = 0 \right\}. \tag{28.25}$$

There are two positive numbers c_1 and c_2 such that the positive eigenvalues μ_k of B, repeated according to multiplicity and ordered by their magnitude, can be estimated by

$$c_1\, k^{-\frac{2\varkappa-n+d}{d}} \le \mu_k \le c_2\, k^{-\frac{2\varkappa-n+d}{d}}, \quad k \in \mathbb{N}. \tag{28.26}$$

Furthermore B is generated by the quadratic form in $\widetilde{H}^\varkappa(\Omega)$

$$\int_\Gamma f(\gamma)\overline{g(\gamma)}\,\mathcal{H}^d(d\gamma) = (B\,f,g)_{\widetilde{H}^\varkappa} \tag{28.27}$$

where $f \in \widetilde{H}^\varkappa(\Omega)$ and $g \in \widetilde{H}^\varkappa(\Omega)$.

Proof.

Step 1. By (28.23) and Theorem 18.6 we have

$$\| \, tr_\Gamma \, f \, | \, L_2(\Gamma) \| \le c \, \|f \, | \, B_{2,1}^{\frac{n-d}{2}}(\Omega)\| \le c' \, \|f \, | \, H^{\varkappa}(\Omega)\| \tag{28.28}$$

for any $f \in H^{\varkappa}(\Omega)$. Since $\Gamma \subset \Omega$, the boundary $\partial\Omega$ does not play any role in (28.28) and we can replace $H^{\varkappa}(\Omega)$ by $\widetilde{H}^{\varkappa}(\Omega)$. Hence the left-hand side of (28.27) is a non-negative bounded quadratic form in $\widetilde{H}^{\varkappa}(\Omega)$. It is well-known that there is a uniquely determined non-negative self-adjoint bounded operator B such that (28.27) holds, see e. g. [Tri92*], p. 91. Furthermore we have

$$\|\sqrt{B} \, f \, | \, \widetilde{H}^{\varkappa}(\Omega)\| = \| \, tr_\Gamma \, f \, | \, L_2(\Gamma) \|. \tag{28.29}$$

This proves (28.25).

Step 2. We prove that B is the same operator as in (28.24). Let $g \in D(\Omega)$ and $f \in \widetilde{H}^{\varkappa}(\Omega)$. By the same reasoning as in (27.70)–(27.72) we have

$$\int_\Gamma f(\gamma) \overline{g(\gamma)} \, \mathcal{H}^d(d\gamma) = \left(A^{\frac{\varkappa}{2m}} B f, A^{\frac{\varkappa}{2m}} g \right)_{L_2(\Omega)} \tag{28.30}$$

$$= \left(A^{\frac{\varkappa}{m}} B f, g \right)_{L_2(\Omega)}.$$

Considered as a dual pairing in $(D(\Omega), D'(\Omega))$ we obtain

$$A^{\frac{\varkappa}{m}} B f = tr^\Gamma \, f \tag{28.31}$$

and (28.24) by the same arguments as in Step 1 of the proof of Theorem 27.15.

Step 3. By 25.1, in particular (25.6), and (28.25) we have

$$\widetilde{H}^{\varkappa}(\Omega) = N(B) \oplus H^{\varkappa - \frac{n-d}{2}}(\Gamma), \tag{28.32}$$

where we used the notation introduced there. Hence the eigenvalues μ_k of B coincide with the eigenvalues of the restriction of B on $H^{\varkappa - \frac{n-d}{2}}(\Gamma)$, denoted by B^Γ. We factorize B^Γ by

$$B^\Gamma = id_3 \circ tr_\Gamma \circ A^{-\frac{\varkappa}{m}} \circ id_2 \circ id_1 \tag{28.33}$$

with

$$id_1 \; : \; H^{\varkappa - \frac{n-d}{2}}(\Gamma) \to L_2(\Gamma)$$

$$id_2 \; : \; L_2(\Gamma) \to B_{2,\infty}^{-\frac{n-d}{2}}(\Omega)$$

$$A^{-\frac{\varkappa}{m}} \; : \; B_{2,\infty}^{-\frac{n-d}{2}}(\Omega) \to B_{2,\infty}^{2\varkappa - \frac{n-d}{2}}(\Omega) \tag{28.34}$$

$$tr_\Gamma \; : \; B_{2,\infty}^{2\varkappa - \frac{n-d}{2}}(\Omega) \to \mathbb{B}_{2,\infty}^{2\varkappa - n + d}(\Gamma)$$

$$id_3 \; : \; \mathbb{B}_{2,\infty}^{2\varkappa - n + d}(\Gamma) \to H^{\varkappa - \frac{n-d}{2}}(\Gamma).$$

By (28.23) and Theorem 20.6 the embedding id_1 is compact and

$$e_k(id_1) \sim k^{-\frac{2\varkappa - n + d}{2d}}, \quad k \in \mathbb{N}. \tag{28.35}$$

The mapping id_2 comes from (18.7). Sine $m + \frac{1}{2} > \frac{n-d}{2}$ the operator $A^{-\frac{\varkappa}{m}}$ is at least well-defined in the space $B_{2,\infty}^{-\frac{n-d}{2}}(\Omega)$ by our discussion in 27.11 and 27.12. However the claimed mapping property in the third line in (28.34) is not completely covered by what had been said at the end of 27.11. It follows from 27.11 and the additional assertion

$$A^{-\tau} \quad : \quad B_{2,\infty}^{0}(\Omega) \to B_{2,\infty}^{2m\tau}(\Omega), \quad \tau > 0. \tag{28.36}$$

In order to prove (28.36) we first remark that iterative application of (27.48), (27.49) yields

$$A^{-k} \quad : \quad L_2(\Omega) \to H^{2mk}(\Omega), \quad k \in \mathbb{N}. \tag{28.37}$$

Now (28.36) can be obtained from (28.37), complex and real interpolation, combined with the assertions at the end of 27.11. For details we refer to the relevant pages in [Tri78] quoted in 27.11 and 27.12. Hence the third line in (28.34) is at least a continuous map into (not necessarily onto). The continuity of tr_Γ comes from $2\varkappa > n - d$ and Definition 20.2. Finally by Theorem 20.6 the embedding id_3 is compact and

$$e_k(id_3) \sim k^{-\frac{2\varkappa - n + d}{2d}}, \quad k \in \mathbb{N}. \tag{28.38}$$

By (6.8) we obtain

$$e_k(B^\Gamma) \le c\, k^{-\frac{2\varkappa - n + d}{d}}, \quad k \in \mathbb{N}, \tag{28.39}$$

where c is independent of $k \in \mathbb{N}$. The right-hand side of (28.26) is now a consequence of (6.13).

Step 4. It remains to prove the estimate from below in (28.26). By Definition 3.1 and, say, $\mathcal{H}^d(\Gamma) = 1$ there is a number $c > 0$ such that for all $j \in \mathbb{N}$ there are $N_j \ge 2^{jd}$ disjoint balls centred at $x^{j,l} \in \Gamma$ and of radius $c2^{-j}$. We may assume that these balls are subsets of Ω (maybe one has to replace $j \in \mathbb{N}$ by $j \ge j_0$ for some $j_0 \in \mathbb{N}$). Let φ be a non-trivial C^∞ function supported near the origin. Let

$$\varphi_{jl}(x) = \varphi\left(2^j(x - x^{j,l})\right), \quad j \in \mathbb{N} \quad \text{and} \quad l = 1, \dots, N_j. \tag{28.40}$$

We may assume that for fixed j the functions $\varphi_{jl}(x)$ have disjoint supports. By the localization property in [ET96], 2.3.2, pp. 35–36, we have

$$\left\| \sum_{l=1}^{N_j} c_{jl}\, \varphi_{jl}(x) \mid \tilde{H}^\varkappa(\Omega) \right\| \sim \left(\sum_{l=1}^{N_j} |c_{jl}|^2 \right)^{\frac{1}{2}} 2^{j(\varkappa - \frac{n}{2})}, \tag{28.41}$$

where «∼» means that the corresponding equivalence constants may be chosen independently of j (and, of course, of $c_{jl} \in \mathbb{C}$). Furthermore we may also assume that

$$\left(\int_\Gamma \left| \sum_{l=1}^{N_j} c_{jl} \, \varphi_{jl}(\gamma) \right|^2 \mathcal{H}^d(d\gamma) \right)^{\frac{1}{2}} \sim 2^{-j\frac{d}{2}} \left(\sum_{l=1}^{N_j} |c_{jl}|^2 \right)^{\frac{1}{2}}. \tag{28.42}$$

Then we obtain by (28.29) and (28.41)

$$\left\| \sqrt{B} \left(\sum_{l=1}^{N_j} c_{jl} \, \varphi_{jl} \right) \mid \widetilde{H}^\varkappa(\Omega) \right\| \sim 2^{-j\frac{d}{2}} \left(\sum_{l=1}^{N_j} |c_{jl}|^2 \right)^{\frac{1}{2}}$$

$$\sim 2^{-j\frac{2\varkappa-n+d}{2}} \left\| \sum_{l=1}^{N_j} c_{jl} \, \varphi_{jl} \mid \widetilde{H}^\varkappa(\Omega) \right\|. \tag{28.43}$$

But now we are in the same position as in (27.84). By the same arguments as there we obtain for the approximation numbers $a_k(\sqrt{B})$ the estimate

$$a_k(\sqrt{B}) \geq c \, k^{-\frac{2\varkappa-n+d}{2d}}, \quad k \in \mathbb{N}, \tag{28.44}$$

where $c > 0$ is independent of k. Here we used $N_j \geq 2^{jd}$. Since B is a non-negative compact self-adjoint operator the left-hand side of (28.26) follows from (28.44) and (24.16).

28.7 The fractal defect revisited
In 27.16 we discussed the fractal defect in connection with (27.65). Now we are in a better situation. In contrast to (27.65) the exponent in (28.26) is sharp. But besides the fractal defect $n - d$, we have now the dimension d in the denominator of the exponent in place of n as in (27.65). In 28.1 we discussed this effect in somewhat philosophical, and hence vague, terms. Further considerations will be given in 28.9 and 28.10.

28.8 L_p-setting
The theorem and its proof depend on Hilbert space techniques. In particular, (24.16) has, in general, no counterpart in Banach spaces. On the other hand, B in (28.24) can be considered also in more general spaces. We did so in Corollary 28.4 or in a slightly different context in Theorem 28.2. The outcome (28.20) and (28.10), respectively, is just the same as the right-hand side of (28.26). Furthermore comparing the above theorem on the one hand and Theorem 28.2 and Corollary 28.4 on the other hand it is now quite clear that the two interpretations (i) and (ii) of B in (28.1) given afterwards in 28.1 are not so different as it might seem at first glance. Dealing with operators of these types in different spaces $L_p(\Gamma)$,

or $\widetilde{H}_p^\varkappa(\Omega)$, $\widetilde{B}_{pq}^\varkappa(\Omega)$, according to (27.43), (27.42), or in other admissible spaces, one is confronted with the question of the *spectral invariance*, that is the independence of the eigenvalues of the underlying space, and of the smoothness of the related *root spaces*, which is the collection of eigenfunctions and associated eigenfunctions. In case of degenerate (elliptic) operators we dealt with problems of this type in some detail in [ET96], Ch. 5. There is hardly any doubt that the techniques developed there can be extended to the fractal cases treated here. It is of interest to find out how smooth are the (associated) eigenfunctions of fractal operators, in particular if these operators are related to fractal drums. We do not go into detail here, but we return to these questions briefly in 30.9.

28.9 Weylian, sub-Weylian, and super-Weylian behaviour
We continue   our previous discussions about fractal defects and deviations from the Weylian behaviour. Assume that the (degenerate) elliptic operator B of order $-\varkappa$ in (26.11) is self-adjoint in a suitable Hilbert space. Then one can expect that (26.17) can be extended to the two-sided estimate

$$|\mu_k| \sim k^{-\frac{\varkappa}{n}}, \quad k \in \mathbb{N}. \tag{28.45}$$

Generalizing this expectation we say that a *compact* (self-adjoint, elliptic) *degenerate or fractal operator* in $\mathbb{R}^n$ of order $-\varkappa < 0$ is *Weylian, sub-Weylian, or super-Weylian* if

$$|\mu_k| \sim k^{-\varrho}, \quad k \in \mathbb{N}, \tag{28.46}$$

with

$$\varrho = \frac{\varkappa}{n}, \quad \varrho < \frac{\varkappa}{n}, \quad \text{or} \quad \varrho > \frac{\varkappa}{n},$$

respectively. In case of the operator B of order $-2\varkappa$ given by (28.24) we have

$$\varrho = \frac{2\varkappa - n + d}{d} \tag{28.47}$$

which must be compared with the *Weyl exponent* $\varkappa_W = \frac{2\varkappa}{n}$.

28.10 Proposition *The operator B given by* (28.24) *is*
  (i) *Weylian if, and only if,* $\varkappa_W = 1$,
 (ii) *sub-Weylian if, and only if,* $\varkappa_W < 1$, *and*
(iii) *super-Weylian if, and only if,* $\varkappa_W > 1$.

Proof. The number ϱ in (28.47) is equal to, less than, or larger than $\varkappa_W$ if, and only if, $2\varkappa(n-d)$ is equal to, less than, or larger than $n(n-d)$. But this is the desired assertion since $0 < d < n$.

28.11 Remark As so often in mathematics the number 1 plays an unexpected role, in our case that it is independent of d.

29 Fractal pseudodifferential operators: limiting cases

29.1 Introduction

We are now half-way in this chapter. As announced in 26.6 we seize the opportunity of looking briefly backwards and forwards. Sections 27 and 28 are devoted to pseudodifferential operators with fractal coefficients and fractal pseudodifferential operators where the distinction between these two classes of operators had been discussed in 26.5. Of interest for us with respect to our later considerations is the special case $\varkappa = m = 1$ in Theorem 28.6, which means

$$B = (-\Delta)^{-1} \circ tr^{\Gamma}, \tag{29.1}$$

where Δ stands for the Dirichlet Laplacian. This is the isotropic fractal drum in the version (iii) in 26.2. We return to this question in Section 30 discussing there also anisotropic and nonisotropic fractal drums. As in the preceding sections we again rely on related estimates for entropy numbers as provided in this case by Section 22. Using these results and methods we discuss in Section 31 the problem of the negative spectrum, which is related to quantum mechanics. The final Section 32, devoted to fractal nonlinearities, is a little bit outside of the outlined main stream. Returning to Sections 27 and 28 the overwhelming importance of estimates for entropy numbers of compact embeddings between related function spaces is now quite clear. The typical assumption ensuring the desired compactness is, for example, in case of Theorem 28.2,

$$\varkappa - n + d > d \left(\frac{1}{r_1} + \frac{1}{r_2} \right). \tag{29.2}$$

What happens if both sides in (29.2) are equal? We dealt with these limiting embeddings in Section 21. It is the main aim of this short section to use Theorem 21.7 to prove the limiting counterpart of Theorem 28.2. One word seems to be in order. In the context of (degenerate) elliptic differential operators limiting cases have attracted much attention. This has its origin in quantum mechanics and the problem of bound states. In connection with degenerate pseudodifferential operators we discussed in [ET96], Ch. 5, also these limiting situations in some detail. There one can find also the necessary references to the quite substantial literature about this topic, see in particular [ET96], 5.2.1 and 5.3. The theorem below is the fractal version of some results obtained there. Finally, in the Sections 20–22 we hinted at further possibilities for estimating entropy numbers. It is quite clear now that this results in estimates of corresponding eigenvalues. Of peculiar interest might be the incorporation of weighted spaces in connection with unbounded diluted d-sets as indicated in 20.11. But we do not go into detail for a simple reason: Nothing has been done so far.

The following theorem is the limiting version of Theorem 28.2, based on Theorem 21.7. We use the notation introduced there. In particular, $L_p(\log L)_a(\Gamma)$ has the same meaning as in Definition 21.2.

29.2 Theorem *Let* Γ *be a compact d-set in* $\mathbb{R}^n$ *with* $0 < d < n$ *according to Definition 3.1. Suppose*

$$1 \leq r_1 \leq \infty, \ 1 \leq r_2 \leq \infty, \ 1 \leq p \leq \infty, \ \varkappa > 0, \ 0 \leq \varrho \leq 1, \qquad (29.3)$$

where

$$0 < \frac{1}{r_2} \leq \frac{1}{p} < 1 - \frac{1}{r_1} \quad \text{and} \quad \varkappa - n + d = d\left(\frac{1}{r_1} + \frac{1}{r_2}\right). \qquad (29.4)$$

Let

$$b_1 \in L_{r_1}(\Gamma), \ b_2 \in L_{r_2}(\log L)_a(\Gamma), \ b(x,D) \in \Psi_{1,\varrho}^{-\varkappa}, \qquad (29.5)$$

with

$$a > 1 + 2\frac{\varkappa - n + d}{d}. \qquad (29.6)$$

Then

$$B = b_2\, b(\cdot, D)\, b_1 \qquad (29.7)$$

is compact in $L_p(\Gamma)$ *and its non-vanishing eigenvalues* μ_k, *repeated according to algebraic multiplicity, and ordered so that*

$$|\mu_1| \geq |\mu_2| \geq \cdots \qquad (29.8)$$

satisfy

$$|\mu_k| \leq c\, \|b_1 \mid L_{r_1}(\Gamma)\|\, \|b_2 \mid L_{r_2}(\log L)_a(\Omega)\|\, k^{-\frac{\varkappa - n + d}{d}}, \quad k \in \mathbb{N}, \qquad (29.9)$$

where c is independent of b_1, b_2 *and k.*

Proof. We modify the proof of Theorem 28.2 and rely as there on Fig. 28.1. As in (28.11) we factorize B by

$$B = b_2 \circ id_2 \circ tr_\Gamma \circ b(\cdot, D) \circ id_1 \circ b_1 \qquad (29.10)$$

with

$$b_1 \ : \ L_p(\Gamma) \to L_q(\Gamma) \quad \text{where} \quad \frac{1}{q} = \frac{1}{p} + \frac{1}{r_1}$$

$$id_1 \ : \ L_q(\Gamma) \to B_{q,\infty}^{-\frac{n-d}{q'}}(\mathbb{R}^n)$$

$$b(\cdot, D) \ : \ B_{q,\infty}^{-\frac{n-d}{q'}}(\mathbb{R}^n) \to B_{q,\infty}^{\varkappa - \frac{n-d}{q'}}(\mathbb{R}^n) \qquad (29.11)$$

$$tr_\Gamma \ : \ B_{q,\infty}^{\varkappa - \frac{n-d}{q'}}(\mathbb{R}^n) \to \mathbb{B}_{q,\infty}^{\varkappa - n + d}(\Gamma)$$

$$id_2 \ : \ \mathbb{B}_{q,\infty}^{\varkappa - n + d}(\Gamma) \to L_t(\log L)_{-a+\varepsilon}(\Gamma) \quad \text{where} \quad \frac{1}{t} = \frac{1}{p} - \frac{1}{r_2}$$

$$b_2 \ : \ L_t(\log L)_{-a+\varepsilon}(\Gamma) \to L_p(\Gamma).$$

The arguments concerning b_1, id_1, $b(\cdot, D)$, and tr_Γ are the same as in (28.12) and covered by the above assumptions. We choose $\varepsilon > 0$ such that also

$$a - \varepsilon > 1 + 2\frac{\varkappa - n + d}{d}. \tag{29.12}$$

We wish to apply Theorem 21.7 to id_2. We have $1 < t \le \infty$ and for $s = \varkappa - n + d > 0$,

$$\frac{1}{t^s} = \frac{1}{t} + \frac{s}{d} = \frac{1}{q}, \tag{29.13}$$

where we used (29.4). Then we obtain by (21.16)

$$e_k(id_2) \sim k^{-\frac{\varkappa - n + d}{d}}, \quad k \in \mathbb{N}. \tag{29.14}$$

Finally the continuity of the mapping b_2 is covered by the arguments in [ET96], formula (7) on p. 214. The rest is the same as in (28.14) and after (27.11).

29.3 Remark The theorem is the fractal counterpart of part (i) of Theorem 5.3.2/1 in [ET96], pp. 207–208. Comparison shows the *fractal defect $n - d$*. Of course, there are also counterparts of the parts (ii) and (iii) of this theorem, but we do not go into detail. Comparison with Theorem 5.3.3/1 in [ET96], pp. 213–214 shows that the results obtained there in the degenerate case are slightly better. We discussed this point in Remark 5.3.3/1 in [ET96], p. 215. One can expect that the arguments provided there apply also to the fractal case.

30 Fractal drums

30.1 Introduction

Let Ω be a bounded domain in the plane $\mathbb{R}^2$ with C^∞ boundary $\partial\Omega$, interpreted as a membrane fixed at its boundary. Vibrations of such a membrane in $\mathbb{R}^3$ are measured by the deflection $u(x,t)$, where $x = (x_1, x_2) \in \Omega$, and $t \ge 0$ stands for the time. In other words, the point $(x_1, x_2, 0)$ in $\mathbb{R}^3$ with $(x_1, x_2) \in \Omega$ of the membrane at rest, is deflected to $(x_1, x_2, u(x,t))$. Up to constants the usual physical description is given by

$$\Delta u(x,t) = m(x)\frac{\partial^2 u(x,t)}{\partial t^2}, \quad x \in \Omega, \quad t \ge 0, \tag{30.1}$$

and

$$u(y,t) = 0 \quad \text{if} \quad y \in \partial\Omega, \quad t \ge 0, \tag{30.2}$$

where the right-hand side of (30.1) is Newton's law with the mass-density $m(x)$. Of course, $\Delta = \frac{\partial^2}{\partial x_1^2} + \frac{\partial^2}{\partial x_2^2}$ stands for the Laplacian. To find the eigenfrequencies one has to insert $u(x,t) = e^{i\lambda t} v(x)$ with $\lambda \in \mathbb{R}$ in (30.1) and obtains

$$-\Delta v(x) = \lambda^2 m(x) v(x), \quad x \in \Omega; \quad v(y) = 0 \quad \text{if} \quad y \in \partial\Omega, \tag{30.3}$$

where one is interested in non-trivial solutions $v(x)$. Hence one asks for the eigen-functions and the eigenvalues of the operator

$$B = (-\Delta)^{-1} \circ m(\cdot), \tag{30.4}$$

where $-\Delta$ stands for the Dirichlet Laplacian. If μ is such a positive eigenvalue then $\lambda = \mu^{-\frac{1}{2}}$ is the related eigenfrequency. Assume that the mass is concentrated in some fractal compact set Γ with $\Gamma \subset \Omega$. Then we arrive at the situation we discussed so far twice in (26.8), (26.9) on the one hand and (29.1) on the other hand. If Γ is an (isotropic) d-set and if the mass is concentrated at Γ and evenly distributed there, then we obtain

$$B = (-\Delta)^{-1} \circ tr^{\Gamma}, \tag{30.5}$$

what coincides with (29.1). These are the *isotropic fractal drums* we have in mind. We refer to 26.2 where we discussed three possible interpretations of what is meant by a fractal drum. Here we are exclusively interested in interpretation (iii), that means in the study of the operator B in (30.5), now in n dimensions. As indicated in 29.1 this is simply a specification of the operator in Theorem 28.6. We formulate the result again and add a few discussions and references. New effects occur if one replaces isotropic d-sets by more general compact fractals Γ with $\Gamma \subset \Omega$. This is the point at which the anisotropic and the nonisotropic d-sets in the plane enter. Based on our previous considerations in the Sections 5, 18, and 22, we play not only the isotropic fractal drum but also the anisotropic and, somewhat faintly, the nonisotropic drum.

Recall that $\overset{\circ}{W}{}_2^1(\Omega)$ is the classical Sobolev space, see 25.3 and 27.11.

30.2 Theorem (Isotropic fractal drum) *Let Ω be a bounded domain in $\mathbb{R}^n$ with C^∞ boundary. Let Γ be a compact d-set in $\mathbb{R}^n$ according to Definition 3.1 such that $\Gamma \subset \Omega$ and $n - 2 < d < n$ (with $0 < d < 1$ in case of $n = 1$). Let tr^{Γ} be the trace operator in the interpretation (28.3), (28.4), whereas tr_Γ stands for the trace on Γ according to (28.2). Then B, given by (30.5), is a non-negative compact self-adjoint operator in $\overset{\circ}{W}{}_2^1(\Omega)$ with null-space*

$$N(B) = \left\{ f \in \overset{\circ}{W}{}_2^1(\Omega), \quad tr_\Gamma f = 0 \right\}. \tag{30.6}$$

There are two positive numbers c_1 and c_2 such that the positive eigenvalues μ_k of B, repeated according to multiplicity and ordered by their magnitude, can be estimated by

$$c_1 \, k^{-\frac{2-n+d}{d}} \le \mu_k \le c_2 \, k^{-\frac{2-n+d}{d}}, \quad k \in \mathbb{N}. \tag{30.7}$$

Furthermore, B is generated by the quadratic form in $\overset{\circ}{W}{}_2^1(\Omega)$,

$$\int_\Gamma f(\gamma) \overline{g(\gamma)} \, \mathcal{H}^d(d\gamma) = (Bf, g)_{W_2^1}, \tag{30.8}$$

where $f \in \overset{\circ}{W}{}_2^1(\Omega)$ and $g \in \overset{\circ}{W}{}_2^1(\Omega)$.

Proof. Let $\varkappa = m = 1$ in Theorem 28.6. Then $A^{-1} = (-\Delta)^{-1}$ is by (27.48), (27.49) the Dirichlet Laplacian. Furthermore (28.23) coincides with $d > n - 2$. Finally, by (27.41), (27.45) we have

$$\widetilde{H}^1(\Omega) = \overset{\circ}{W}{}^1_2(\Omega).$$

Hence the above theorem is a special case of Theorem 28.6.

30.3 Interpretation By 30.1,

$$\lambda_k \sim k^{\frac{2-n+d}{2d}}, \qquad k \in \mathbb{N}, \tag{30.9}$$

are the eigenfrequencies of the n-dimensional isotropic fractal drum, where the respective layer is an (isotropic) d-set with $n - 2 < d < n$ (that means $0 < d < 1$ in case of $n = 1$). Modifying the famous question, *Can one hear the shape of a drum*, [Kac66], one can ask

Can one hear the fractality of a drum?

The answer depends on the dimension if one interpretes this question as the possibility to recover d from the exponent in (30.9) or (what is the same) in (30.7). If $n = 2$, what corresponds to the vibrating (isotropic fractal) drum, then we have by (30.7)

$$\mu_k \sim k^{-1}, \qquad k \in \mathbb{N}, \tag{30.10}$$

and the answer is *no*. This is the classical Weyl exponent as discussed in 26.2, in particular (26.5). Hence *the Weyl exponent 1 in the plane is very resilient and survives even fractal distortions.* But this is not a surprise. In our case we have $\varkappa_W = 1$ in Proposition 28.10, and in this case the operator B is Weylian. By the same proposition,

$$\begin{array}{ll}
B & \text{is super-Weylian if } n = 1, \\
B & \text{is Weylian if } n = 2, \\
B & \text{is sub-Weylian if } n \geq 3, \tag{30.11}
\end{array}$$

where $n = 1$ refers to the vibrating string.

30.4 Discussion and references
The fractal drum in the version of 26.2(i) refers to a membrane Ω with constant mass density where the boundary $\partial\Omega$ might be a fractal, say an (isotropic) d-set with $n - 1 < d < n$. It is the battle about the remainder term in (26.4) and there is quite a substantial literature now. We gave in 26.2(i) a few references. On the other hand, as far as isotropic fractal drums according to Theorem 30.2 are concerned, nothing is known to the author in the case of d-sets as treated here (not to speak about our anisotropic ambitions). The situation is somewhat

different if the d-set in question is in addition self-similar. The vibrating string where Γ is a Cantor set has been considered in [Fuj87]. Higher-dimensional cases may be found in [NaS94] and [NaS95]. Of interest is also the paper [SoV95]. In this paper one finds a discussion whether λ_k in (30.9) with $n = 1$ can be refined by an asymptotic assertion of type

$$\lambda_k = c\, k^{\frac{1+d}{2d}} \left(1 + o(1)\right), \quad k \to \infty. \tag{30.12}$$

The answer is *negative*, the quotient $\lambda_k\, k^{-\frac{1+d}{2d}}$ might oscillate. This shows that

even under the restriction to self-similarity there is no hope to find something similar as in (26.4).

Since the days of H. Weyl much attention has been paid to investigate especially the second term on the right-hand side of (26.4). Nowadays this phenomenon is well understood, including a thorough study of the behaviour of this second term for «non-Weylian» boundaries where some oscillation may occur, see [SaV96].

It is remarkable that in the above fractal setting possible oscillations jump from the second term to the main term.

30.5 Anisotropic fractal drums: the set-up
Theorem 30.2 applies to compact d-sets in $\mathbb{R}^n$ according to Definition 3.1 which we call now *isotropic d-sets*, in order to distinguish them from *anisotropic d-sets* and *nonisotropic d-sets* introduced in Definitions 5.2 and 5.3, respectively. By Proposition 5.6 the PXT-fractals and the OF-fractals are illuminating constructive examples. In what follows we wish to identify Γ in (30.5) with anisotropic or non-isotropic d-sets. A final assertion as in (30.7) cannot be expected for the following reason. By Definition 5.2 an anisotropic d-set is governed by the parameters d and a with

$$0 < d < 2 \quad \text{and} \quad 0 \le a \le 1, \tag{30.13}$$

where the latter is the *deviation* from the isotropicity. As remarked in 5.7 with a reference to 4.24 these two numbers do not characterize an anisotropic d-set to such an extent that an equivalence of type (30.7) can be expected. Just on the contrary. By the discussions quoted and also by the outcome in 30.6 and 30.7 below one should try to make the deviation of a given anisotropic d-set as small as possible (this reminds of the structure theory in functional analysis where one asks how near a given Banach space to a Hilbert space is). Nevertheless under these circumstances the estimate in Theorem 30.7 below is rather satisfactory, where one needs for the estimate from below the additional assumption that the anisotropic d-set in question is *proper* according to Definition 5.11. By 5.12–5.14 this additional assumption is not very restrictive.

To prepare what follows we recall that now $-\Delta$ stands for the Dirichlet Laplacian in a bounded domain Ω in the plane $\mathbb{R}^2$ with a C^∞ boundary $\partial\Omega$.

Then $-\Delta$ maps any space

$$B_{pq,0}^{\sigma}(\Omega) = \{g \in B_{pq}^{\sigma}(\Omega), \quad tr_{\partial\Omega} \, g = 0\} \qquad (30.14)$$

onto $B_{pq}^{\sigma-2}(\Omega)$ provided that

$$1 \leq p \leq \infty, \quad 1 \leq q \leq \infty \quad and \quad \sigma > \frac{1}{p}. \qquad (30.15)$$

Here $B_{pq}^{\sigma}(\Omega)$ has the usual meaning, see 23.1. Mapping properties of $-\Delta$ had been discussed several times in this book. The above version may be found in [Tri78], 5.7.1, Remark 1 on p. 402, complemented by [Tri83], 4.3.3 and 4.3.4. Furthermore for the above compact anisotropic d-set with (30.13) we have Theorem 18.15. Let $\Gamma \subset \Omega$. With the rather typical combination

$$d_a = 2 - \frac{d}{1+a} \qquad (30.16)$$

we obtain by Theorem 18.15 and with the same interpretation as in (28.3), (28.4)

$$tr^{\Gamma} \quad : \quad B_{pq}^{\frac{d_a}{p}+s}(\Omega) \to L_p(\Gamma) \to B_{p,\infty}^{-\frac{d_a}{p'}}(\Omega), \qquad (30.17)$$

where

$$1 \leq p \leq \infty, \, 0 < q \leq \infty, \, \frac{1}{p} + \frac{1}{p'} = 1 \text{ and } s \geq 0$$

(with $q \leq 1$ in case of $s = 0$).

30.6 Proposition *Let Ω be a bounded domain in the plane $\mathbb{R}^2$ with C^{∞} boundary. Let Γ be an anisotropic d-set with the deviation a according to Definition 5.2 and (30.13) such that $\Gamma \subset \Omega$. Let*

$$1 \leq p \leq \infty, \quad \frac{1}{p} + \frac{1}{p'} = 1, \quad and \quad d_a = 2 - \frac{d}{1+a};$$

then

$$B = (-\Delta)^{-1} \circ tr^{\Gamma} \qquad (30.18)$$

with the trace operator tr^{Γ} in the interpretation (30.17) is compact in $B_{p,\infty}^{2-\frac{d_a}{p'}}(\Omega)$. Furthermore there is a constant $c > 0$ such that for the entropy numbers of B we have

$$e_k(B) \leq c \, k^{-\frac{d}{d+2a}}, \quad k \in \mathbb{N}. \qquad (30.19)$$

Proof. First we remark that the line $\sigma = 2 - \frac{d_a}{p'}$ is above the line $\sigma = \frac{1}{p}$ in Fig. 30.1. Hence $-\Delta$ makes sense on that line. Secondly, with

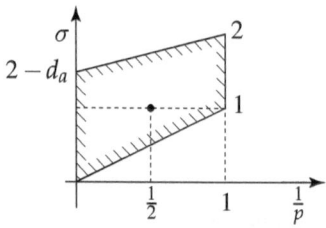

Fig. 30.1

$$s = 2 - d_a = \frac{d}{1+a} > 0, \qquad (30.20)$$

the trace operator tr^Γ in (30.17) makes sense and we have by Theorem 22.2 with $p_1 = p_2 = p$,

$$e_k(tr^\Gamma) \le c \, k^{-\frac{1+a}{d+2a}s} = c \, k^{-\frac{d}{d+2a}}, \qquad k \in \mathbb{N}, \qquad (30.21)$$

for some $c > 0$. The proposition is the consequence of these two observations.

30.7 Theorem (Anisotropic fractal drum) *Let Ω be a bounded domain in the plane $\mathbb{R}^2$ with C^∞ boundary. Let Γ be a proper anisotropic d-set with the deviation a according to the Definitions 5.2 and 5.11, and (30.13) such that $\Gamma \subset \Omega$. Then B given by (30.18) is a non-negative compact self-adjoint operator in $\overset{\circ}{W}{}^{\frac{1}{2}}_2(\Omega)$ with null-space*

$$N(B) = \left\{ f \in \overset{\circ}{W}{}^{\frac{1}{2}}_2(\Omega), \quad tr_\Gamma f = 0 \right\}. \qquad (30.22)$$

There are two positive numbers c_1 and c_2 such that the positive eigenvalues μ_k of B, repeated according to multiplicity and ordered by their magnitude, can be estimated by

$$c_1 \, k^{-\frac{d+2a}{d}} \le \mu_k \le c_2 \, k^{-\frac{d}{d+2a}}, \qquad k \in \mathbb{N}. \qquad (30.23)$$

Furthermore, B is generated by the quadratic form in $\overset{\circ}{W}{}^{\frac{1}{2}}_2(\Omega)$

$$\int_\Gamma f(\gamma) \overline{g(\gamma)} \, \mu(d\gamma) = (Bf, g)_{W^{\frac{1}{2}}_2} \qquad (30.24)$$

where $f \in \overset{\circ}{W}{}^{\frac{1}{2}}_2(\Omega)$ and $g \in \overset{\circ}{W}{}^{\frac{1}{2}}_2(\Omega)$, and μ stands for the Radon measure according to Theorem 5.5.

Proof.

Step 1. By (30.17) and (30.16) we have

$$\| \, tr_\Gamma \, f \, | \, L_2(\Gamma) \| \le c \, \| f \, | \, B_{2,1}^{\frac{d_a}{2}}(\Omega) \| \le c' \, \| f \, | \, W_2^1(\Omega) \|. \tag{30.25}$$

By the same arguments as in Step 1 of the proof of Theorem 28.6 it follows that there is a non-negative self-adjoint operator B in $\overset{\circ}{W}\,{}_2^1(\Omega)$ with (30.24),

$$\| \sqrt{B} \, f \, | \, W_2^1(\Omega) \| = \| \, tr_\Gamma \, f \, | \, L_2(\Gamma) \|, \tag{30.26}$$

and hence (30.22). The proof that B is given by (30.18) is the same as in Step 2 of the proof of Theorem 28.6, where we assume tacitly that $\overset{\circ}{W}\,{}_2^1(\Omega)$ is normed as indicated in the proofs of Theorems 30.2 and 28.6.

Step 2. Since the embedding of $W_2^1(\Omega)$ in $B_{2,1}^{\frac{d_a}{2}}(\Omega)$ is compact it follows by (30.25), (30.26) that B is compact. Let $f \in \overset{\circ}{W}\,{}_2^1(\Omega)$ be an eigenfunction. Then it follows by (30.18), (30.17) that f belongs even to $B_{2,\infty}^{2-\frac{d_a}{2}}(\Omega)$. Hence it is also an eigenfunction of the operator B restricted to this space. Obviously, the converse is also true. Hence the root systems of B considered in $W_2^1(\Omega)$ (or $\overset{\circ}{W}\,{}_2^1(\Omega)$, what is the same in our context) and in $B_{2,\infty}^{2-\frac{d_a}{2}}(\Omega)$ coincide. Then the eigenvalues of B considered in these two spaces also coincide, inclusively their multiplicities. The right-hand side of (30.23) follows now from Proposition 30.6 and (6.13).

Step 3. We prove the left-hand side of (30.23). Let φ be a non-trivial C^∞ function in $\mathbb{R}^2$ with compact support near the origin. Let $x^{j,l}$ be the centres of the rectangles R_l^j according to Definition 5.2. Let

$$\varphi_{jl}(x) = \varphi \left(2^{a_1^{j,l}}(x_1 - x_1^{j,l}), \, 2^{a_2^{j,l}}(x_2 - x_2^{j,l}) \right), \tag{30.27}$$

where all symbols have the same meaning as in Definition 5.2 and where we tacitly assumed, without restriction of generality, that $2^{-a_1^{j,l}}$ and $2^{-a_2^{j,l}}$ are the side-lengths of R_l^j in the directions x_1 and x_2, respectively. We may suppose

$$supp \, \varphi_{jl} \subset R_l^j \tag{30.28}$$

and

$$\int_\Gamma |\varphi_{jl}(\gamma)|^2 \, \mu(d\gamma) \sim 2^{-dj}; \quad l = 1, \dots, N_j; \tag{30.29}$$

where the equivalence constants can be chosen independently of $j \in \mathbb{N}$ (and of l). Here we use for the first time that the anisotropic d-set is proper according to Definition 5.11. By (5.10) we have

$$\|\varphi_{jl} \, | \, W_2^1(\Omega)\| \le c \, 2^{a_2^{j,l}} \, 2^{-j} \le c' \, 2^{(1+a)j-j} = c' \, 2^{aj}. \tag{30.30}$$

Now we are in the same position as in Step 4 of the proof of Theorem 28.6. Using (30.26) we have the following counterparts of (28.41)–(28.43)

$$\left\| \sum_{l=1}^{N_j} c_{jl}\, \varphi_{jl} \mid W_2^1(\Omega) \right\| \le c \left(\sum_{l=1}^{N_j} |c_{jl}|^2 \right)^{\frac{1}{2}} 2^{ja}, \tag{30.31}$$

$$\left(\int_\Gamma \left| \sum_{l=1}^{N_j} c_{jl}\, \varphi_{jl}(\gamma) \right|^2 \mu(d\gamma) \right)^{\frac{1}{2}} \sim 2^{-j\frac{d}{2}} \left(\sum_{l=1}^{N_j} |c_{jl}|^2 \right)^{\frac{1}{2}} \tag{30.32}$$

and

$$\left\| \sqrt{B} \left(\sum_{l=1}^{N_j} c_{jl}\, \varphi_{jl} \right) \mid W_2^1(\Omega) \right\| \ge c'\, 2^{-j\frac{d+2a}{2}} \left\| \sum_{l=1}^{N_j} c_{jl}\, \varphi_{jl} \mid W_2^1(\Omega) \right\| \tag{30.33}$$

for some $c' > 0$. Together with $N_j \sim 2^{jd}$ we can argue as in (27.84)–(27.87) and obtain as in (28.44) for the related approximation numbers

$$a_k(\sqrt{B}) \ge c\, k^{-\frac{d+2a}{2d}}, \quad k \in \mathbb{N}, \tag{30.34}$$

where $c > 0$ is independent of k. Since B is a non-negative compact self-adjoint operator, the left-hand side of (30.23) follows from (30.34) and (24.16).

30.8 Discussion
If the deviation $a = 0$ then we have by (30.23)

$$\mu_k \sim k^{-1}, \quad k \in \mathbb{N}, \tag{30.35}$$

what coincides with (30.10). This means that *the Weyl exponent 1 occurs also in the case of proper anisotropic d-sets in the plane with the deviation $a = 0$ according to Definition 5.11*. This is not a surprise since these fractals are very near to (isotropic) compact d-sets according to Definition 3.1. If $0 < a \le 1$ then we have only the estimate (30.23). But by our discussion at the beginning of 30.5 it is quite clear that one cannot expect an equivalence of type (30.35). Under these circumstances (30.23) looks rather satisfactory: although the methods how the two sides of (30.23) are obtained are rather different, the product of the two exponents involved is 1.

30.9 A comment on 28.4, 28.6, and 28.8 We return briefly to isotropic problems. In Step 3 of the proof of Theorem 28.6 we proved the right-hand side of (28.26) by the decomposition (28.33), (28.34), which results in (28.39). The corresponding assertion in Theorem 30.7 was based on Proposition 30.6 and the independence of the root systems and the eigenvalues of B in (30.18) on the underlying spaces. This independence holds also for the operator B in (28.24). In other words one can try to prove the right-hand side of (28.26) by looking for an appropriate counterpart of Proposition 30.6 for the operator B given by (28.24) which in an L_p-setting reads now as follows:

Let $1 \le p \le \infty$, $\frac{1}{p} + \frac{1}{p'} = 1$ and let

$$m > \frac{n-d-1}{p'} \quad \text{and} \quad \varkappa > \frac{n-d}{2}.$$

Then

$$B = A^{-\frac{\varkappa}{m}} \circ tr^{\Gamma} \tag{30.36}$$

is compact in $B_{p,\infty}^{2\varkappa - \frac{n-d}{p'}}(\Omega)$ and its entropy numbers satisfy

$$e_k(B) \le c\, k^{-\frac{2\varkappa - n + d}{d}}, \quad k \in \mathbb{N}. \tag{30.37}$$

As for the proof we remark first that

$$2\varkappa - \frac{n-d}{p'} = \frac{n-d}{p} + 2\varkappa - n + d. \tag{30.38}$$

Then by Theorems 18.2 and 20.6 complemented by (18.18), and Definition 20.2 we have

$$e_k\left(tr^{\Gamma} \;:\; B_{p,\infty}^{2\varkappa - \frac{n-d}{p'}}(\Omega) \to B_{p,\infty}^{-\frac{n-d}{p'}}(\Omega)\right) \le c\, k^{-\frac{2\varkappa - n + d}{d}}, \quad k \in \mathbb{N}. \tag{30.39}$$

The restriction for m is the same as in (28.17) with $r_1 = \infty$, whereas the additional restriction for $\varkappa$ mentioned there is not necessary as we remarked in 28.5. Then we obtain (30.37) by the same arguments as there in (28.21) and (28.22). Perhaps the proof in Step 3 in 28.6 is more transparent and better adapted to the Hilbert spaces. The above argument can also be applied if B is not related to a Hilbert space. We discussed these questions in 28.8.

30.10 Corollary (Nonisotropic fractal drum) *Let Ω be a bounded domain in the plane $\mathbb{R}^2$ with C^{∞} boundary. Let Γ be a proper nonisotropic d-set with the deviation a according to the Definitions 5.3 and 5.11, and*

$$2 > d > \frac{1+a}{2}, \quad 0 \le a \le 1, \tag{30.40}$$

such that $\Gamma \subset \Omega$. Then B given by (30.18) is a non-negative compact self-adjoint operator in $\overset{\circ}{W}{}^1_2(\Omega)$ with null-space (30.22). There are two positive numbers c_1 and c_2 such that the positive eigenvalues μ_k of B, repeated according to multiplicity and ordered by their magnitude, can be estimated by

$$c_1\, k^{-\frac{d+2a}{d}} \le \mu_k \le c_2\, k^{-\varrho} \quad \text{with} \quad \varrho = \frac{2d - 1 - a}{2d + 1 + 5a}, \tag{30.41}$$

$k \in \mathbb{N}$. Furthermore, B can be generated by the quadratic form (30.24).

Proof.

Step 1. We indicate the necessary modifications compared with the proofs of Proposition 30.6 and Theorem 30.7. First we modify (30.16) by

$$d'_a = \frac{5}{2} - \frac{d}{1+a}. \qquad (30.42)$$

Then we have by Theorems 22.5 and 18.17 with $s > 0$

$$tr^\Gamma \quad : \quad B^{\frac{d'_a}{p}+s}_{p,\infty}(\Omega) \to B^{-\frac{d'_a}{p}}_{p,\infty}(\Omega) \qquad (30.43)$$

and

$$e_k(tr^\Gamma) \leq c\, k^{-\lambda} \quad \text{with} \quad \lambda = \frac{2s(1+a)}{2d+1+5a}. \qquad (30.44)$$

Of interest is the case $p = 2$. We wish to apply $(-\Delta)^{-1}$ to the right-hand space in (30.43) and to arrive at the space where we started from. That means

$$\frac{d'_a}{2} + s = 2 - \frac{d'_a}{2} = \frac{3}{4} + \frac{d}{2(1+a)} > 1, \qquad (30.45)$$

by (30.40), and

$$s = 2 - d'_a = \frac{d}{1+a} - \frac{1}{2} > 0. \qquad (30.46)$$

Inserting s in (30.44) we obtain the exponent ϱ in (30.41). Since the number 1 in (30.45) stands for $\overset{\circ}{W}{}^1_2(\Omega)$ we have now the same situation as in Step 2 of the proof of Theorem 30.7. In Step 1 of this proof one has to replace $\frac{d_a}{2}$ in (30.25) by $\frac{d'_a}{2}$. By (30.42) and (30.40) we have $\frac{d'_a}{2} < 1$. Hence B, generated by (30.24), is given by (30.18) and we have the right-hand side of (30.41).

Step 2. The proof of the left-hand side of (30.41) is the same as in Step 3 of the proof of Theorem 30.7, where $\varphi_{jl}(x)$ in (30.27) must be modified appropriately.

30.11 The music of the ferns

The most interesting anisotropic and nonisotropic d-sets are the self-affine fractals as considered in Section 4. By Proposition 5.6 the PXT-fractals introduced in 4.17 are anisotropic d-sets and the OF-fractals introduced in 4.19 are nonisotropic d-sets. The additional assumption to be *proper* excludes by Proposition 5.13 only pathological cases, where the whole fractal retreats in the boundary of the square or circle. In other words, the lovely ferns created in this way are admitted fractals Γ in Theorem 30.7 and Corollary 30.10. Surely (30.23) is more satisfactory than (30.41). We complained about these shortcomings in 18.18 and 22.8. The interesting question arises whether additional geometric reasoning in connection with these self-affine fractals provides the possibility to improve these estimates or even to calculate the correct exponent ϱ in $\mu_k \sim k^{-\varrho}$ (if there is any).

31 Schrödinger operators with fractal potentials

31.1 Introduction

Based on [HaT94b] we studied in [ET96], 5.4.1 and in the following subsections, operators of the type

$$H_\beta = a(x, D) + \beta a(x) v(x, D) a(x) = a(x, D) + \beta V, \quad \beta \in \mathbb{R}, \qquad (31.1)$$

where $a(x, D) \in \Psi_{1,\varrho}^{\varkappa}$, with $\varkappa > 0$, $0 \leq \varrho \leq 1$, is assumed to be a positive-definite, self-adjoint pseudodifferential operator in $L_2(\mathbb{R}^n)$, perturbed by the symmetric operator $v(x, D) \in \Psi_{1,\delta}^{\lambda}$ with $\lambda < \varkappa$, $0 \leq \delta \leq 1$, whereas the real function $a(x)$ belongs typically to some $L_r(\mathbb{R}^n)$ or $H_r^s(\mathbb{R}^n)$, say, with $1 \leq r \leq \infty$ and $s \geq 0$, such that

$$a(x, D)^{-1} V \qquad (31.2)$$

is compact in $L_2(\mathbb{R}^n)$. Then the essential spectra of the self-adjoint operators $a(x, D)$ and H_β coincide. They are located in $\mathbb{R}_+ = (0, \infty)$. Of interest is the number of eigenvalues, counted with respect to their multiplicities, in $(-\infty, 0]$ in dependence on $\beta \to \infty$. Recall that $\#M$ denotes the number of elements of a finite set M. Let $\sigma(H_\beta)$ be the spectrum of the self-adjoint operator H_β in $L_2(\mathbb{R}^n)$; then

$$\# \left\{ \sigma(H_\beta) \cap (-\infty, 0] \right\} \leq \# \left\{ k \in \mathbb{N} : \sqrt{2}\,\beta\,e_k \left(a(x, D)^{-1} V \right) \geq 1 \right\}. \qquad (31.3)$$

This follows from the *Birman-Schwinger principle* in the version in [HaT94b] and [ET96], 5.4.1, p. 223. There one finds also more detailed explanations. As always in this book, e_k stands for the entropy numbers of the related compact operators, introduced in 6.2, and used extensively throughout this chapter, with Corollary 6.10 as the link to spectral theory. In literature this is sometimes called the problem of the *negative spectrum*.

The interest in these questions comes from quantum mechanics with the typical hydrogen-like operator in $L_2(\mathbb{R}^3)$,

$$H = -\frac{\hbar^2}{2m} \Delta + V(x), \qquad (31.4)$$

where $\hbar$ is Planck's constant, m is the mass of the electron and $V(x)$ is the potential. Asking for (negative) eigenvalues one can divide H by $\frac{\hbar^2}{2m}$ and obtains

$$H_\beta = -\Delta + \beta V(x), \qquad (31.5)$$

where β is proportional to $\hbar^{-2}$. As complained after (26.30) of interest is the semi-classical limit $\hbar \to 0$, or better, $\beta \to \infty$, hence the behaviour of the cardinal number in (31.3) in dependence on β. In [HaT94b] and [ET96], 5.4, we studied problems of a type somewhat between (31.1) and (31.5), that means preferably with $v(x, D) = 1$ in (31.1). Here we restrict our attention to (31.5), considered in

$\mathbb{R}^n$. As throughout in this chapter, all our arguments are qualitative and there is no problem to replace $-\Delta$ in (31.5) by $(-\Delta)^l$ with $l \in \mathbb{N}$, as done in literature, or by symmetric (exotic) pseudodifferential operators. But we stick for simplicity at $-\Delta$. In [ET96] and [HaT94b] we dealt with the case where the potential $V(x)$ is a function typically belonging to some $L_r(\mathbb{R}^n)$. In (31.3) this results in degenerate elliptic operators in $L_2(\mathbb{R}^n)$. Now we wish to replace $V(x)$ by fractals or related finite Radon measures. Of course, the domain of definition of H_β in (31.5) or in (31.3) is now questionable. Nevertheless such singular potentials attracted quite a lot of attention, going back to Fermi in 1936. This interest stems from *short range potentials* and the attempt to idealize them. We refer to [BEKS94] where one finds more details and the relevant references, see also [Bra95], Ch. 11.

31.2 The set-up

By our previous considerations in this chapter it is quite clear how to deal with H_β in (31.5) or better $(-\Delta + id)^{-1} V$ in (31.3) if V is a fractal in analogy to Theorems 28.6, 30.2, and 30.7. We drew all the desired information from the interplay of the explicit representation of the operator in question and its quadratic form. We will do here the same. We follow 28.1 and 28.2. Let Γ be a compact d-set in $\mathbb{R}^n$ with $0 < d < n$ according to Definition 3.1. Let $\mu = \mathcal{H}^d | \Gamma$ be the respective Hausdorff measure restricted to Γ. Let $b \in L_r(\Gamma)$ be real and

$$0 < \frac{1}{p} \leq \frac{1}{p} + \frac{1}{r} = \frac{1}{q} < 1; \tag{31.6}$$

then we have in analogy to (28.2)–(28.4), based on (18.23),

$$tr_\Gamma \; : \; B_{p,1}^{\frac{n-d}{p}} (\mathbb{R}^n) \to L_p(\Gamma), \tag{31.7}$$

$$b \, tr_\Gamma \; : \; B_{p,1}^{\frac{n-d}{p}} (\mathbb{R}^n) \to L_q(\Gamma), \tag{31.8}$$

$$tr_b^\Gamma \; : \; B_{p,1}^{\frac{n-d}{p}} (\mathbb{R}^n) \to B_{q,\infty}^{-\frac{n-d}{q'}} (\mathbb{R}^n), \tag{31.9}$$

where in the last line tr_b^Γ is the interpretation of $b \, tr_\Gamma$ according to (18.7). One can formalize the understanding of tr_b^Γ as in (28.4). In slight modification of (31.5) we are interested in the negative spectrum of

$$H_\beta = id - \Delta + \beta \, tr_b^\Gamma . \tag{31.10}$$

The related quadratic form is given by

$$Q[f] = \|f \mid W_2^1(\mathbb{R}^n)\|^2 + \beta \int_\Gamma b(\gamma) \, |\, tr_\Gamma \, f(\gamma)|^2 \, \mu(d\gamma), \tag{31.11}$$

where $W_2^1(\mathbb{R}^n)$ is the very classical Sobolev space as in (10.10). One word seems to be in order. In Section 24 we discussed the relation between quadratic forms and corresponding operators. But not only from the mathematical point of view, but also from the physical point of view there might be a preference of quadratic forms compared with the generated operators. This comes from the fact that quadratic forms, representing the energy of the underlying system, have a direct physical meaning, somewhat in contrat to the related operators. For a discussion of this point of view we refer to [Sim71], p. 33. However we avoid the occasionally somewhat delicate problem about the domain of definition of operators generated by unbounded quadratic forms. We follow the scheme developed in 28.6 and 30.2 and interpret $Q[f]$ in $W_2^1(\mathbb{R}^n)$, where it becomes a bounded quadratic form.

31.3 Theorem *Let Γ be a compact d-set in $\mathbb{R}^n$ according to Definition 3.1 with $n - 2 < d < n$ (where $0 < d < 1$ in case of $n = 1$). Let $b \in L_r(\Gamma)$ be real with*

$$0 \le \frac{1}{r} < \frac{2 - n + d}{d} = 1 - \frac{n - 2}{d} \tag{31.12}$$

(where $0 \le \frac{1}{r} < 1$ in case of $n = 1$), and let tr_b^Γ be the operator in the interpretation of (31.9) with $\frac{2}{p} = 1 - \frac{1}{r}$ and $\frac{2}{q} = 1 + \frac{1}{r}$. Then B,

$$B = (id - \Delta)^{-1} \circ tr_b^\Gamma, \tag{31.13}$$

is a compact self-adjoint operator in $W_2^1(\mathbb{R}^n)$ with the null-space

$$N(B) \supset \left\{ f \in W_2^1(\mathbb{R}^n) , \ tr_\Gamma f = 0 \right\}, \tag{31.14}$$

where both sides in (31.14) coincide if $b(\gamma) \ne 0$ a.e. on Γ. It is generated by the quadratic form in $W_2^1(\mathbb{R}^n)$,

$$\int_\Gamma b(\gamma) f(\gamma) \overline{g(\gamma)} \, \mu(d\gamma) = (Bf, g)_{W_2^1}. \tag{31.15}$$

Let

$$G_\beta = id + \beta B \quad with \quad \beta > 0 \tag{31.16}$$

and let

$$N_\beta = \# \left\{ \sigma(G_\beta) \cap (-\infty, 0] \right\} \tag{31.17}$$

be the number of non-positive eigenvalues of G_β. Then there is a number $c > 0$ such that

$$N_\beta \le c \, (\|b \mid L_r(\Gamma)\| \, \beta)^{\frac{d}{2 - n + d}} \quad for \ all \quad \beta > 0 \tag{31.18}$$

and all $b \in L_r(\Gamma)$.

Proof.

Step 1. First we prove that the quadratic form on the left-hand side of (31.15) makes sense in $W_2^1(\mathbb{R}^n) = H^1(\mathbb{R}^n)$. If $f \in H^1(\mathbb{R}^n)$, then by (20.3) we have

$$f(\gamma) = (tr_\Gamma \, f)(\gamma) \in H^{1-\frac{n-d}{2}}(\Gamma) = \mathbb{B}_{2,2}^{1-\frac{n-d}{2}}(\Gamma). \qquad (31.19)$$

Because

$$0 < \frac{2}{p} = 1 - \frac{1}{r} \le 1 \qquad (31.20)$$

we obtain by Proposition 20.5 the compact embedding

$$H^{1-\frac{n-d}{2}}(\Gamma) \subset L_p(\Gamma) \qquad (31.21)$$

since

$$1 - \frac{n-d}{2} - \frac{d}{2} = 1 - \frac{n}{2} > -\frac{d}{2} + \frac{d}{2r} = -\frac{d}{p}. \qquad (31.22)$$

Then it follows by (31.20) and Hölder's inequaltiy that (31.15) is a real bounded quadratic form on $W_2^1(\mathbb{R}^n)$ and that the generated self-adjoint bounded operator B is compact. Recall that we specified the norming in $W_2^1(\mathbb{R}^n)$ by

$$\|f \mid W_2^1(\mathbb{R}^n)\| = \left\| (id - \Delta)^{\frac{1}{2}} f \mid L_2(\mathbb{R}^n) \right\|. \qquad (31.23)$$

Then we obtain (31.13) for this operator B by the same reasoning as in Step 2 of the proof of Theorem 28.6 and as in (27.70). Furthermore, by Theorem 3.8 and (31.15) we have $f \in N(B)$ if, and only if,

$$b(\gamma) \, (tr_\Gamma \, f)(\gamma) = 0 \quad \text{a. e. on} \quad \Gamma. \qquad (31.24)$$

This proves the assertions about $N(B)$.

Step 2. To estimate the eigenvalues of B by their magnitude we decompose

$$W_2^1(\mathbb{R}^n) = H^1(\mathbb{R}^n) = \{f \in W_2^1(\mathbb{R}^n), \, tr_\Gamma \, f = 0\} \oplus H^{1-\frac{n-d}{2}}(\Gamma) \qquad (31.25)$$

analogously to (28.32). By (31.14) the eigenvalues of B coincide with the eigenvalues of the restriction of B on $H^{1-\frac{n-d}{2}}(\Gamma)$, denoted by B^Γ. By (6.13) we have to estimate the corresponding entropy numbers. As indicated in Fig. 31.1 we factorize B^Γ by

$$B^\Gamma = id_3 \circ tr_\Gamma \circ (id - \Delta)^{-1} \circ id_2 \circ b \circ id_1 \qquad (31.26)$$

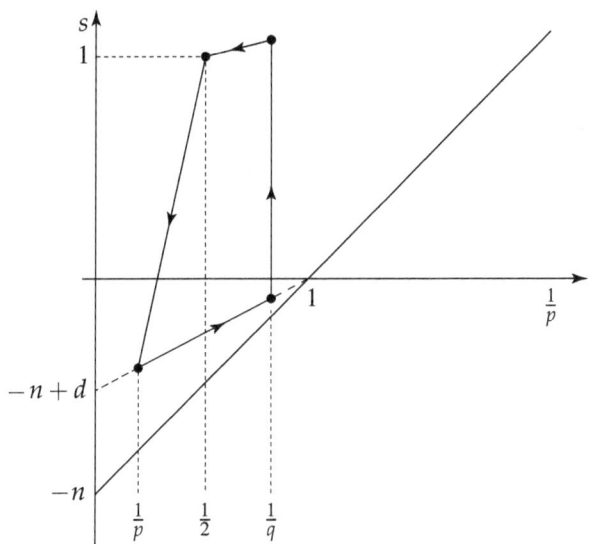

Fig. 31.1

with

$$id_1 \; : \; H^{1-\frac{n-d}{2}}(\Gamma) \to L_p(\Gamma), \quad \text{where} \quad \frac{1}{p} = \frac{1}{2}\left(1 - \frac{1}{r}\right)$$

$$b \; : \; L_p(\Gamma) \to L_q(\Gamma) \quad \text{where} \quad \frac{1}{q} = \frac{1}{p} + \frac{1}{r}$$

$$id_2 \; : \; L_q(\Gamma) \to B_{q,\infty}^{-\frac{n-d}{q'}}(\mathbb{R}^n)$$

$$(id-\Delta)^{-1} \; : \; B_{q,\infty}^{-\frac{n-d}{q'}}(\mathbb{R}^n) \to B_{q,\infty}^{2-\frac{n-d}{q'}}(\mathbb{R}^n)$$

$$tr_\Gamma \; : \; B_{q,\infty}^{2-\frac{n-d}{q'}}(\mathbb{R}^n) \to \mathbb{B}_{q,\infty}^{2-n+d}(\Gamma)$$

$$id_3 \; : \; \mathbb{B}_{q,\infty}^{2-n+d}(\Gamma) \to H^{1-\frac{n-d}{2}}(\Gamma). \tag{31.27}$$

By Theorem 20.6 and (31.21), (31.22) we have

$$e_k(id_1) \leq c\, k^{-\frac{2-n+d}{2d}}, \quad k \in \mathbb{N}. \tag{31.28}$$

The multiplication with b is simply Hölder's inequality, where

$$\frac{1}{q} = \frac{1}{2}\left(1 + \frac{1}{r}\right) < 1. \tag{31.29}$$

The continuity of id_2 and tr_Γ follows from Theorem 18.2 and Definition 20.2 since $d > n - 2$. Application of $(id - \Delta)^{-1}$ is simply the lift (10.12). Finally to show that id_3 is compact we need

$$2 - n + d - \frac{d}{q} > 1 - \frac{n}{2}, \qquad (31.30)$$

where we used (31.22). This follows from $\frac{1}{q} = \frac{1}{2}(1 + \frac{1}{r})$, (31.12) and

$$2 - n + d - \frac{d}{2}\left(1 + \frac{2 - n + d}{d}\right) = \frac{2 - n + d}{2} - \frac{d}{2} = 1 - \frac{n}{2}. \qquad (31.31)$$

By Theorem 20.6 we have

$$e_k(id_3) \le c\, k^{-\frac{2-n+d}{2d}}, \quad k \in \mathbb{N}. \qquad (31.32)$$

By (6.8) we obtain

$$e_k(B^\Gamma) \le c\, k^{-\frac{2-n+d}{d}}, \quad k \in \mathbb{N}. \qquad (31.33)$$

Finally we arrive at

$$e_k(B) \le c\, \|b \,|\, L_r(\Gamma)\|\, k^{-\frac{2-n+d}{d}}, \quad k \in \mathbb{N}, \qquad (31.34)$$

where c is independent of b and k.

Step 3. We estimate N_β given by (31.17). We apply (31.3) now with id and B in place of $a(x,D)^{-1}$ and V. Hence, by (31.34) we ask for the largest $k \in \mathbb{N}$ such that

$$\beta \,\|b \,|\, L_r(\Gamma)\|\, k^{-\frac{2-n+d}{d}} \ge c > 0, \qquad (31.35)$$

and obtain (31.18).

31.4 Discussion

We add a remark about the *quantum-mechanical relevance* of our considerations so far. Let

$$H_\beta = id - \Delta + \beta V \qquad (31.36)$$

as a slight modification of (31.5). Assume that an L_2-setting makes sense (maybe $V = V(x)$ is a suitable degenerate function) and that $V(id - \Delta)^{-1}$ and $(id - \Delta)^{-1}V$ are compact, say, in $L_2(\mathbb{R}^n)$. Then we have (31.3) with $a(x,D) = id - \Delta$. But it comes out that one can replace $e_k\left((id - \Delta)^{-1}V\right)$ by

$$e_k\left(V(id - \Delta)^{-1}\right) \quad \text{or} \quad e_k\left((id - \Delta)^{-\frac{1}{2}}V(id - \Delta)^{-\frac{1}{2}}\right).$$

We discussed these symmetric and non-symmetric versions of the Birman-Schwinger principle in [ET96], 5.2.1 and 5.4.1, pp. 186 and 223. Furthermore we proved in

[ET96], Ch. 5, that one has *spectral invariance*, at least in non-limiting situations. Applied to our situation it comes out that the root systems and the eigenvalues of, say, the compact operator $(id - \Delta)^{-1} V$ are independent of the underlying space L_2, L_p or some H_p^s, maybe W_2^1, as far as these operators make sense. In other words, in case of good potentials V, one may choose one of the above operators

$$(id - \Delta)^{-1} V, \quad V (id - \Delta)^{-1}, \quad \text{or} \quad (id - \Delta)^{-\frac{1}{2}} V (id - \Delta)^{-\frac{1}{2}} \quad (31.37)$$

in an arbitrary admitted space to estimate the related entropy numbers and to get via the Birman-Schwinger principle an assertion about the negative spectrum of H_β in (31.36). If V degenerates strongly, maybe to tr_b^Γ in the above meaning, then the situation is more delicate. It may happen that some of the operators in (31.37) make sense in a certain space, but others do not. By the above discussion it is reasonable to ask for a suitable version of the operators, say, of type (31.37), in an appropriate space in order to discuss the negative spectrum of H_β in (31.36). This is just what we did in the above theorem. Such a point of view is also supported by the above remarks about quadratic forms in connection with (31.10), (31.11). One begins with the quadratic form $Q[f]$ in (31.11) and asks for a suitable space where it makes sense. In our case this is $W_2^1(\mathbb{R}^n)$. Afterwards one determines the generated operator. In our case this is G_β in (31.16). Hence, H_β in (31.10) and G_β in (31.16), are generated by the same quadratic form $Q[f]$ in L_2 and in W_2^1 (where in the case H_β this must be understood only in a formal way, since it does not make sense rigorously).

31.5 Theorem *Let Γ be a compact d-set in $\mathbb{R}^n$ according to Definition 3.1 with $n - 2 < d < n$ (where $0 < d < 1$ in case of $n = 1$). Let tr^Γ be the operator*

$$tr^\Gamma \quad : \quad B_{2,1}^{\frac{n-d}{2}}(\mathbb{R}^n) \to B_{2,\infty}^{-\frac{n-d}{2}}(\mathbb{R}^n) \quad (31.38)$$

(that means $b = 1$ and $p = q = 2$ in (31.9)). Then B,

$$B = (id - \Delta)^{-1} \circ tr^\Gamma, \quad (31.39)$$

is a compact non-negative self-adjoint operator in $W_2^1(\mathbb{R}^n)$ with null-space

$$N(B) = \left\{ f \in W_2^1(\mathbb{R}^n), \ tr_\Gamma f = 0 \right\}. \quad (31.40)$$

It is generated by the quadratic form in $W_2^1(\mathbb{R}^n)$,

$$\int_\Gamma f(\gamma) \overline{g(\gamma)} \, \mu(d\gamma) = (Bf, g)_{W_2^1}. \quad (31.41)$$

Let

$$G_\beta = id - \beta B \quad \text{with} \quad \beta > 0, \quad (31.42)$$

and let N_β, given by (31.17), be the number of non-positive eigenvalues of G_β. There are three positive numbers c_1, c_2 and β_0 such that

$$c_1 \, \beta^{\frac{d}{2-n+d}} \leq N_\beta \leq c_2 \, \beta^{\frac{d}{2-n+d}} \quad \text{for all} \quad \beta \geq \beta_0. \quad (31.43)$$

Proof. All assertions are covered by Theorem 31.3 with exception of the left-hand side of (31.43) which we shall prove now. The quadratic form belonging to the self-adjoint operator G_β in (31.42) in $W_2^1(\mathbb{R}^n)$ is given by

$$Q_\beta[f] = \|f \mid W_2^1(\mathbb{R}^n)\|^2 - \beta \int_\Gamma |\, tr_\Gamma \, f(\gamma)|^2 \, \mu(d\gamma). \qquad (31.44)$$

Let φ be a non-trivial C^∞ function in $\mathbb{R}^n$ with compact support near the origin. Let

$$\varphi_{jl}(x) = \varphi\left(2^j(x - x^{j,l})\right) ; \quad j \in \mathbb{N}, \quad l = 1, \ldots, M_j, \qquad (31.45)$$

be the same functions as in (28.40): If $j \in \mathbb{N}$ is fixed then the functions $\varphi_{jl}(x)$ have disjoint supports, $M_j \geq 2^{jd}$, and $x^{j,l} \in \Gamma$. We may assume that for any fixed $j \in \mathbb{N}$,

$$\left\{ 2^{-j(1-\frac{n}{2})} \varphi_{jl}(x), \quad l = 1, \ldots, M_j \right\} \qquad (31.46)$$

is an orthonormal system in $W_2^1(\mathbb{R}^n)$ and that

$$\int_\Gamma 2^{-j(2-n)} |\varphi_{jl}(\gamma)|^2 \, \mu(d\gamma) \sim 2^{-j(2-n)} \, 2^{-jd} = 2^{-j(2-n+d)}. \qquad (31.47)$$

Let $\beta = c \, 2^{j(2-n+d)}$ then we have for a suitably chosen $c > 0$, and all $j \in \mathbb{N}$ and $l = 1, \ldots, M_j$,

$$Q_\beta\left[2^{-j(1-\frac{n}{2})} \varphi_{jl} \right] \leq 1 - c' < 0. \qquad (31.48)$$

Hence the quadratic form Q_β with $\beta = c \, 2^{j(2-n+d)}$ is negative on the span of the $M_j \geq 2^{jd}$ functions $\varphi_{jl}(x)$. Again by the Max-Min principle, see [EEv87], pp. 489–492, it follows

$$N_\beta \geq 2^{jd} \quad \text{with} \quad \beta = c \, 2^{j(2-n+d)}. \qquad (31.49)$$

Now the estimate from below in (31.43) is an easy consequence of (31.49).

31.6 Remark First one might ask whether there is a constant $c > 0$ such that

$$N_\beta \, \beta^{-\frac{d}{2-n+d}} \to c \quad \text{if} \quad \beta \to \infty. \qquad (31.50)$$

But, in general, this is unlikely. We refer to the discussion in 30.4 and the paper [SoV95] quoted there.

31.7 Generalizations

We reduced the problem of the negative spectrum in the Theorems 31.3 and 31.5 via the Birman-Schwinger principle to the estimate of entropy numbers of the operators B in (31.13) and (31.39). However we obtained such estimates in this chapter for more general operators and there is no problem to replace $-\Delta$ in (31.13) and (31.39) by suitable formally self-adjoint elliptic pseudodifferential operators $a(x, D)$ as described at the beginning of this section in connection with (31.1), (31.2), and possible replacements of V by, say, tr_b^{Γ} or tr^{Γ}. The preference of $-\Delta$ comes from the quantum-mechanical background. In case of an (isotropic) compact d-set one has the rather satisfactory two-sided estimate (31.43) which is the direct counterpart of (30.7) connected with the isotropic fractal drum. Rephrasing the interpretation 30.3 one may ask whether the

fractality d of the potential Γ can be recovered by counting the negative eigenvalues of G_{β} in the semi-classical limit $\beta \to \infty$.

Furthermore one can replace the isotropic d-set by anisotropic or nonisotropic d-sets. Then one obtains the counterparts of (30.23) and (30.41). Instead of the music of the ferns according to 30.11 one obtains now estimates of the negative energy levels for fern-like potentials.

32 Nonlinear elliptic equations related to fractals

32.1 Two nonlinear problems

It is not surprising that the techniques developed in this book can also be applied to nonlinear problems related to fractals. We restrict ourselves to two examples. All our arguments are qualitative and it is quite clear that these prototypes can be generalized in several directions. Let Ω be a bounded domain in $\mathbb{R}^n$ with C^{∞} boundary and let Γ be a compact d-set in $\mathbb{R}^n$ according to Definition 3.1 with $\Gamma \subset \Omega$. Let Δ be the Dirichlet Laplacian in Ω, let $1 < p < \infty$ and $0 < \varkappa < 1$.

First, we ask for solutions of the nonlinear Dirichlet problem

$$\Delta u + |tr_{\Gamma} u|^{\varkappa} = f \in L_p(\Gamma). \tag{32.1}$$

One has to explain what is meant by (32.1). This will be done in 32.2.

Secondly, under the same conditions for Ω and Γ we ask e. g. for the solvability of

$$\Delta u + \sum_{j=1}^{n} b_j(x) u \frac{\partial u}{\partial x_j} = f \in L_q(\Gamma). \tag{32.2}$$

The interest in (32.2) comes from the *stationary Navier-Stokes equations*, where the (vector version) of the involved nonlinearities is just of that type. To study these nonlinear equations is the first step to deal with (stationary) Navier-Stokes equations.

First we discuss (32.1), whereas the following longer part of this section is devoted to (32.2).

32.2 First problem: The set-up
Let Ω be a bounded C^∞ domain in $\mathbb{R}^n$ and let Γ be a compact d-set in $\mathbb{R}^n$ according to Definition 3.1 with $\Gamma \subset \Omega$ and $n - 2 < d < n$ (where $0 < d < 1$ if $n = 1$). By 20.2 and 18.6 we have

$$tr_\Gamma \, u \in L_p(\Gamma) \quad \text{if} \quad u \in B^s_{pq}(\Omega), \; 1 < p < \infty, \; s > \frac{n-d}{p}, \tag{32.3}$$

and $0 < q \le \infty$. Let $0 < \varkappa \le 1$, then we have also $|\, tr_\Gamma \, u|^\varkappa \in L_p(\Gamma)$ by Hölder's inequality. According to Theorem 18.2 the outcome will be interpreted as

$$|tr_\Gamma \, u|^\varkappa \in B^{-\frac{n-d}{p'}}_{p,\infty}(\Omega), \quad \frac{1}{p} + \frac{1}{p'} = 1. \tag{32.4}$$

This interpretation applies also to f in (32.1). Hence the first candidate for the Dirichlet problem (32.1) is the space

$$\left\{ g \in B^{2-\frac{n-d}{p'}}_{p,\infty}(\Omega) \, , \; g|\partial\Omega = 0 \right\}. \tag{32.5}$$

Since

$$2 - \frac{n-d}{p'} = \frac{1}{p'}(2 - n + d) + \frac{2}{p} > \frac{1}{p} \tag{32.6}$$

and

$$2 - \frac{n-d}{p'} = 2 - n + d + \frac{n-d}{p} > \frac{n-d}{p}, \tag{32.7}$$

both $g|\partial\Omega$ in (32.5) and (32.3) make sense. On the other hand, at least in this context, (32.5) is the largest space which meets the above requirements.

32.3 Theorem *Let Ω be a bounded C^∞ domain in $\mathbb{R}^n$ and let Γ be a compact d-set in $\mathbb{R}^n$ according to Definition 3.1 with $\Gamma \subset \Omega$ and $n - 2 < d < n$ (where $0 < d < 1$ if $n = 1$). Let $1 < p < \infty$ and $0 < \varkappa < 1$. For $f \in L_p(\Gamma)$ the Dirichlet problem*

$$\Delta u + |\, tr_\Gamma \, u|^\varkappa = f \tag{32.8}$$

in the set-up of 32.2 has a solution in the space (32.5).

Proof.
Step 1. By 32.2 the problem makes sense. Let

$$T = (-\Delta)^{-1} \circ |\, tr_\Gamma \, |^\varkappa \tag{32.9}$$

in the decomposition

$$T = (-\Delta)^{-1} \circ id_2 \circ |id_1|^\varkappa \circ tr_\Gamma \tag{31.10}$$

and its explanation

$$tr_\Gamma \; : \; B_{p,\infty}^{2-\frac{n-d}{p'}}(\Omega) \to L_p(\Gamma)$$

$$|id_1|^\varkappa \; : \; L_p(\Gamma) \to L_p(\Gamma)$$

$$id_2 \; : \; L_p(\Gamma) \to B_{p,\infty}^{-\frac{n-d}{p'}}(\Omega)$$

$$(-\Delta)^{-1} \; : \; B_{p,\infty}^{-\frac{n-d}{p'}}(\Omega) \to B_{p,\infty}^{2-\frac{n-d}{p'}}(\Omega). \tag{32.11}$$

The trace operator tr_Γ is compact by (32.7), 20.2, and 20.6. Furthermore,

$$|id_1|^\varkappa \quad : \quad u \mapsto |u|^\varkappa \tag{32.12}$$

with

$$\| \, |id_1|^\varkappa u \, | \, L_p(\Gamma)\| = \| \, |u|^\varkappa \, | \, L_p(\Gamma)\| \le c \, \|u \, | \, L_p(\Gamma)\|^\varkappa \tag{32.13}$$

comes from Hölder's inequality. The operator id_2 stands for the extension accord-
ing to Theorem 18.2. Finally $(-\Delta)^{-1}$ is the isomorphic map of $B_{p,\infty}^{-\frac{n-d}{p'}}(\Omega)$ onto
the space in (32.5): We used assertions of this type several times. An explicit
formulation may be found in [Tri78], 5.7.1, p. 402 (in the correction of the sec.
edition in 1995). Hence T is a compact operator in the space (32.5).

Step 2. By (32.8) we have to find a fixed point of the compact operator

$$V u = T u - (-\Delta)^{-1} f \tag{32.14}$$

in the space (32.5). By (32.13) we have

$$\|V u \, | \, B_{p,\infty}^{2-\frac{n-d}{p'}}(\Omega)\| \le c \, \|f \, | \, L_p(\Gamma)\| + c \, \|u \, | \, B_{p,\infty}^{2-\frac{n-d}{p'}}(\Omega)\|^\varkappa \tag{32.15}$$

and

$$\|V u_1 - V u_2 \, | \, B_{p,\infty}^{2-\frac{n-d}{p'}}(\Omega)\|$$

$$\le c \, \| \, |tr_\Gamma \, u_1|^\varkappa - |tr_\Gamma \, u_2|^\varkappa \, | \, L_p(\Gamma)\|$$

$$\le c \, \| \, |tr_\Gamma \, (u_1 - u_2)|^\varkappa \, | \, L_p(\Gamma)\| \tag{32.16}$$

$$\le c' \, \|u_1 - u_2 \, | \, B_{p,\infty}^{2-\frac{n-d}{p'}}(\Omega)\|^\varkappa,$$

where the middle estimate comes from $0 < \varkappa < 1$ and $(a+b)^\varkappa \le a^\varkappa + b^\varkappa$ if
$a > 0$, $b > 0$, which results in $a^\varkappa - b^\varkappa \le |a - b|^\varkappa$. Hence, V is continuous and
compact, and maps a sufficiently large ball of the space in (32.5) into itself. By
Schauder's fixed point theorem, see e. g. [Sma74], 4.1, p. 25, the map V has a
fixed point in that ball.

32.4 Remark Crucial for our arguments is the assumption $\varkappa < 1$. Otherwise the above theorem can be extended in many directions. One can generalize $|tr_\Gamma|^\varkappa$ by $b\,|tr_\Gamma|^\varkappa$ with some $b \in L_r(\Gamma)$. Furthermore instead of Δ one can deal with operators $A^{\frac{\lambda}{m}}$ as discussed in 27.11 and used in 28.6. Also the replacement of Δ by some pseudodifferential operators $a(x, D) \in \Psi^\mu_{1,\varrho}$ with $0 \le \varrho \le 1$ is possible. Finally instead of (isotropic) d-sets one can consider anisotropic and nonisotropic d-sets according to Definitions 5.2 and 5.3 in the same way as in 30.5–30.11. We do not go into detail.

32.5 Second problem: The set-up
Let Ω be a bounded domain in $\mathbb{R}^n$ with C^∞ boundary $\partial\Omega$. The *stationary Navier-Stokes equations* are given by

$$-\Delta u + \sum_{j=1}^{n} u_j \frac{\partial u}{\partial x_j} + \nabla P = f \quad \text{in} \quad \Omega, \quad u|\partial\Omega = 0, \qquad (32.17)$$

$$div\, u = 0 \quad \text{in} \quad \Omega, \qquad (32.18)$$

where $u = (u_1, \ldots, u_n)$ is a vector-function, and P is a scalar function. Navier-Stokes equations and the above stationary Navier-Stokes equations have attracted much attention. We refer to [Tem83], [Tem84], and [Lio96]. In any case function spaces play a decisive role. Nearest to our intentions here are [Joh93], [Joh95], and [JoR96]. We refer in particular to [Joh93], Theorem 5.4.2. A first step to handle (stationary) Navier-Stokes equations is to put $P = 0$, to omit (32.18), and to study the resulting equations, where the scalar case is sufficient now. We generalize the outcome by studying the (scalar) equation

$$\Delta u + \sum_{j=1}^{n} b_j(x)\, u \, \frac{\partial u}{\partial x_j} = f, \qquad (32.19)$$

where Δ again is the Dirichlet Laplacian in Ω. We are interested in strong solutions in worst cases as far as the coefficients $b_j(x)$ and (fractal) f's are concerned. To avoid awkward writing we introduce some notation. Recall that $H^s_p(\mathbb{R}^n)$ are the Sobolev spaces as introduced in 10.5, where $s \in \mathbb{R}$ and $1 < p < \infty$, and that $H^s_p(\Omega)$ stands for the restriction of $H^s_p(\mathbb{R}^n)$ on Ω similarly as in 23.1. As in 27.11 the completion of $C^\infty_0(\Omega)$ in $H^s_p(\Omega)$ is denoted by $\overset{\circ}{H}{}^s_p(\Omega)$. Recall

$$\overset{\circ}{H}{}^s_p(\Omega) = \{g \in H^s_p(\Omega)\,, \; g|\partial\Omega = 0\}, \quad 1 < p < \infty, \quad \frac{1}{p} < s \le 1 + \frac{1}{p}, \quad (32.20)$$

see [Tri78], 4.7.1, p. 330. Let $1 < p < \infty$ and again let

$$\frac{1}{p^s} = \frac{1}{p} + \frac{s}{n} \ge 0, \quad s \in \mathbb{R}, \qquad (32.21)$$

as in 11.5. We are interested here only in the three values $s = \frac{1}{2}$, $s = \frac{3}{2}$, and $s = -\frac{1}{2}$. Instead of (32.21) we put

$$\frac{1}{p(1)} = \frac{1}{p} + \frac{1}{2n}, \quad \frac{1}{p(2)} = \frac{1}{p} + \frac{3}{2n}, \quad \frac{1}{p(3)} = \frac{1}{p} - \frac{1}{2n}, \tag{32.22}$$

and similarly, of course, $q(1)$, etc.

32.6 Proposition *Let $n \geq 3$ and let Ω be a bounded domain in $\mathbb{R}^n$ with C^∞ boundary. Let $p < \infty$,*

$$0 < \frac{1}{q} \leq \frac{1}{n} - \frac{1}{p}, \tag{32.23}$$

and

$$b_j(x) \in H^{\frac{1}{2}}_{q(1)}(\Omega). \tag{32.24}$$

Let $(-\Delta)^{-1}$ be the inverse of the Dirichlet Laplacian. Then B,

$$B u = (-\Delta)^{-1} \left(\sum_{j=1}^{n} b_j(x) u \frac{\partial u}{\partial x_j} \right), \tag{32.25}$$

is a continuous and bounded map in $\overset{\circ}{H}{}^{\frac{1}{2}}_{p(1)}(\Omega)$ and there is a constant c such that

$$\left\| B u_1 - B u_2 \,|\, H^{\frac{1}{2}}_{p(1)}(\Omega) \right\|$$
$$\leq c \left(\|u_1 \,|\, H^{\frac{1}{2}}_{p(1)}(\Omega)\| + \|u_2 \,|\, H^{\frac{1}{2}}_{p(1)}(\Omega)\| \right) \|u_1 - u_2 \,|\, H^{\frac{1}{2}}_{p(1)}(\Omega)\| \tag{32.26}$$

for all $u_1 \in \overset{\circ}{H}{}^{\frac{1}{2}}_{p(1)}(\Omega)$ and $u_2 \in \overset{\circ}{H}{}^{\frac{1}{2}}_{p(1)}(\Omega)$.

Proof.
Step 1. We put

$$\frac{1}{r_1} = \frac{1}{p} + \frac{1}{q} \leq \frac{1}{n}, \tag{32.27}$$

$$\frac{1}{r_2} = \frac{1}{p} + \frac{1}{n} < \frac{2}{n} < 1, \tag{32.28}$$

and

$$\frac{1}{r_3} = \frac{1}{r_1} + \frac{1}{r_2} \leq \frac{1}{p} + \frac{2}{n} < 1. \tag{32.29}$$

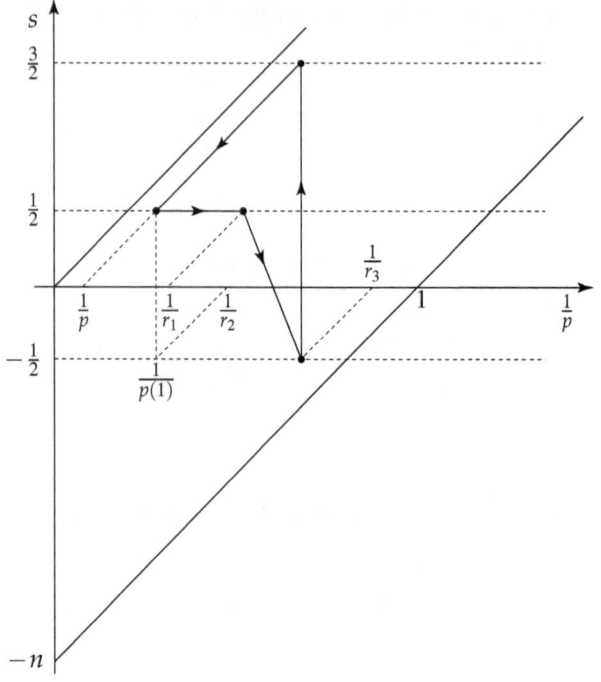

Fig. 32.1

Let $u \in H^{\frac{1}{2}}_{p(1)}(\Omega)$ then it follows by (11.33), (32.22), and (32.27)

$$b_j u \in H^{\frac{1}{2}}_{r_1(1)}(\Omega) \tag{32.30}$$

as indicated in Fig. 32.1. By (32.28) and (32.22) we have

$$\frac{\partial u}{\partial x_j} \in H^{-\frac{1}{2}}_{p(1)}(\Omega) = H^{-\frac{1}{2}}_{r_2(3)}(\Omega) \tag{32.31}$$

and by (11.33), and (32.29), (32.30),

$$b_j u \frac{\partial u}{\partial x_j} \in H^{-\frac{1}{2}}_{r_3(3)}(\Omega). \tag{32.32}$$

We apply $(-\Delta)^{-1}$ to (32.32) and arrive at

$$\{v \in H^{\frac{3}{2}}_{r_3(3)}(\Omega), \ v|\partial\Omega = 0\}.$$

For justification see the references at the end of Step 1 of the proof of Theorem 32.3. Since $\frac{1}{p(2)} \geq \frac{1}{r_3(3)}$ we obtain by continuous embedding

$$B\, u \in \overset{\circ}{H}{}^{\frac{1}{2}}_{p(1)}(\Omega), \qquad (32.33)$$

where Fig. 32.1 stands for the limiting case. Here we used that $-\Delta$ is the Dirichlet Laplacian and that (32.20) with $s = \frac{1}{2}$ can be applied since $\frac{1}{p(1)} < \frac{3}{2n} \leq \frac{1}{2}$. By construction

$$\|B\, u \,|\, H^{\frac{1}{2}}_{p(1)}(\Omega)\|$$

$$\leq c \,\|u \,|\, H^{\frac{1}{2}}_{p(1)}(\Omega)\| \left(\sum_{j=1}^{n} \left\| b_j \,|\, H^{\frac{1}{2}}_{q(1)}(\Omega) \right\| \, \left\| \frac{\partial u}{\partial x_j} \,|\, H^{-\frac{1}{2}}_{p(1)}(\Omega) \right\| \right) \qquad (32.34)$$

$$\leq c \,\left\| u \,|\, H^{\frac{1}{2}}_{p(1)}(\Omega) \right\|^2 \, \sum_{j=1}^{n} \left\| b_j \,|\, H^{\frac{1}{2}}_{q(1)}(\Omega) \right\|.$$

Hence B is bounded.

Step 2. To prove the continuity of B we apply the above estimate to

$$B\, u_1 - B\, u_2 = (-\Delta)^{-1} \sum_{j=1}^{n} b_j(x) \left[u_1 \left(\frac{\partial u_1}{\partial x_j} - \frac{\partial u_2}{\partial x_j} \right) + \frac{\partial u_2}{\partial x_j} (u_1 - u_2) \right]$$

$$(32.35)$$

and obtain (32.26). This implies the continuity.

32.7 Remark The assumptions (32.24) for $b_j(x)$ are rather weak. Since Ω is bounded we have

$$\mathcal{C}^{\alpha}(\Omega) \subset H^{\frac{1}{2}}_{q(1)}(\Omega) \quad \text{for any} \quad \alpha > \frac{1}{2}, \qquad (32.36)$$

where $\mathcal{C}^{\alpha} = B^{\alpha}_{\infty,\infty}$ are the usual Hölder-Zygmund spaces. Hence $b_j \in \mathcal{C}^{\alpha}(\Omega)$, $\alpha > \frac{1}{2}$, are admitted coefficients in the above proposition.

32.8 Theorem *Let $n \geq 3$ and let Ω be a bounded domain in $\mathbb{R}^n$ with C^{∞} boundary. Let $p < \infty$,*

$$0 < \frac{1}{q} \leq \frac{1}{n} - \frac{1}{p}, \qquad (32.37)$$

and

$$b_j(x) \in H^{\frac{1}{2}}_{q(1)}(\Omega) \quad \text{with} \quad \frac{1}{q(1)} = \frac{1}{q} + \frac{1}{2n}. \qquad (32.38)$$

There is a number $\varepsilon > 0$ such that for all

$$g \in H_{p(1)}^{-\frac{3}{2}}(\Omega) \quad \text{with} \quad \|g \,|\, H_{p(1)}^{-\frac{3}{2}}(\Omega)\| \leq \varepsilon \quad \text{and} \quad \frac{1}{p(1)} = \frac{1}{p} + \frac{1}{2n}, \qquad (32.39)$$

the equation

$$\Delta u + \sum_{j=1}^{n} b_j(x)\, u\, \frac{\partial u}{\partial x_j} = g \qquad (32.40)$$

has a solution in $\overset{\circ}{H}{}_{p(1)}^{\frac{1}{2}}(\Omega)$.

Proof. First we repeat that $\frac{1}{p(1)} < \frac{3}{2n} \leq \frac{1}{2}$ and hence $\overset{\circ}{H}{}_{p(1)}^{\frac{1}{2}}(\Omega)$ makes sense according to (32.20). In other words, (32.40) is a nonlinear Dirichlet problem with vanishing boundary data. Let B be given by (32.25). By Proposition 32.6,

$$A u = B u - (-\Delta)^{-1} g \qquad (32.41)$$

is a continuous and bounded map in $\overset{\circ}{H}{}_{p(1)}^{\frac{1}{2}}(\Omega)$ and we have (32.26) with A in place of B. A solution of (32.40) is a fixed point of A and vice versa. Let

$$U_\tau = \left\{ u \in \overset{\circ}{H}{}_{p(1)}^{\frac{1}{2}}(\Omega) \;:\; \|u \,|\, H_{p(1)}^{\frac{1}{2}}(\Omega)\| \leq \tau \right\}, \qquad (32.42)$$

where $\tau > 0$. Let u, u_1, u_2 be elements of U_τ and let g be given by (32.39), then we obtain by (32.26)

$$\|A u \,|\, H_{p(1)}^{\frac{1}{2}}(\Omega)\| \leq c\varepsilon + c\tau^2 \qquad (32.43)$$

and

$$\|A u_1 - A u_2 \,|\, H_{p(1)}^{\frac{1}{2}}(\Omega)\| \leq c\tau \, \|u_1 - u_2 \,|\, H_{p(1)}^{\frac{1}{2}}(\Omega)\|, \qquad (32.44)$$

where $c > 0$ is independent of τ and ε. We choose first $\tau > 0$ and then $\varepsilon > 0$ such that $c\tau < 1$ in (32.44) and

$$c\varepsilon + c\tau^2 < \tau \qquad (32.45)$$

in (32.43). Then A is a contraction which maps U_τ into itself. Hence A has in U_τ a uniquely determined fixed point. As had been said above this fixed point is a solution of (32.40).

32.9 Remark We proved that (32.40) with (32.39) has a unique solution in U_τ. Global uniqueness cannot be expected. As we mentioned in 32.5 a thorough study of stationary Navier-Stokes equations and related problems in function spaces of the above type may be found in the papers J. Johnsen and T. Runst quoted there. For corresponding nonlinear problems in a more general context we refer also to [RuS96].

32.10 The fractal case

Let Γ be a compact d-set in $\mathbb{R}^n$ according to Definition 3.1 with $\Gamma \subset \Omega$, where Ω has the same meaning as in Theorem 32.8. The question is under which conditions g in (32.40) can be identified with the restriction of the Hausdorff measure $\mathcal{H}^d$ on Γ or, more general, $g \in L_r(\Gamma)$ for a suitable r. If $g \in L_r(\Gamma)$ with $1 < r \leq \infty$ then we have by Theorem 18.2

$$g \in B_{r,\infty}^{-\frac{n-d}{r'}}(\Omega), \quad \frac{1}{r} + \frac{1}{r'} = 1. \tag{32.46}$$

We ask for a solution of (32.40) with respect to this interpretation. For sake of simplicity we assume $b_j(x) = 1$. Then we do not need q in (32.37) and (32.38).

32.11 Corollary *Let $n \geq 3$, let Ω be a bounded domain in $\mathbb{R}^n$ with C^∞ boundary, and let Γ be a compact d-set in $\mathbb{R}^n$ according to Definition 3.1 with $\Gamma \subset \Omega$ and*

$$n > d > 2n\frac{n-3}{2n-3}. \tag{32.47}$$

There are numbers r and r' with $\frac{1}{r} + \frac{1}{r'} = 1$ and

$$\frac{2n}{3} < r < 2n \quad \text{and} \quad r' > \frac{2}{3}(n-d). \tag{32.48}$$

For fixed r there is an $\varepsilon > 0$ such that for any $g \in L_r(\Gamma)$ with $\|g \,|\, L_r(\Gamma)\| \leq \varepsilon$ the equation

$$\Delta u + \sum_{j=1}^{n} u \frac{\partial u}{\partial x_j} = g \tag{32.49}$$

has a solution in $\overset{\circ}{H}{}_{r}^{\frac{1}{2}}(\Omega)$.

Proof. We reduce this assertion to Theorem 32.8 and put $r = p(1)$. This explains the first inequality in (32.48). By (32.47) we have

$$\frac{2}{3}(n-d) < \frac{2n}{3}\left(1 - \frac{2n-6}{2n-3}\right) = \frac{2n}{2n-3}. \tag{32.50}$$

This is the conjugate number to $\frac{2n}{3}$. Hence if r is near $\frac{2n}{3}$ then r' satisfies the second inequality in (32.48). Together with (32.46) we have

$$g \in H_r^{-\frac{3}{2}}(\Omega) \quad \text{and} \quad \|g \,|\, H_r^{-\frac{3}{2}}(\Omega)\| \leq c\varepsilon. \tag{32.51}$$

The corollary is now a consequence of the above theorem.

32.12 Remark Since (32.40) is the nonlinearity in the stationary Navier-Stokes equations, the *case $n = 3$* is of peculiar interest. In that case

(32.47) *reduces to $0 < d < 3$ and r in* (32.48) *must be chosen sufficiently near to 2.*

This is a rather satisfactory situation. Furthermore our considerations are qualitative. They can be generalized in several directions as indicated in 32.4.

References

[AdH96]   Adams, D.R. and Hedberg, L.I., Function spaces and potential theory. Berlin, Springer, 1996

[Ama95]   Amann, H., Linear and quasilinear parabolic problems. I. Basel, Birkhäuser, 1995

[Bar88]   Barnsley, M., Fractals everywhere. Boston, Academic Press, 1988

[BeS88]   Bennett, C. and Sharpley, R., Interpolation of operators. Boston, Academic Press, 1988

[BEKS94] Brasche, J.F., Exner, P., Kuperin, Yu.A. and Šeba, P., Schrödinger operators with singular interactions. J. Math. Analysis Applications **184** (1994), 112–133

[BPT96a] Bui, H.- Q., Paluszyński, M. and Taibleson, M.H., A maximal function characterization of weighted Besov-Lipschitz and Triebel-Lizorkin spaces. Studia Math. **111** (1996), 219–246

[BPT96b] Bui, H.-Q., Paluszyński, M. and Taibleson, M.H., Characterization of the Besov-Lipschitz and Triebel-Lizorkin spaces. The case $q < 1$. Preprint 1996

[Bra95]   Brasche, J.F., On spectra of self-adjoint extensions and generalized Schrödinger operators. Habilitationsschrift, Frankfurt (Main), 1995

[Carl81] Carl, B., Entropy numbers, s-numbers and eigenvalue problems. J. Funct. Analysis **41** (1981), 290–306

[CaS90]   Carl, B. and Stephani, I., Entropy, compactness and approximation of operators. Cambridge Univ. Press, 1990

[CaT80]   Carl, B. and Triebel, H., Inequalities between eigenvalues, entropy numbers, and related quantities of compact operators in Banach spaces. Math. Ann. **251** (1980), 129–133

[DaS93]   David, G. and Semmes, S., Analysis of and on uniformly rectifiable sets. Math. Surveys and Monographs **38**, Providence, AMS, 1993

[Dau92]   Daubechies, I., Ten lectures on wavelets. Philadelphia, SIAM, 1992

[Dav95]   Davies, E.B., Spectral theory and differential operators. Cambridge Univ. Press, 1995

[DeJ92]   Deliu, A. and Jawerth, B., Geometrical dimension versus smoothness. Constr. Approx. **8** (1992), 211–222

[DeVL93] DeVore, R.A. and Lorentz, G.G., Constructive approximation. Berlin, Springer, 1993

H. Triebel, *Fractals and Spectra*, Modern Birkhäuser Classics,
DOI 10.1007/978-3-0348-0034-1, © Birkhäuser Verlag 1997

[DiU77]   Diestel, J. and Uhl, J.J., Vector measures. Math. Surveys and Mono-graphs **15**. Providence, AMS, 1977

[DuS58]   Dunford, N. and Schwartz, J.T., Linear operators, I. New York, Inter-science Publ., 1958

[Edg90]   Edgar, G.A., Measure, topology, and fractal geometry. New York, Springer 1990

[EEv87]   Edmunds, D.E. and Evans, W.D., Spectral theory and differential op-erators. Oxford Univ. Press, 1987

[ET96]   Edmunds, D.E. and Triebel, H., Function spaces, entropy numbers, differential operators. Cambridge Univ. Press 1996

[Fal85]   Falconer, K.J., The geometry of fractal sets. Cambridge Univ. Press, 1985

[Fal90]   Falconer, K.J., Fractal geometry. Chichester, Wiley, 1990

[Far97]   Farkas, W., Atomic and subatomic decompositions in anisotropic func-tion spaces. Preprint, Jena, 1997

[Fed96]   Federer, H., Geometric measure theory. Berlin, Springer, 1996 (Reprint of the 1969 ed.)

[FJW91]   Frazier, M., Jawerth, B. and Weiss, G., Littlewood-Paley theory and the study of function spaces. CBMS-AMS Regional Conf. Ser. **79**, 1991

[FlV93]   Fleckinger-Pellé, J. and Vassiliev, D.G., An example of a two- term asymptotics for the «counting function» of a fractal drum. Trans. AMS **337** (1993), 99–116

[FLV95]   Fleckinger, J., Levitin, M. and Vassiliev, D., Heat equation on the triadic von Koch snowflake: asymptotic and numerical analysis. Proc. London Math. Soc. **71** (1995), 372–396

[FrJ85]   Frazier, M. and Jawerth, B., Decomposition of Besov spaces. Indiana Univ. Math. J. **34** (1985), 777–799

[FrJ90]   Frazier, M. and Jawerth, B., A discrete transform and decomposition of distribution spaces. J. Funct. Analysis **93** (1990), 34–170

[Fuj87]   Fujita, T., A fractal dimension, self-similarity and generalized diffu-sion operators. In «Probabilistic methods in mathematical physics», Boston, Academic Press, 1987, 83–90

[Har95]   Haroske, D., Approximation numbers in some weighted function spaces. J. Approx. Theory **83** (1995), 104–136

[HaS94]   Han, Y.S. and Sawyer, E.T., Littlewood-Paley theory on spaces of homogeneous type and the classical function spaces. Memoirs AMS **110**, 530. Providence, AMS, 1994

[HaT94a]   Haroske, D. and Triebel, H., Entropy numbers in weighted function spaces and eigenvalue distributions of some degenerate pseudodifferential operators I. Math. Nachr. **167** (1994), 131–156

[HaT94b]   Haroske, D. and Triebel, H., Entropy numbers in weighted function spaces and eigenvalue distributions of some degenerate pseudodifferential operators II. Math. Nachr. **168** (1994), 109–137

[HKM93]   Heinonen, J., Kilpeläinen, T. and Martio, O., Nonlinear potential theory of degenerate elliptic equations. Oxford Univ. Press, 1993

[Hör85]   Hörmander, L., The analysis of linear partial differential operators III. Berlin, Springer, 1985

[Hut81]   Hutchinson, J.E., Fractals and self similarity. Indiana Univ. Math. J. **30** (1981), 713–747

[Joh93]   Johnsen, J., The stationary Navier-Stokes equations in L_p-related spaces. Ph.D-thesis, Copenhagen, 1993

[Joh95]   Johnsen, J., Regularity properties of semilinear boundary problems in Besov and Triebel-Lizorkin spaces. Journeés «Équations aux Dériveés Partielles» (Saint-Jean-de-Monts, 1995), Exp. **14**, École Polytech., Palaiseau, 1995

[Jon93]   Jonsson, A., Atomic decomposition of Besov spaces on closed sets. In: Function spaces, differential operators and non-linear analysis. Teubner-Texte Math. **133**, Leipzig, Teubner, 1993, 285–289

[JoR96]   Johnsen, J. and Runst, Th., Boundary problems with composition-type non-linearities in L_p-related spaces. Preprint, Copenhagen, 1996

[JoW84]   Jonsson, A. and Wallin, H., Function spaces on subsets of $\mathbb{R}^n$. Math. reports **2**, 1, London, Harwood acad. publ., 1984

[JoW95]   Jonsson, A. and Wallin, H., The dual of Besov spaces on fractals. Studia Math. **112** (1995), 285–300

[Kac66]   Kac, M., Can one hear the shape of a drum? Amer. Math. Monthly **73** (1966), 1–23

[Kig93]   Kigami, J., Harmonic metric and Dirichlet form on the Sierpinski gasket. In: Asymptotic problems in probabilistic theory: stochastic models and diffusion on fractals. Pitman Research Notes in Math. **283**, London, Longman Sci. Techn., 1993, 201–218

[Kig95]   Kigami, J., Laplacians on self-similar sets and their spectral distributions. In: Fractal geometry and stochastics. Basel, Birkhäuser, 1995, 221–238

[Kön86]   König, H., Eigenvalue distribution of compact operators. Basel, Birkhäuser, 1986

[Kühn84] Kühn, T., Entropy numbers of matrix operators in Besov sequence spaces. Math. Nachr. **119** (1984), 165–175

[Lap91] Lapidus, M.L., Fractal drums, inverse spectral problems for elliptic operators and a partial resolution of the Weyl-Berry conjecture. Trans. AMS **325** (1991), 465–529

[Lap93] Lapidus, M.L., Vibrations of fractal drums. The Riemann hypothesis, waves in fractal media, and the Weyl-Berry conjecture. In: Ordinary and partial differential equations, vol. IV. Pitman Research Notes in Math. **289**, London, Longman Sci. Techn. 1993, 126–209

[Lau95] Lau, K.-S., Self-similarity, L^p-spectrum and multifractal formalism. In: Fractal geometry and stochastics. Basel, Birkhäuser, 1995, 55–90

[LeV96] Levitin, M. and Vassiliev, D., Spectral asymptotics, renewal theorem, and the Berry conjecture for a class of fractals. Proc. London Math. Soc. **72** (1996), 188–214

[LGM96] Lorentz, G.G., Golitschek, M.v. and Makovoz, Y., Constructive approximation, advanced problems. Berlin, Springer, 1996

[Lio96] Lions, P.-L., Mathematical topics in fluid mechanics. I. Oxford, Clarendon Press, 1996

[Man77] Mandelbrot, B.B., Fractals: form, chance, and dimension. San Francisco, Freeman, 1977

[Mat95] Mattila, P., Geometry of sets and measures in euclidean spaces. Cambridge Univ. Press, 1995

[Maz85] Maz'ja, V.G., Sobolev spaces. Berlin, Springer, 1985

[Mey92] Meyer, Y., Wavelets and operators. Cambridge Univ. Press, 1992

[NaS94] Naimark, K. and Solomyak, M., On the eigenvalue behaviour for a class of operators related to self-similar measures on $\mathbb{R}^d$. C. R. Acad. Sci. Paris **319 (I)** (1994), 837–842

[NaS95] Naimark, K. and Solomyak, M., The eigenvalue behaviour for the boundary value problems related to self-similar measures on $\mathbb{R}^d$. Math. Research Letters **2** (1995), 279–298

[Net88] Netrusov, Yu.V., Embedding theorems for traces of Besov spaces and Lizorkin-Triebel spaces. Soviet Math. Dokl. **37** (1988), 270–273

[Net90] Netrusov, Yu.V., Sets of singularities of functions in spaces of Besov and Lizorkin-Triebel type. Proc. Steklov Inst. Math. **187** (1990), 185–203

[Net92a] Netrusov, Yu.V., Metric estimates of the capacities of sets in Besov spaces. Proc. Steklov Inst. Math. **190** (1992), 167–192

264   *References*

[Net92b]   Netrusov, Yu.V., Estimates of capacities associated with Besov spaces
           (Russian). Zap. Nauchn. Sem. St.-Peterburg, Otdel. Mat. Inst. Steklov
           (POMI) **201** (1992), 124–156

[Päi83]   Päivärinta, L., Pseudo differential operators in Hardy-Triebel spaces.
           Z. Anal. Anwendungen **2** (1983), 235–242

[Per93]   Peruggia, M., Discrete iterated function systems. Wellesley, Peters,
           1993

[Pie80]   Pietsch, A., Operator ideals. Amsterdam, North-Holland, 1980

[Pie87]   Pietsch, A., Eigenvalues and s-numbers. Cambridge Univ. Press, 1987

[Rud91]   Rudin, W., Functional analysis. Sec. ed., New York, McGraw-Hill,
           Inc., 1991

[Run85]   Runst, Th., Pseudo-differential operators of the «exotic» class $L_{1,1}^0$ in
           spaces of Besov and Triebel-Lizorkin type. Ann. Global Anal. Geom.
           3 (1985), 13–28

[RuS96]   Runst, Th. and Sickel, W., Sobolev spaces of fractional order, Ne-
           mytzkij operators and nonlinear partial differential equations. Berlin,
           de Gruyter, 1996

[SaV96]   Safarov, Yu. and Vassiliev, D., The asymptotic distribution of eigen-
           values of partial differential operators. Translations Math. Monographs
           155, Providence, AMS, 1996

[Schm87]   Schmeisser, H.-J., Vector-valued Sobolev and Besov spaces. Teubner-
           Text. Math. **96**, Leipzig, Teubner, 1987, 4–44

[Schott96a] Schott, Th., Function spaces with exponential weights I. Math. Nachr.

[Schott96b] Schott, Th., Function spaces with exponential weights II. Math. Nachr.

[SchT87]   Schmeisser, H.-J. and Triebel, H., Topics in Fourier analysis and func-
           tion spaces. Chichester, Wiley, 1987

[Schü84]   Schütt, C., Entropy numbers of diagonal operators between symmetric
           Banach spaces. J. Approx. Theory **40** (1984), 121–128

[Schw57/8] Schwartz, L., Distributions a valeurs vectorielles. I. Ann. Inst. Fourier
           7 (1957), 1–142. II. Ann. Inst. Fourier **8** (1958), 1–209

[Shi93]   Shima, T., The eigenvalue problem for the Laplacian on the Sierpin-
           ski gasket. In: Asymptotic problems in probability theory: stochastic
           models and diffusions on fractals. Pitman Research Notes in Math.
           283, London, Longman Sci. Techn., 1993, 279–288

[Sic90]   Sickel, W., Spline representations of functions in Besov-Triebel-Li-
           zorkin spaces on $\mathbb{R}^n$. Forum Math. **2** (1990), 451–475

[SicT95]   Sickel, W. and Triebel, H., Hölder inequalities and sharp embeddings in function spaces of B_{pq}^s and F_{pq}^s type. Z. Anal. Anwendungen **14** (1995), 105–140

[Sim71]   Simon, B., Quantum mechanics for Hamiltonians defined as quadratic forms. Princeton Univ. Press, 1971

[Sma74]   Smart, D.R., Fixed point theorems. Cambridge Univ. Press, 1974

[Sol96]   Solomyak, M., Piecewise-polynomial approximation of functions from Sobolev spaces, revisited. Preprint, 1996

[SoV95]   Solomyak, M. and Verbitsky, E., On a spectral problem related to self-similar measures. Bull. London Math. Soc. **27** (1995), 242–248

[Ste70]   Stein, E.M., Singular integrals and differentiability properties of functions. Princeton Univ. Press, 1970

[Str90a]   Strichartz, R.S., Fourier asymptotics of fractal measure. J. Funct. Analysis **89** (1990), 154–187

[Str90b]   Strichartz, R.S., Self-similar measures and their Fourier transforms I. Indiana Univ. Math. J. **39** (1990), 797–817

[Str91]   Strichartz, R.S., Wavelet expansions of fractal measures. J. Geom. Analysis **1** (1991), 269–289

[Str93a]   Strichartz, R.S., Self-similar measures and their Fourier transforms II. Trans. AMS **336** (1993), 335–361

[Str93b]   Strichartz, R.S., Self-similar measures and their Fourier transforms III. Indiana Univ. Math. J. **42** (1993). 367–411

[Str94]   Strichartz, R.S., Self-similarity in harmonic analysis. J. Fourier Anal. Appl. **1** (1994), 1–37

[Tay81]   Taylor, M.E., Pseudodifferential operators. Princeton Univ. Press, 1981

[Tay91]   Taylor, M.E., Pseudodifferential operators and nonlinear PDE. Boston, Birkhäuser, 1991

[Tem83]   Temam, R., Navier-Stokes equations and nonlinear functional analysis. Philadelphia, SIAM, 1983

[Tem84]   Temam, R., Navier-Stokes equations. Amsterdam, North-Holland, 1984

[Tor86]   Torchinsky, A., Real-variable methods in harmonic analysis. San Diego, Academic Press, 1986

[Torr90]   Torres, R.H., Continuity properties of pseudodifferential operators of type 1,1. Comm. Part. Diff. Equations **15(9)** (1990), 1313–1328

[Torr91]   Torres, R.H., Boundedness results for operators with singular kernels on distribution spaces. Memoirs AMS **90**, 442. Providence, AMS, 1991

[Tri78]   Triebel, H., Interpolation theory, function spaces, differential opera-
           tors. Amsterdam, North-Holland, 1978 (Sec. ed. Heidelberg, Barth,
           1995)

[Tri83]   Triebel, H., Theory of function spaces. Basel, Birkhäuser, 1983

[Tri92]   Triebel, H., Theory of function spaces II. Basel, Birkhäuser, 1992

[Tri92*]   Triebel, H., Higher analysis. Leipzig, Barth, 1992

[TrW95]   Triebel, H. and Winkelvoss, H., The dimension of a closed subset
           of $\mathbb{R}^n$ and related function spaces. Acta Math. Hungarica **68** (1995),
           117–133

[TrW96a]   Triebel, H. and Winkelvoss, H., Intrinsic atomic characterizations of
           function spaces on domains. Math. Z. **221** (1996), 647–673

[TrW96b]   Triebel, H. and Winkelvoss, H., A Fourier analytical characteriza-
           tion of the Hausdorff dimension of a closed set and related Lebesgue
           spaces. Studia Math. **121** (1996), 149–166

[Vas91]   Vassiliev, D., One can hear the dimension of a connected fractal in
           $\mathbb{R}^2$. In: Integral equations and inverse problems, New York, Wiley,
           1991, 270–273

[Wal88]   Wallin, H., New and old function spaces. In: Lect. Notes Math. **1302**,
           Berlin, Springer, 1988, 99–114

[Win95]   Winkelvoss, H., Function spaces related to fractals. Intrinsic atomic
           characterizations of function spaces on domains. PhD-thesis, Jena,
           1995

Symbols

Index